Metamaterials and Metasurfaces

Basics and trends

Online at: https://doi.org/10.1088/978-0-7503-5532-2

Series in Electromagnetics and Metamaterials

Other titles in this Series:

Ricardo A Depine, *Graphene Optics: Electromagnetic Solution of Canonical Problems*

Tom G Mackay and Akhlesh Lakhtakia, *Modern Analytical Electromagnetic Homogenization*

Christopher M Collins, *Electromagnetics in Magnetic Resonance Imaging: Physical Principles, Applications, and Ongoing Developments*

Didier Felbacq, *Advanced Numerical and Theoretical Methods for Photonic Crystals and Metamaterials*

V Kesari and B N Basu, *High Power Microwave Tubes: Basics and Trends, Volume 1*

V Kesari and B N Basu, *High Power Microwave Tubes: Basics and Trends, Volume 2*

Igor I Smolyaninov, *Hyperbolic Metamaterials*

Muhammad Faryad and Akhlesh Lakhtakia, *Infinite-Space Dyadic Green Functions in Electromagnetism*

Syed H Murshid, *Optical Fiber Multiplexing and Emerging Techniques: SDM and OAM*

Series in Electromagnetics and Metamaterials

The Series on Electromagnetics and Metamaterials, published by IOP, is an innovative and authoritative source of information on a fundamental science that has been enabling a multitude of transformative technologies for two centuries and more. The electromagnetic spectrum extends from millihertz waves to microwaves to terahertz radiation to ultraviolet light and even soft x-rays. In each spectral regime, different classes of materials have different kinds of electromagnetic response characteristics. Much has been discovered and much has been technologically exploited, but even more remains to be discovered and even more remains to be put to use for diverse applications.

Electromagnetics is an evermore vibrant arena of techno-scientific research. This is amply exemplified by the huge current interest in metamaterials. By virtue of carefully designed and engineered morphology, metamaterials exhibit response characteristics that are either completely absent or muted in their constituent materials.

Each book in the series offers an extended essay on a foundational topic: an emerging topic; a currently hot topic; and/or a tool for metrology, design, and application. Ranging from 60 to 120 pages, books are written by internationally renowned experts who have been charged with making the content not only authoritative, but also easy to understand, thereby offering more synthesis and depth than a typical review article in a journal.

Illustrated in full color for both ebook and printed copies, these short books are easily searchable in the ebook format. The series is thus more modular and dynamic than traditional handbooks and more coherent than contributed volumes.

This series is edited by Akhlesh Lakhtakia, the Charles Godfrey Binder (Endowed) Professor of Engineering Science and Mechanics at the Pennsylvania State University. Initial topics targeted in the series include symmetries of Maxwell equations, homogenization of bianisotropic materials, metamaterials and metasurfaces, transformation optics, nanophotonics for medicine and biology, single photons, radiation sources, optical bolometry, magnetic resonance imaging, and detection and imaging of buried objects. Additional topic suggestions are welcomed and will be promptly considered and decided upon. As part of the IOP digital library, students and professors at purchased institutions will have unlimited access to the ebooks for classroom and research usage.

Do you have an idea of a book that you would like to explore?

For further information and details of submitting book proposals, contact Akhlesh Lakhtakia at akhlesh314@icloud.com.

A full list of the titles published in this series can be found at: https://iopscience.iop.org/bookListInfo/iop-series-on-electromagnetics-and-metamaterials#series.

Metamaterials and Metasurfaces

Basics and trends

Subal Kar

Former Professor and Head,
Institute of Radio Physics and Electronics,
University of Calcutta, Kolkata, India

IOP Publishing, Bristol, UK

ISBN 978-0-7503-5532-2 (ebook)
ISBN 978-0-7503-5530-8 (print)
ISBN 978-0-7503-5533-9 (myPrint)
ISBN 978-0-7503-5531-5 (mobi)

DOI 10.1088/978-0-7503-5532-2

Version: 20230401

IOP ebooks

British Library Cataloguing-in-Publication Data: A catalogue record for this book is available from the British Library.

Published by IOP Publishing, wholly owned by The Institute of Physics, London

IOP Publishing, No.2 The Distillery, Glassfields, Avon Street, Bristol, BS2 0GR, UK

US Office: IOP Publishing, Inc., 190 North Independence Mall West, Suite 601, Philadelphia, PA 19106, USA

Dedicated to all the students and young researchers of the world.

Contents

Preface

The subject of metamaterials and metasurfaces demands the knowledge of both physics and engineering in the electromagnetic spectrum from RF through THz to optical frequency domain. I am privileged in this regard as I have BSc Physics (Hons.) and also post-BSc, BTech, MTech, and PhD (Tech) degrees in radio physics and electronics. This has given me a sound base for learning the subject of metamaterials and metasurfaces both from the physics and engineering perspective.

My research on metamaterials began in 2006 when I received a request from a senior scientist from Bhabha Atomic Research Centre (BARC), Mumbai, India to take up a project on the development of a metamaterial research facility in India as a part of 'Technology Vision 2020'. This ultimately took shape in late 2006 as a collaborative research of the University of Calcutta, SAMEER (Society for Applied Microwave Electronics Engineering and Research) Kolkata Centre and BARC. We completed a BRNS (Board of Research in Nuclear Sciences) project during 2010–14. A metamaterial measurement facility has been developed at SAMEER Kolkata Centre where students and researchers work with great success. The First successful plasmonic metamaterial in India was designed, fabricated and tested indigenously in 2009 by our group of the University of Calcutta in collaboration with SAMEER Kolkata Centre and BARC, Mumbai.

I had the occasion to deliver invited talks on the 'Progress with Metamaterial Research' in Indian Science Congress (2012, 2017), Oxford University, UK (2013), Cockcroft Institute, UK (2013), EPFL, Switzerland (2014) and many other universities and institutes in India and abroad. With all these exposures and experiences gained with my mission of life as a 'curious and inquisitive learner' I have planned to write this book.

The purpose of my writing this book stems from the feeling and observation that there may be a number of good edited/specialized books/monographs that help with advanced research in metamaterials and metasurfaces but are primarily useful for experts in the subject. However, the students at UG/PG level and young and new entrants in research in this emerging topic need a suitable textbook that deals in a systematic and lucid way covering the basics and trends of metamaterial and metasurface developments from its inception till today. A holistic scenario of this development placed before them will possibly give them a stepping platform for initial learning and also enable them to look forward with confidence to indulge in innovative ideas for their future research.

Following on from this background of my learning on metmaterials/metasurfaces and the purpose of writing this book, I will now discuss in brief the chapter-wise coverage in this book.

Chapter **1** primarily deals with the various counter-intuitive properties like reversal of Snell's law, reversed Doppler effect and reversed Čerenkov effect exhibited by metamaterials, or left-handed materials (LHMs). Metamaterials, being manmade, i.e. a kind of artificial material, need to be differentiated from other artificial materials like artificial dielectrics, photonic crystals, and chiral materials; a brief discussion in this regard has also been included in this chapter. This is followed by the mathematical basis of the counter-intuitiveness of metamaterial including the boundary conditions. Next, the transmission properties of electromagnetic signal via material are discussed in detail, including plane-slab focussing and the so-called 'evanescent wave growth' and 'sub-wavelength imaging' by metamaterial plane slab. Detailed mathematical analysis for characterization of transmission properties of electromagnetic signal via metamaterial has also been discussed in this chapter. Metamaterials are known to make things invisible with the so-called cloaking device; the physical principle of metamaterial-based cloaking and the cloaking devices at microwave and optical frequencies have been discussed. Finally, the chapter ends with the effect of losses in metamaterial performance with some interesting results.

A detailed discussion of two types of metamaterials—the plasmonic type and transmission-line type—has been covered in chapter **2**. The first successful metamaterial was of plasmonic type made of thin wire (TW) and split-ring resonator (SRR) in 2001 followed by the transmission-line metamaterial that was first reported in 2002. Negative permittivity is realizable with TW (and also with cut-wire, CW), which acts as an artificial 'electric atom', and negative permeability is realized with SRR, which acts as an artificial 'magnetic atom'. There are a number of variants of the magnetic inclusion structure SRR, like MSRR (multiple inclusion split-ring resonators), SR (spiral resonator) and LR (labyrinth resonator), and many others too each having relative advantages and disadvantages. The physical principle, mathematical analysis and design techniques of some of these plasmonic magnetic inclusion structures are discussed in this chapter. This is followed by transmission-line metamaterial which are basically periodically loaded transmission line (PLTL) in which the host RH (right-handed) transmission line is periodically loaded with series capacitance and shunt inductance (the LH or left-handed circuit elements). PLTL in practice behaves as a *composite right/left-handed* (CRLH) transmission line. Finally, the basic characteristics of LH and CRLH transmission lines have been discussed and the use of CRLH as zeroth-order resonator is also discussed with dispersion characteristics of pure right-handed (PRH), pure left-handed (PLH) and CRLH transmission line.

Chapter **3** discusses metamaterial-inspired passive components, antennas, and active devices. Firstly the design of CRLH based microwave filter is given, followed by CSRR (complimentary split-ring resonator) loaded microstrip patch antenna design. This is followed by the review of the development of metamaterial-inspired various other passive components and antennas. Metamaterial-based antennas including zero-order antenna, dual-band ring antenna, leaky-wave metamaterial antenna and also other microstrip antennas with metamaterial-inspiration are discussed in this chapter. Next, different types of metamaterial couplers like symmetric impedance coupler, asymmetric phase coupler including other types of metamaterial-inspired couplers are discussed. Thereafter, phase compensator and its applications and metamaterial-inspired phase shifter are discussed. Some innovative trends for passive components and antenna developments are also discussed; specifically the textured electromagnetic surfaces or high impedance surface are covered in this respect. Finally, the chapter ends with a discussion that covers the metamaterial-inspired microwave active devices, both solid-state and tube devices.

Optical and terahertz (THz) metamaterials are very important from the real-world application point of view. Thus, chapter **4** deals with THz and optical metamaterials. As far as the THz metamaterials are concerned, the electric and magnetic inclusions used at microwave frequency can be scaled down so as to operate at THz frequency. But at optical frequency this does not work, and also the fabrication of nanometre dimension SRRs etc are not possible due to technical limitations of their size miniaturization. But the nanofabrication facility possible these days allows the design of optical metmamaterials with nanowire and fishnet structures that has made widespread use of optical metamaterials for various applications. Under the heading of THz metamaterial we have discussed metamaterial-based THz absorbers and sensors. For optical metamaterials, we have discussed the basic optical metamaterial structures and different fabrication technologies used in the case of the fabrication of optical metamaterials. This is followed by applications of optical metamaterials like hyperlens and metamaterial absorbers at optical frequency. Other emerging applications of optical metamaterial like laser SPASER, metmaterial-assisted carbon nanotube etc are also discussed in brief along with the sensors and light harvesting applications at optical frequency. Finally, the so-called transparent metamaterial useful for photonic integrated circuits and single-photon generation is discussed.

Chapter **5** is dedicated for modelling and characterization of metamaterials. Beginning with a brief comparative discussion on various analytical/numerical techniques of modelling of metamaterials, the popularly used *S*-parameter-based retrieval method for modelling of metamaterials is discussed in detail. The simulation based results of eight of the magnetic inclusion structures are given and their performance comparison is provided. A few of these structures have been fabricated and experimentally characterized with free-space focussed beam LHM characterization setup at X band (designed and developed indigenously) and the experimental results are compared with simulation results. Further, the experimental characterization of metamaterial prism/wedge structure and plane-slab focussing structure is also given.

Metasurfaces or 2D/surface versions of metamaterials have emerged as a very important alternative of 3D and bulky metamaterials covering microwave, THz and optical frequencies with excitingly new and varied applications. Thus, chapter **6** is fully devoted to metasurfaces and begins with a brief comparison of them with metamaterials and frequency selective surfaces (FSSs). This is followed by a detailed discussion on various physical aspects of metasurfaces related to application viewpoints—where the anomalous reflection and refraction, polarization conversion, dielectric metasurfaces, active, tunable/reconfigurable and non-linear metasurfaces are discussed in brief. Modelling and characterization of metasurfaces are discussed next. Finally, various applications of metasurfaces are discussed, some of which are already in place while others are indicating the future trends. This includes metasurface-based absorbers, metasurface in antenna design, metalens with metasurface, metasurface-based wave guidance and coupling, miniature resonators based on metasurface, RF to electricity with metasurface and so forth. A brief scenario of the developments and challenges/possibilities ahead for meta-research has also been included in the concluding part of chapter 6.

As an author it is now my honest duty to acknowledge the help and cooperation from all those who have made my endeavour for writing this book fruitful and rewarding.

At the very outset I would like to acknowledge the help and support I have received from my wife Rina and son Sauradeep for writing this book. The quality time they have provided me to write this book with a concentrated mind has actually made it possible to translate my dream to reality. Without their active support and encouragement this effort of mine would not have been successful.

I am also thankful to all my students of the University of Calcutta, India, who worked with me on metamaterials since 2006. The list of these students is not very short but I would like to mention the names of some of them who have contributed significantly in my metamaterial research endeavour. The list starts with T Roy and D Banerjee, with whom I started my first journey of metamaterial research in 2006. Then subsequently, P Ganguli, S Pal, S S Sikdar, T K Saha, S Das, R R Day, K Sadhukhan, D Roy, M Ghosh, S K Ghosh, A Saha and many more diligent and hardworking students have helped me to take forward my metamaterial research.

I would further like to offer my special thanks to the collaborators from BARC, Shantanu Das, Scientist G, and SAMEER Kolkata Centre, Arijit Majumder, Scientist E, who worked with our group of the University of Calcutta in the BRNS project entitled ‘Left Handed Maxwell’s Systems—Experimental Studies’ implemented at SAMEER, Kolkata Centre during 2010–14. The success of the project was made possible with active participation of my research student T Roy and two scientists of SAMEER, Kolkata Centre, A Kumar and S Chatterjee. However, I should not miss this opportunity to appreciate the help and cooperation that was extended by Dr A L Das, the then Program Director of SAMEER, Kolkata Centre.

My whole-hearted and sincere gratitude obviously goes to Professor Sir John Pendry FRS of the Department of Physics, The Blackett Laboratory, Imperial College London, UK, the father-figure of metamaterial research in the world, who has gladly accepted my request to write the foreword for this book. This is definitely a great reward I have received on writing this dream book of mine.

Finally, I must thank Ms Emily Tapp, Commissioning Editor of IOP Publishing Limited, UK and her team including Mr Chris Benson, Production Editor of IOP Publishing, and the typesetters for doing their job excellently with patience and meticulousness. I am also thankful to SRS Publishing Services, Puducherry, India, for their assistance in helping me to draw professionally most of the illustrations in this book.

If this book is liked by the inquisitive readers, for whom it has been written, then all my humble efforts will be truly rewarded.

January 1, 2023
Kolkata: 700059, India

Subal Kar
Former Professor and Head
Institute of Radio Physics and Electronics
University of Calcutta, India

Foreword by Sir J B Pendry

Professor Sir John Brain Pendry, FRS.
Department of Physics,
The Blackett Laboratory,
Imperial College London, SW7 2AZ, UK.

Awards and Distinctions: FRS (1984), Dirac prize (1996), Knighted (2004), Royal Society Bakerian Medal (2005), Royal Medal (2006), Isaac Newton Medal (2013), Kavli Prize in Nanoscience (2014); and many more.

Most fashionable topics in science have a lifespan of about 10 years, and then fade away as a new fad arrives. A flash in the pan one might say. In the case of metamaterials there were two flashes, but they lit a fire that continues to the present day.

The first flash was the realisation that metamaterials could realise negative refraction. Half a century ago it was postulated theoretically that materials with a negative refractive index should be possible if only the right materials could be found, but none were to be found. Then much later through work I did with the Marconi Company came the concept of designing the microstructure of a material to dramatically change its properties. This sometimes happens in Nature: think of silver which can be highly reflecting as in a mirror, or very black as in the colloidal silver nanoparticles found in photographic negatives—same metal, different microstructure. Metamaterials can generate a huge range of different magnetic and electrical properties amongst which is the ability to reverse the response of a material to a field. It was just this effect which was needed to fill the gap in materials technology that had existed for more than thirty years. Then followed the realisation that a slab of negative material could in fact be designed to be a 'perfect' lens in the sense that there is no theoretical obstacle to its resolution, only the limitation of how accurately it could be constructed. This violation of the Abbe limit caused some consternation at the time until it was realised experimentally both for microwaves and at optical wavelengths. Although metamaterials are not confined to negative index materials, this concept of negative refraction and perfect lenses created the first spark interest in the field.

The second flash came with the design of a cloak of invisibility. This required values of the permittivity and permeability to vary continuously within the cloak and was only practical if metamaterials were employed. Many realisations of the basic concept have been made and tested starting with a design by the team at Duke University. That attracted a considerable amount of attention not only from scientists and engineers but also from school children saturated with the Harry Potter stories. Once the excitement of invisibility died down scientists started to ask if metamaterials could make something as remarkable as a cloak maybe there were many other applications to be found. The fire was taking hold.

This was the beginning, but the continued interest in metamaterials stems from mundane but more important considerations: their application spans numerous areas of engineering ranging from 5G technology to ultra-thin lenses. Many of these applications can be found documented in this book. To keep the metamaterial fire burning, each successive generation of scientists and engineers needs to learn the ropes. That is the aim of this book which draws on the author's considerable experience to give an up-to-date account of the basics of metamaterials/metasurfaces and their application in the real world.

J B Pendry
London
2023

Author biography

Subal Kar

Subal Kar is former Professor and Head of the Institute of Radio Physics and Electronics, University of Calcutta, India. His field of specialization is microwave and terahertz (THz) engineering, metamaterials, and high-energy physics. Dr Kar has three patents to his credit and has published a large number of research papers in peer-reviewed international journals. He was visiting scientist to various universities and institutes in the United States, Europe, and Asia, including Lawrence Berkeley National Laboratory, Oxford University (https://vimeo.com/87101068), Cockcroft Institute, EPFL, and Kyoto University. Dr Kar has authored a textbook *Microwave Engineering—Fundamentals, Design and Applications* published by Universities Press, Orient Blackswan, India (1st edition 2016 and 2nd edition 2022) and has also authored another book entitled *Physics and Astrophysics—Glimpses of the Progress* published by CRC Press of Taylor and Francis Group, USA. He has contributed a number of chapters in edited books published by Elsevier, Springer Nature, and CRC Press. He is the recipient of the young scientist award of URSI and IEEE MTT, and the Fulbright award from the US Government.

An interview of the author with a science journalist

Gardening to invisibility cloak

Posted on December 9, 2012

This is an exciting tale of a lad more obsessed with gardening than study who went on to become one of India's finest physicists and now dealing with exotic subjects like high-energy physics and metamaterials. Meet **Professor Subal Kar** who shares his journey through the mazes of scientific riddles, puzzles and solutions tracing back to his school days with science journalist **Biplab Das**[1]

(1) **What triggered you to become a physicist? Can you mention any life-changing incident that set the stage?**

There is no recipe that can make a physicist. However, milieu does shape a tender mind. I grew up in Shillong, the capital of Meghalaya, and one of the smallest states in India. The hilly regions with lush forest cover fascinated me since childhood. Lavish nature with rich tapestry of life piqued my curiosity and broadened my mind. To begin with, I was not a bright student. The stuff of textbooks seemed dull. Instead, I was obsessed with building scientific models. I preferred gardening to textbooks and grew vegetables, flowers and fruit in our backyard. I studied at Jail Road Boys' High School, Shillong (https://jrbhss.in/).

This happy-go-lucky life took an astonishing turn after the final school exam of ninth standard. One day, seeing my apathy towards study, the headmaster of our school took me to a deserted corner in the school campus and asked why I was not as good as my elder brother in study. He said 'gardening wouldn't take you closer to your goals'. He was aware of my penchant for making scientific models. This remark shook me to a great extent. I immersed myself in study. And gardening took a back seat.

A year later I could find my name among the toppers. I soon found some like-minded peers with common interest in space and astronomy. With peers, I made a prototype space vehicle mimicking the Apollo 9, the third manned spacecraft that blasted off in 1969. With great hope, we launched the tiny spacecraft into the air only to see it crash to the ground. This failure didn't shatter our spirit, however. With a positive mind, I entered St. Edmunds' College in Shillong (https://sec.edu.in/).

(2) **Did the educational ambience in college shape your mind?**

My college life was a special one. It was peppered with memorable events. My obsession with physics grew stronger and it was greatly shaped

[1] ***Biplab Das*** *is a Science Journalist associated with Nature India and other reputed science news media.*

by Professor N Gangooly, the then head of the physics department of this college. And I felt an urge to study physics beyond college. We had a state-of-the-art physics laboratory with a rich library. This set the stage to launch my career in physics. I also dabbled in philosophy and literature (particularly Western).

I shared my passions with Brother De Mellow, a fellow classmate, and I paid occasional visits to the serene atmosphere of our College Church. My forays into making scientific models continued. And I made a 10-inch telescope and explored the star-studded night sky with a childlike curiosity. My stargazing took off on 29 August 1972. On that day, I was awestruck by the immensity of the night sky dotted with glittering stars. I felt as if I was Tycho Brahe, the experimental astronomer of 16th century.

(3) **At what point of your life did you decide that you wanted to be a scientist?**

I left college and reached Calcutta (now Kolkata) in 1974 to pursue higher study. I inched closer to my dreams taking admission at Institute of Radio Physics and Electronics (http://www.irpel.in/) of University College of Science and Technology and Agriculture (popularly known as Science College), University of Calcutta (https://www.caluniv.ac.in/), popularly known as Calcutta University (CU). The university already had a glorious past.

It touched several milestones by developing the vacuum tube (Types 80 and 6C5) [1949], analog computer [1956] and magnetron (CW S-band) [1959] for the first time in India. As a student I basked in the glory of such achievements and listened to thought-provoking lectures of great teachers. Besides poring over the great theories, I honed my skills in experiments. Armed with a mind prone to discover novel things, I embarked on a research career under the guidance of respected teacher Professor S K Roy, who pioneered the research on IMPATT (Impact Avalanche Transit Time) device in India. Although I was absorbed in research, I didn't banish the study of Indian philosophy, listening to music and the passion for reading Bengali literature.

(4) **Which area of physics did you pursue during the early days of your research career?**

My research career started with the development of a new type of microwave IMPATT oscillators and amplifiers. In a short period of time, I successfully invented a broadband and high power IMPATT oscillator and amplifier under the guidance of Professor. S K Roy. These devices were granted patents in the late 1980s. We also propounded a new physics for the device-circuit interaction at microwave frequency for oscillators and amplifiers.

After obtaining a PhD in 1989, I finished post-doctoral research at the same institute and presented a number of invited papers at international conferences in France, US, and Japan. By 1999, I moved to the US on a prestigious Fulbright Fellowship. On my return to India, I continued research at my alma mater. In 2003, I tasted success by inventing a new type of microwave power combiner which earned a patent.

(5) **Tell how you landed up in Berkeley Lab superconducting collider project in US?**

I was one of the seven scientists of Berkeley Lab actively involved in the aforesaid project. The other organizations in the collaboration were Fermi Lab, Chicago, University of Mississippi and Illinois Institute of Technology, US. My research at Berkeley lab was extremely fruitful and this gave me a spirit and training for taking up any challenging, front-line research activity in a big way.

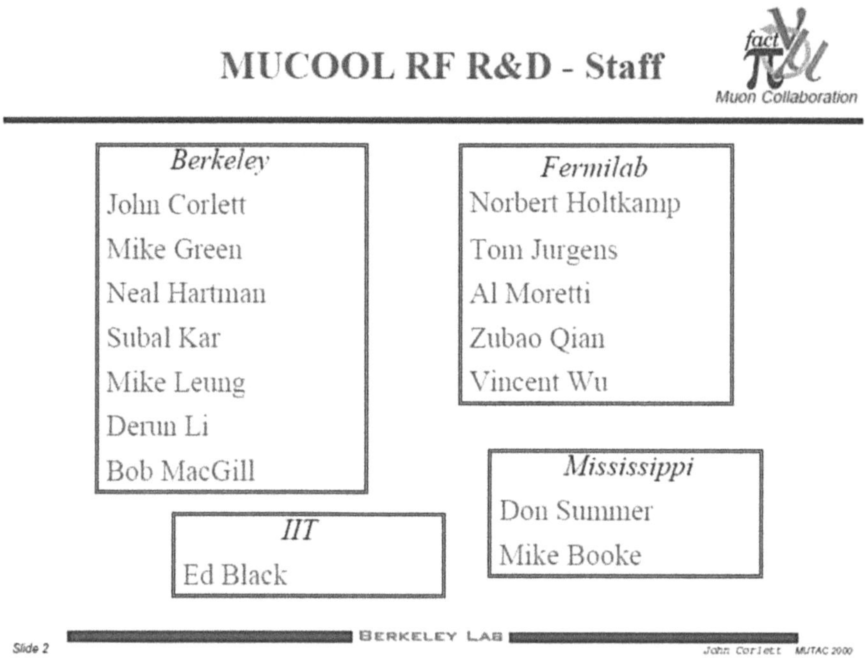

Berkeley Lab, US, presentation document, 2000.

(6) **Please share something about your research experience during your stay in Berkeley, US?**

My research stint as visiting Fulbright scientist at Berkeley Lab (https://www.lbl.gov/) was in the Accelerator and Fusion Research Division in connection with the radio frequency (RF) system development required for the design of the Muon Collider. Particle accelerators or colliders are used to produce energetic electrons (leptons) or protons (hadrons) or muons with energy reaching to the order of a million or even trillion electron volts that are made to collide nearly at the speed of light in a collider ring, spawning exotic subatomic particles setting the stage to explore the quirky world of subatomic particles.

For example, the Large Hadron Collider of CERN, Geneva is a hadron collider in which protons and antiprotons collide in the collider ring in the hope of glimpsing the most fundamental particle: the Higgs Boson, the so-called 'God particle'. The Muon Collider project for which I have worked at Berkeley lab uses muon (a subatomic particle that is generated when high energy protons hit a liquid jet of platinum oxide target popping up pions that decay very quickly into muons). Muons are leptons like electrons but each muon has a mass 207 times that of an electron. When positive and negative muons collide in the collider ring they generate copious numbers of neutrinos (another subatomic particle weighing less than a billionth of a hydrogen atom).

By the time you finish reading this line, a huge number of neutrinos are passing through us. The study of a neutrino is extremely important as it is believed to be a vital constituent of so-called 'dark matter' and 'dark energy', which are supposed to make up 95% of the invisible matter of the Universe.

Well, I am not a nuclear scientist. My role in the muon collider research was to design the RF systems associated with it. To energize the muon particles to the order of a million electron volts, they are to be passed through a large chain of RF cavities. These are very specialized cavities called interleaved cavity of pill-box type with beryllium windows. The cavities are to be operated at liquid nitrogen temperature to realize very high quality factor of the cavities.

In 2004, I began to collaborate with the Berkeley Lab for a new research project. This time the project aimed to design and develop a Laser-based Ultra-fast X-ray (LUX) source (an international user facility) useful for nanosized diagnostic facilities cutting across all branches of science. Our contribution from the University of Calcutta in this project is to design and develop the 'timing and synchronization' systems for the LUX.

Dr Kar at work in his laboratory (IRPEL, CU), 2005.

(7) **Of late, you are exploring the realms of metamaterials. How challenging is this field and what is its potential?**

Yes, it is my latest research venture. We embarked on metamaterial research at University of Calcutta in 2006. We made the first successful metamaterial crystal in India in 2009 as a collaborative effort of the University of Calcutta, SAMEER, Kolkata, and BARC, Mumbai. The metamaterial crystal was unveiled at a National Meet at BARC, Mumbai on 17 August, 2009. It was highlighted as a feature article in *Nature India* (https://www.nature.com/articles/nindia.2009.273).

Metamaterials are popularly known to make things 'invisible'. Such materials reverse several known physical phenomena and bend light in the wrong directions. They can make things invisible and may even catch such minute details of objects that no other sophisticated optical/electron microscope can. Metamaterials are already showing promise that can help science leap from the pages of science fiction to the fringes of reality.

The hallmark of a metamaterial crystal is that it does not absorb or scatter any part of the microwave radiation or any other electromagnetic radiation. Light could also be made to flow smoothly around the metamaterial like water flowing past a smooth pebble. Such an outcome has far-reaching implications.

When light hits a small object, it generates waves that vanish without a trace. Such waves carry minute details of a small object. Metamaterials could be designed to capture such vanishing waves giving rise to a super-microscope that can image even an individual strand of DNA. Besides conjuring up the possibility of an invisibility cloak, metamaterials are on the brink of revolutionizing the interaction of electromagnetic waves with objects, from modems and MRI to radios and radar.

(8) **What are your missions and visions as a physicist?**

My mission embodies two goals. One of them is to motivate bright young students towards a research path. The other is to keep myself busy in breaking new ground in physics. For the past 30 years of my research and teaching, I have been sincerely trying to instill a zeal for research into students' minds to carve their niches and identities.

Regarding vision, I am always optimistic and thus I believe new discoveries and inventions are due in the current and the next decade. The process has already started. For example, metamaterials research is sure to bring a revolution with remarkable technological fruit and challenging understanding of new dimensions in physics. Many exotic applications are possible with metamaterials which will usher in a new paradigm in applied physics.

(9) **What are the special qualities that a scientist should have to crack puzzles leading us to new horizons?**

The thought process of a scientist should be that of a child: inquisitiveness with single-minded quest for truth. Knowledge and skills are necessary, but that are only to sail through one's career, but that should not influence or limit the independent thinking power of a scientist. A good research guide for a young scientist is a much-needed thing in the arena of scientific research. However, the guide shouldn't block the free thinking of the research scholar. Besides having a novelty in thinking, honesty, sincerity, and dedication are qualities that drive a scientist towards their goals.

Chapter 1

Metamaterials or left-handed materials—a counter-intuitive artificial material

Anything counter-intuitive to an average mind is not so to an insightful mind

1.1 Introduction

Human history is often defined in terms of the materials that shape the era and the way in which materials enable us to manipulate our needs and demand of the age concerned. Starting from the crude realization of the physical world around us from the 'stone' age, humankind discovered the ways and means to manipulate metals in the 'bronze' and 'iron' ages. From the middle of the 20th century we have mastered silicon and other advanced material technologies leading to the finest of the technologies of today—nanotechnology. Entering into the 21st century, we have found ourselves on the cusp of a new era of the so-called metamaterials, which can artificially mimic the atomic structure of natural materials but respond to electromagnetic signal exhibiting a range of unique and tailorable electromagnetic properties that are not observed with naturally occurring materials.

The first successful experimental realization of negative refractive index with bi-dimensional metamaterials was reported in 2001 [1] by the scientists of UCSD, USA (though the uni-dimensional structure was reported by the same group in 2000, see chapter 2) using a combined structure of thin wires (TWs) and split-ring resonators (SRRs). In 2009, our group at the University of Calcutta, India, was successful in demonstrating, for the first time in India, the negative refractive index properties of metamaterials using a combination of TW and labyrinth resonator (LR) structure [2]. Metamaterials are known to exhibit counter-intuitive properties like negative refractive index; reversed Doppler effect and reversed Čerenkov effect (see figure 1.1).

doi:10.1088/978-0-7503-5532-2ch1

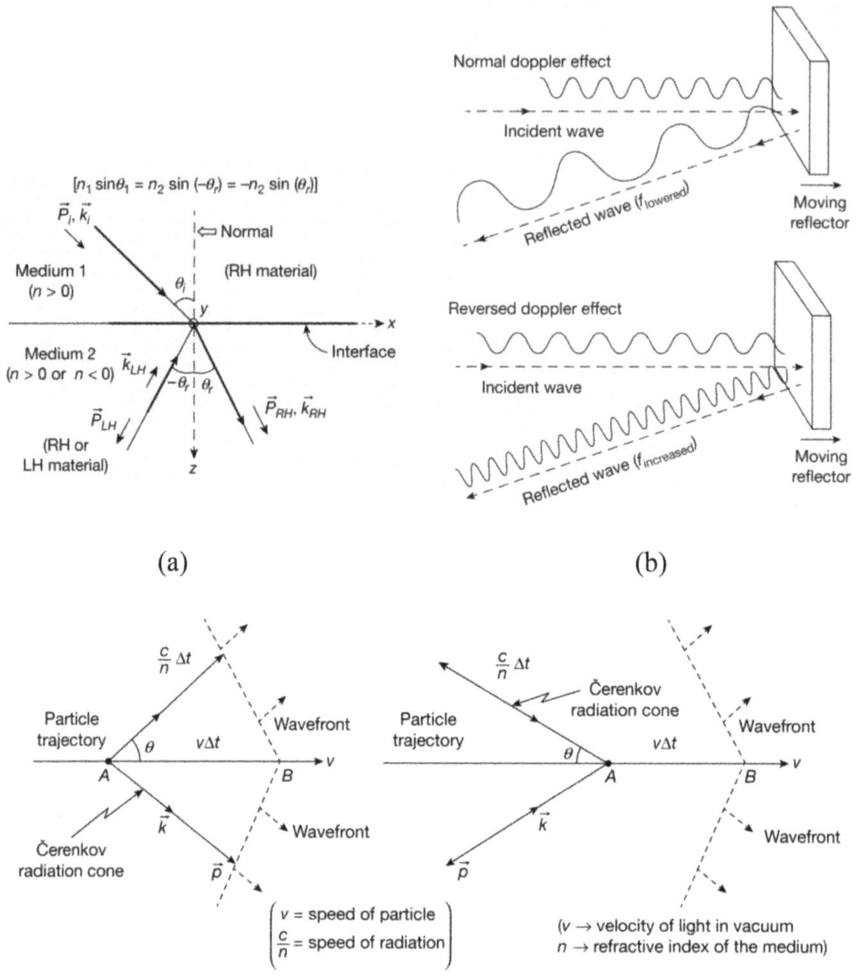

(c)

Figure 1.1. Counter-intuitive phenomena with metamaterial. (a) Reversal of Snell's law of refraction. (b) Reversed Doppler effect. (c) Reversal of Čerenkov radiation.

From our school days we are familiar with *Snell's law of refraction* propounded by the Dutch mathematician Willebrord Snellius in 1621 which passed down from generation to generation unquestioned. According to this law, when a ray of light travels obliquely from a lighter medium (say air) to a denser medium (say glass) it is bent or refracted at the surface separating the two media. The refraction occurs because light travels at slightly different velocities in different media—slows down when it travels from air into water or other substances having higher refractive index than air (the speed of propagation of light in water is only 0.75*c*, where *c* is the velocity of light in air, more specifically in vacuum)—this happens because of frequent interaction of photons (a particle representing quantum of light or other electromagnetic radiation) with matter. Thus in the two media there will be a slight change of wavelength of the propagating light wave (longer in lighter medium and shorter in denser medium). A similar phenomenon also applies to wave motions for all electromagnetic waves other than light. It may be mentioned that the index of refraction/refractive index provides a measure of the speed of electromagnetic wave as it propagates within a material.

Doppler effect is a fundamental frequency-shift phenomenon, which was proposed by Johann Christian Doppler in 1843, that occurs whenever a wave source and an observer are moving with respect to one another. The Doppler effect explains why the pitch of the whistle of a train or the siren of a police van changes from high to low when they pass by us. The use of Doppler effect is numerous: the police use CW (continuous wave) Doppler radar to catch speeders, it is used in weather and aircraft radar systems, satellites use this effect to track space debris, even the measurement of blood flow in unborn fetal vessels can be done by using the Doppler effect.

The *Čerenkov effect*, named after the Soviet scientist Pavlov Čerenkov who observed this radiation in 1934 along with his research student Sergey Vavilov (that is why it is also known as Vavilov–Čerenkov radiation), is a common term with particle physicists and accelerator physicists but the common public is less familiar with it. Čerenkov radiation is extensively used by experimental particle physicists for identification, detection and characterization of charged particles moving very close to the relativistic speed in accelerators. The most advanced type of Čerenkov radiation detector is RICH (ring imaging Čerenkov detector) developed in the 1980s and is in use in the Large Hadron Collider of CERN, Geneva, in which the last missing fundamental particle of the Standard Model of Physics, the Higgs boson, was detected in 2012.

Anyway, it is known from Snell's law of refraction that when an incident electromagnetic signal goes from a less dense natural medium (i.e. the so-called right-handed or RH medium) to a more dense RH medium, the refracted ray turns away from the normal to the interfacing surface with a $+\theta_r$ angle of refraction. But when the less dense medium is RH-type (with positive refractive index) and the more dense medium is LH-type (LH stands for left-handed)—the refracted ray is bent in the $-\theta$ direction (i.e. on the same side of the normal to the interfacing surface as the incident ray is) with a $-\theta_r$ angle of refraction (see figure 1.1(a)); and we say that the LH medium exhibits a *negative refractive index*. It may be noted that a negative refractive index implicitly means that the phase velocity (v_p), which describes the propagation of individual wavefronts in a wave group, represented by $\vec{k}_{LH}$, of a propagating wave is opposite to the movement of the energy flux of the wave, represented by the Poynting vector ($\vec{P}_{LH}$). This somewhat counter-intuitive (anti-parallel) propagation phenomenon that takes place within an LH media is essentially the characteristics of a negative refractive index material that leads to a variety of intriguing effects.

The consequence of this *negative refraction* and realization of the so-called negative refractive index possible with a bizarre class of artificial/man-made material, i.e. LH material or metamaterial, is phenomenal and has led to a number of exotic applications from microwave through terahertz to optical frequency domain in recent years. In fact, negative refraction has generated an enormous potential for a new generation of lenses, the 'superlens' or 'perfect lens', with plane metamaterial slab. For instance, air or a vacuum has a refractive index of $n = +1$, so a piece of natural or RH material with refractive index $n = +1$ does not refract rays, and cannot form a lens. But metamaterial or LH material using plane-slab (and not curved surface) with $n = -1$ is able to refract rays through it, making the possibility of the so-called superlens (details of which will be discussed in section 1.5.1). However, curved LH material can also behave as a lens but a converging lens made of negative-index material should have a concave surface rather than a convex one (unlike RH material lens). It may be mentioned that negative index material (NIM) possible with metamaterial is the subject of constant capacity for surprise: innocent assumptions led to unexpected and sometimes profound consequences. This has generated great enthusiasm during the last two decades but also had its controversies; that ultimately were settled, enriching the critical concepts with deep scrutiny. A lot of experimental data has now been produced that has validated the concept with confidence.

Again we know from our common experience that the whistle heard from a passing train/police van/ambulance car becomes shrill (i.e., its pitch increases) when the train/vehicle is approaching us as its frequency (f) reaching us increases by Δf. But the sound of the same whistle flattens out (i.e., its pitch decreases) when the source of sound is receding from us as its frequency reaching us decreases by Δf due to the well-known Doppler effect. This is the normal Doppler effect we are familiar with for wave propagation in natural (RH) media. But the case just reverses and we can observe reversed/inverse Doppler effect for waves propagating in LH material (see figure 1.1(b)). The reversed Doppler effect realizable for propagation of electromagnetic signal in LH media might lead to exotic applications in radar and possibly in other domains too not yet envisioned.

Further, in particle accelerators, it is known that when elementary particles like electrons, protons and so on travel close to the velocity of light an electromagnetic radiation cone travels along with the fast moving particle and this radiation cone is known as Čerenkov radiation. Particle physicists identify/discover new particles by studying the signature of the Čerenkov radiation emitted by the elementary particle. When the particle moves in RH media, both the particle and the associated Čerenkov radiation moves in the same direction (i.e., in the $+z$-direction). But if the particle moves in LH media then the generated Čerenkov radiation will move in the $-z$-direction (see figure 1.1(c)). The reversal of Čerenkov radiation via LH media might make the detectors of Čerenkov radiation provide more sensitivity of detection as the spurious signal from high-speed particle movement will not load the detectors used for sensing the Čerenkov radiation in accelerators.

It was Victor Georgievich Veselago in 1967 [3] (original Russian paper was published in 1967 while the English translated paper was published in 1968) who made theoretical investigations on the solutions to Maxwell's equations in hypothetical media having simultaneously negative isotropic permittivity and permeability and observed that such material has the possibility of exhibiting the above-mentioned three counter-intuitive properties. Because no naturally occurring

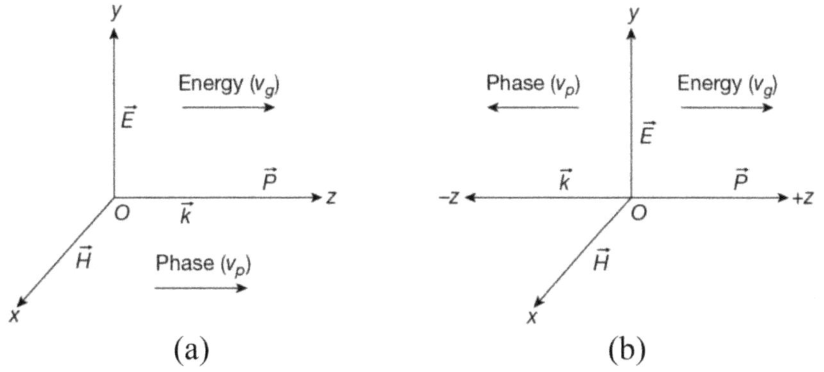

Figure 1.2. The vector triplet *E*, *H*, and *k* in (a) right-handed material (RHM) and (b) left-handed material (LHM).

material or compound has ever been demonstrated with simultaneous negative ε and μ, Veselago wondered whether such hypothetical material was just happenstance or perhaps had a more fundamental origin. He, however, believed that if such material becomes possible by any chance (artificially fabricated?) it would exhibit remarkable properties unlike those of any known natural material and would give a twist to or change the landscape of virtually all electromagnetic phenomena. Veselago always referred to such exotic material with simultaneous negative ε and μ as 'left-handed', because the wave propagation vector $\vec{k}$ is antiparallel to the usual RH cross product of the intensity of the electric field $\vec{E}$ and magnetic intensity $\vec{H}$.

Henceforth, such hypothetical material was known as left-handed material (LHM) in contrast to the natural or right-handed material (RHM)—because RHM forms a right-handed triplet with the *E*, *H*, and *k* vectors of the propagating electromagnetic wave while the LHM forms a left-handed triplet with the *E*, *H*, and *k* vectors (see figure 1.2). It may be noted from this figure that although the phase velocity (v_p) of electromagnetic signal is counter directed in RHM and LHM but the group velocity (v_g) that carries the energy of the electromagnetic signal, in other words the direction of Poynting vector, remains in the +ve *z*-direction—otherwise the causality will be violated.

It may be noted that the formation of RH triplet with *E*, *H*, and *k* vectors in natural material or RHM and LH triplet by LHM follows from Maxwell's equations. According to Maxwell's equations for a plane monochromatic wave: $\vec{k} \times \vec{E} = \omega\mu\vec{H}$ and $\vec{k} \times \vec{H} = -\omega\varepsilon\vec{E}$. Thus for RHM with ε and μ both positive: $\vec{E}$,$\vec{H}$ and $\vec{k}$ forms an RH triplet but in LHM since ε and μ both are simultaneously negative $\vec{E}$,$\vec{H}$ and $\vec{k}$ would form an LH triplet. But, the direction of energy flow, given by the Poynting vector: $\vec{P} = \vec{E} \times \vec{H}$, does not depend on the sign of permittivity (ε) and permeability (μ) of the medium and hence it remains in the same (+*z*) direction for both RHM and LHM. Detailed mathematical treatment of this issue is given in section 1.3.

The term 'metamaterial' is derived from the Greek word *meta* which means 'beyond' and the Latin word *materia*, meaning 'matter' or 'material'—that is, the characteristics exhibited by such materials are beyond what we see in the natural materials. The term 'mctamaterial' was coined by Roger Walser, for the so-called left-handed or the Veselago material mentioned above, at a 1999 DARPA (Defense Advanced Research Projects Agency) Workshop on composite materials, where the prefix 'meta' was chosen to convey that such composites transcend the properties of natural materials. Although definitions abound and continue to evolve, metamaterials may generally be described as (typically periodic) structures consisting of subwavelength arrangements of metallic and/or dielectric inclusions engineered to effect exotic or otherwise inaccessible macroscopic properties not found in the natural materials that comprise them. Metamaterials or LHMs are a novel functional material that are designed around unique micro- and nano-scale patterns or structures, mimicking the natural material, see figure 1.3, which cause them to interact with light and other forms of energy in ways not found in Nature.

These artificially engineered composite materials derive their properties from internal micro- and nanostructures, rather than the chemical composition of natural materials of which they are made. In other words, whereas the conventional or natural materials derive their electromagnetic characteristics from the properties of atoms and molecules, the metamaterial or these man-made artificial materials enable us to design our own atoms ('electric atom' and 'magnetic atom') and thus access new functionalities, such as 'invisibility cloak', 'imaging with unlimited resolution' and so forth.

The negative refractive index materials are better designated with 'metamaterial' rather than LHM to get rid of the confusion with chirality, i.e. chiral material, which is an entirely different phenomenon (as discussed in section 1.2.4). It may further be noted that the metamaterials are also known as double-negative (DNG) material as for such materials the effective permittivity and effective permeability are simultaneously negative in a given frequency band. They are also known as negative-index materials (NIMs) as they are capable of exhibiting negative refractive index, backward-wave (BW) media as the propagation vector (k) goes in the negative z-direction (backward compared to natural or RH material).

All known natural materials have a positive refractive index so that light that crosses from one medium to another gets bent on the positive side of the normal to the interface between two media in the direction of propagation. For example, air at standard conditions has the lowest index of refraction, i.e. refractive index, hovering just above 1. The refractive index of water is 1.33, for glass it is 1.52 and that for diamond is about 2.417. The higher a material's refractive index, the more is the

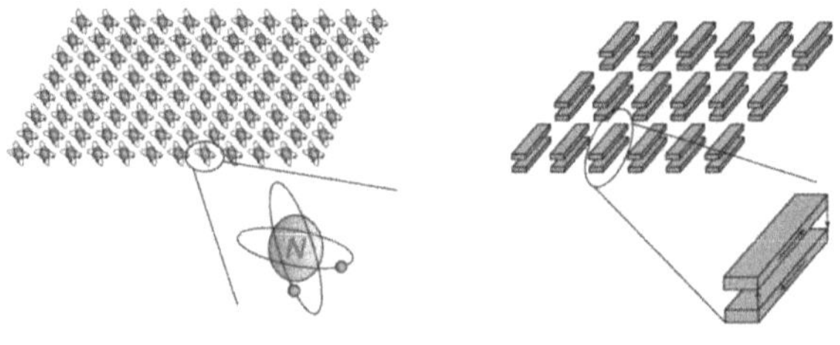

(a) (b)

Figure 1.3. A comparison of natural material with metamaterial: (a) natural material and (b) metamaterial.

angle of bending ($+\theta_r$). With metamaterial, one can tune the refractive index of the metamaterial to positive, near-zero or negative values; the negative refraction causes the incident light and other radiation to get bent in the 'wrong direction', i.e. on the negative side of the normal (with angle of refraction being $-\theta_r$). This counter-intuitive characteristic realizable with metamaterial might make peculiar/exotic things happen. When we toss a rock into a hypothetical metamaterial pond the ripples might be seen to flow inward, toward the point of impact; rather than flowing outward as we are familiar with in our day-to- day experience. Consider a fish in a tank of water. If water had a negative refractive index—which it doesn't—the fish would appear to be swimming upside down above the water, in the sky!

Ideally, with a view to applications, we would want unlimited power to control the refractive index. A computer-chip maker, for example, will be thrilled to have a lens of huge refractive index in their lithographic machine, because such a lens would allow chips to be made that are much smaller and perform better than those currently available. But Nature cannot always supply our ideals: naturally occurring materials only have a limited range of optical refractive index, typically between 1 and 3. However, metamaterial research has given us good news that we can have metamaterials with unnaturally high refractive index [4].

The ε–μ diagram, see figure 1.4, depicts the refraction and reflection (with no transmission) characteristics of different types of materials found in nature or artificially designed. The material with ε and μ both positive is the conventional material also called double positive (DPS) material that exhibits the usual Snell's law in which the incident ray is refracted in the positive θ direction of the normal to the interface as may be seen in the first quadrant of the ε–μ plot in figure 1.4. But if either of ε or μ becomes negative, then we have epsilon (ε) negative (ENG) or mu (μ) negative (MNG) materials (as in the second and fourth quadrant of ε–μ diagram). Under such a condition the refractive index becomes imaginary and the incident ray is reflected back to the incident medium (with no transmission via the medium). However, when both ε and μ are simultaneously negative for a medium as in a DNG material or the so-called metamaterial (which occurs in the third quadrant of ε–μ plot) the incident ray gets refracted in the negative θ-direction of the normal to the interface thereby reversing the Snell's law of refraction, see figure 1.4. It may be mentioned that the noble metals gold and silver exhibit ENG characteristics in the infrared and visible spectrum. The so-called gyrotropic or gyromagnetic materials (one whose material property has been altered by the presence of a quasi-static magnetic field, for e.g., ferrite material) exhibit MNG characteristics.

Although materials with simultaneous negative ε and μ are hard to find in nature, as mentioned earlier, materials having negative ε include metals like gold, silver and aluminum at optical frequencies. Also, it is long known that the ionosphere in the Earth's upper atmosphere (where atoms are ionized by the heat of Sun and behave as a plasma of electrons), which is used for short-wave radio transmission via sky-wave propagation, exhibits negative permittivity below its plasma frequency. Further, materials with negative μ include resonant ferromagnetic or antiferromagnetic systems. However, it was only in 1996 that John Brian Pendry of Imperial College, London demonstrated the realization of negative permittivity with an array of very

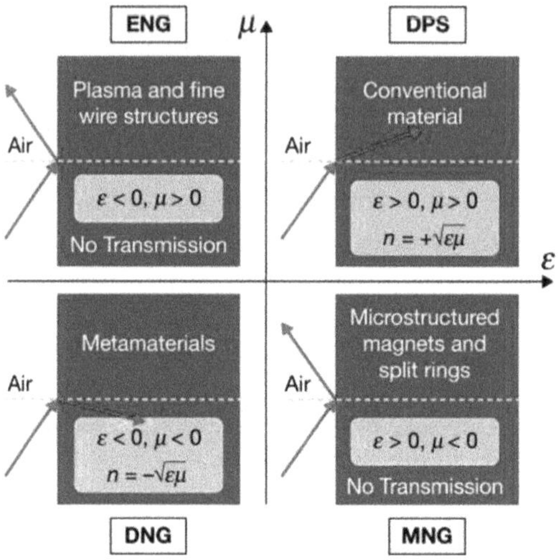

Figure 1.4. The ε–μ diagram depicting the refraction and no-transmission characteristics via different material media.

thin wire (TW) below its electric plasma frequency in the microwave spectrum [5]. Further, in 1999, it was again J B Pendry and his colleagues, who were able to realize negative permeability below the magnetic plasma frequency of split-ring resonator (SRR) at microwave frequency [6]. These two landmark achievements of Pendry led to the development of the first successful bi-dimensional plasmonic metamaterial exhibiting negative refractive index in 2001, as mentioned above. Pendry's seminal work of 1999 on artificial magnetism dramatically extended the palette of realizable electromagnetic properties by a new class of artificial material—the metamaterial—whose response to radiation could be custom-designed. Scientists have long known that they could change the behaviour of a material by altering its chemistry. For instance, one can alter the colour and hardness of glass by adding lead. But the iconoclastic work of J B Pendry could alter the functionality of the metamaterial by changing the artificially made material's internal structure on a very fine scale, less than a wavelength of whatever he was manipulating. His work was at microwave frequency but applicable at optical domain too—a wave of visible light, just about the size of a virus, has a length of a few hundred nanometers. It may be commented here that though the possibility of LHM was hypothesized by V G Vaselago in 1967, it was J B Pendry's work of 1996 and 1999 which made it practically possible to make a left-handed material or metamaterial in the beginning of this century.

In 2002, George V Eleftheriades *et al* [7], and Christophe Caloz *et al* [8] almost simultaneously proposed an alternative way to realize LHM property using transmission lines, known as *periodically loaded transmission line* (PLTL) metamaterial. They recognized that the *dual* of the normal transmission line is capable of exhibiting LH waves that resemble the BW already known (since the 1940s) to exist in periodic structures (like that in helix TWT). Negative refraction at microwave frequency with PLTL was reported by G V Eleftheriades and his group of the

University of Toronto, Canada [7]. Since transmission line is non-resonant, PLTL is capable of exhibiting simultaneously low-loss and broad bandwidth and are thus well suited for RF and microwave circuit applications. In contrast, SRR and its variants are resonant structures while TW exhibits high-pass characteristic but its variant cut-wire (CW) is again a resonant structure.

Metamaterials are popularly known to make things invisible by the so-called cloaking device (reminding us of the Harry Potter's invisibility cloak). Technically speaking, the LHM or metamaterial is artificially structured material (commonly made of metal–dielectric composite, though presently only dielectric based metamaterials are also under development). The structural properties, rather than the chemistry of the material like copper/gold and so on (with which it is designed), determine the characteristics of LH materials. Metamaterials are said to mimic the lattice structure of natural material but unlike natural material the lattice constant (p) of metamaterial is less than the operating wavelength (λ) of the irradiating electromagnetic signal (see figure 1.5); typically $p = 0.25\ \lambda$–0.1λ for metamaterials; thus they are also known as sub-wavelength structures.

Metamaterials have intrinsic inhomogeneity (from a microscopic point of view) but to an incident electromagnetic signal it is extrinsically homogeneous (from a macroscopic point of view). Metamaterials are thus said to exhibit bulk property, i.e., they possess effective negative permittivity ($-\varepsilon_{\text{reff}}$) and effective negative permeability ($-\mu_{\text{reff}}$) exhibiting negative effective refractive index given by: $n_{\text{eff}} = -\sqrt{\varepsilon_{\text{reff}}\mu_{\text{reff}}}$. Since the refractive index n appears as a square power in the

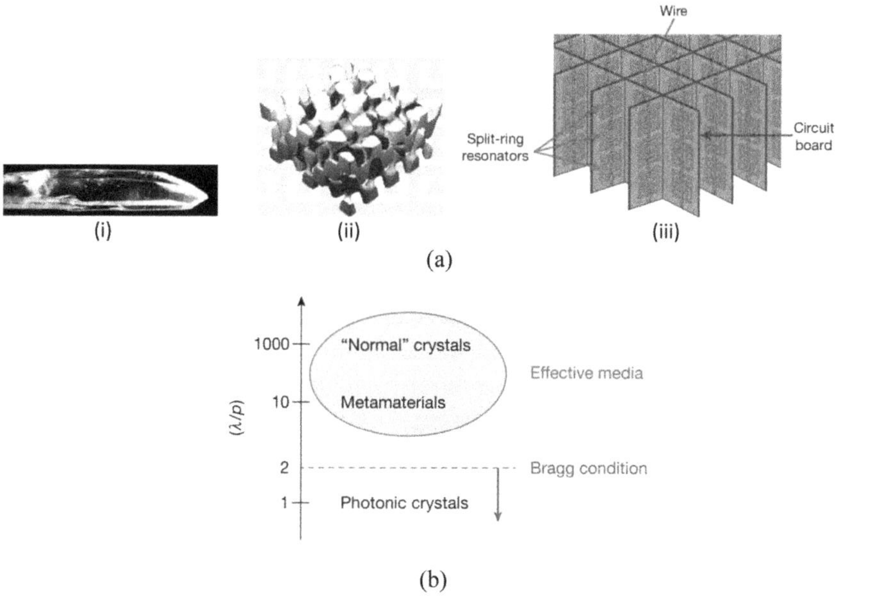

Figure 1.5. (a) Lattice constant (p) compared with the wavelength (λ) of the incident electromagnetic signal for (i) normal crystal ($\lambda \gg p$), (ii) photonic crystal ($\lambda \approx p$), and (iii) metamaterial ($\lambda > p$). (Reprinted/adapted with permission from [1] with permission from American Association for Advancement of Science.) (b) Ratio of wavelength (λ) to lattice constant (p) for different types of crystals/materials.

Maxwell's wave equation: $\nabla^2\Psi + n^2(\omega/c)^2\Psi = 0$ where $n^2 = \varepsilon_r\mu_r$, regardless of the sign of ε and μ, the fundamental Maxwell's equations, and their boundary conditions (with appropriate modification as discussed in section 1.4) for refraction of light and also for wave propagation can be satisfied for LHM too.

The primary research efforts on metamaterials, since the possibility of the realization of negative permittivity with TW [5] and negative permeability with SRR [6] and hence the experimental demonstration of negative refractive index with metamaterial [1], was directed towards various applications of NIM leading to plane-slab focussing and sub-wavelength imaging, cloaking and so forth. However, some theoretical studies and numerical simulation results and experiments regarding Čerenkov radiation in DNG material or metamaterial have also been reported in literature, which have been nicely reviewed in a paper by Duan *et al* [9]. Further, reports are available on metamaterial-based accelerator structures for accelerating, bending and focussing of electrons [10]. Some research reports are also available [11] in published literature on the reversed/inverse Doppler effect with metamaterial. However, practical applications of the phenomenon of reversed Čerenkov radiation and reversal of Doppler effect in metamaterial or DNG material is yet to emerge with reasonable success for practical applications.

1.2 Metamaterial versus other artificial materials

1.2.1 Introduction

In the last section, a brief discussion regarding the counter-intuitive nature of metamaterials has been given. Metamaterials fall in the class of 'artificial material' or man-made material (of course all are made with natural materials but using novel and innovative design approaches) those are capable of exhibiting tailorable properties in the electromagnetic spectrum, sometimes exotic ones too. The research work on artificial materials is understood to have begun from the last decade of the 19th century, with extensive research since the 1940s. A concise discussion regarding different artificial materials is warranted to understand the metamaterials in due perspective within the broad class of artificial materials.

The history of development of artificial materials can be traced back to the work of Jagadish Chandra Bose of Presidency college of the University of Calcutta who published his work on the rotation of the plane of polarization of newly found invisible waves (the so-called radio waves) by man-made twisted jute structures in 1898 [12], which by today's terminology, were artificial chiral structures. Karl Ferdinand Lindman in 1920 published his studies on artificial chiral media formed by a collection of randomly-oriented small wire helices [13]. In 1948 Winston Edward Kock manipulated the effective refractive index of a medium by periodically placing disks, spheres and strips [14]—all recalling artificial materials of some form or the other. Afterwards, there were several other investigators in the first half of the twentieth century who studied various man-made materials. Then in the 1950s and 1960s, artificial dielectrics were explored for lightweight microwave antenna lenses. This was followed by the development of another type of artificial material, known as photonic band-gap (PBG)/electromagnetic band-gap (EBG) materials, in the 1980s.

The interest in artificial chiral materials was also resurrected in the 1980s and 1990s and they were investigated for microwave radar absorber and other applications. The possibility of metamaterials or LHMs, a type of artificially structured material was, however, first proposed by V G Veselago of Lebedev Physics Institute, Academy of Sciences, USSR in 1967 [3] whose practical realization became possible in 2001 [1] with the seminal work of J B Pendry in 1996 [5] and 1999 [6].

1.2.2 Artificial dielectrics

Artificial dielectrics are man-made media that mimic the properties of naturally occurring dielectric media, or even manifest properties that do not generally occur in natural dielectrics. For example, the well-known dielectric property, the refractive index, which usually has a value equal to or greater than unity, can have a value less than unity in an artificial dielectric. This idea was first briefly introduced by the microwave community over half a century ago [14–16]. Artificial dielectrics are fabricated composite materials, often consisting of arrays of conductive strips or particles in a nonconductive support matrix, designed to have specific electromagnetic properties resembling those realizable with natural dielectrics and also beyond. The difference between the natural and artificial dielectric substance is that in the latter the atoms or molecules are artificially (human) constructed. The term artificial dielectric was originated by W E Kock in 1948 when he was employed by Bell Laboratories. The artificial dielectrics were borne out of a need for lightweight and low-loss materials for large and otherwise heavy devices.

In natural dielectrics, or natural materials, the local responses or scattering of incident electromagnetic signal occur on the atomic or molecular level and the macroscopic responses of the material are then described with electric permittivity (ε) and magnetic permeability (μ). However, this macroscopic response demands that the wavelength (λ) of the irradiating electromagnetic signal is larger than the lattice spacing (p). To design artificial dielectrics, that mimics the natural dielectrics, in 1940s, the artificial lattice structures that could be designed with scatterers in place (with the available technology of the time) were at RF and microwave frequencies realizing $p < \lambda$. The man-made/designed scatterers in such artificial material responded to the incident electromagnetic signal similar to atoms and molecules in natural materials, and the media behaved much like natural dielectrics with an effective media response with constitutive parameters ε and μ. The geometric shapes of the scattering elements used in such artificial materials were spheres, disks, conducting strips and so forth that also contributed to the design parameters. Today, with precise fabrication technology, especially with nanofabrication technology, artificial dielectrics are possible to design even at THz and higher frequencies.

Further, the artificial impedance surfaces or high-impedance surfaces (HIS) [15], which is a form of two-dimensional (2D) artificial material, find a lot of applications in the design of electromagnetic absorbers, impedance waveguides, and electromagnetic band-gap structures including planar reflect-arrays and leaky-wave antennas just to name a few. Compared to conventional absorbers, the use of artificial impedance surfaces in their design helps to reduce the electrical thickness of the structure.

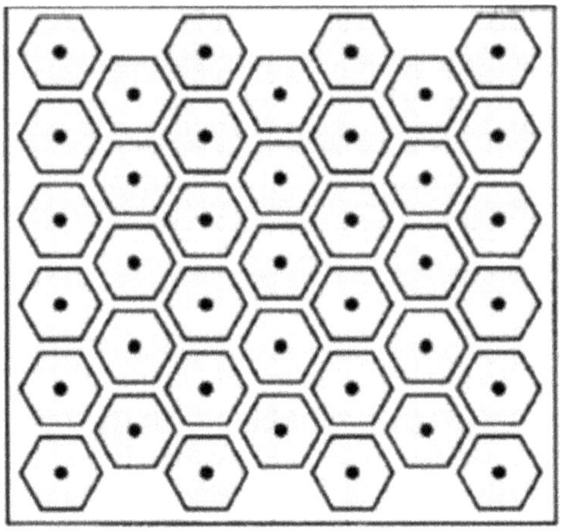

Figure 1.6. Top-view of high-impedance surface showing a lattice of hexagonal metal plates.

A smooth conducting sheet is known to have low surface impedance, but with a specially designed geometry, a textured surface, can have high surface impedance and this is the basic philosophy of the design of high-impedance surfaces/artificial impedance surfaces [17], see figure 1.6. More details about high impedance surfaces and their applications will be discussed in section 3.5.2.2 of chapter 3.

1.2.3 Photonic band-gap (PBG) materials

Photonic crystals [18, 19] are composed of periodic dielectric or metallo-dielectric structures, an engineered nanostructure, that is designed to form energy band structures for photons that either allow or forbid the propagation of electromagnetic waves in the optical range—the band of frequencies in which light propagation in the photonic crystal is forbidden is known as photonic band-gap (PBG). The PBG or more generally termed as EBG (electromagnetic band-gap) material affects the properties of photons in the same way that semiconductor materials affect the properties of electrons with their energy band structure. Thus it may be said that photonic crystals can manipulate light in the same way that a semiconductor like silicon can steer electrical current. Photonic crystals have been the subject of numerous investigations since the original work of Eli Yablonoviteh [20] and Sajeev John [21] in 1987. Because of their unique characteristics, the potential applications of photonic crystals are highly prospective. They are now used in data processing and in waveguides for laser surgery including those in inkless printing, reflective flat displays and also in the design of optical filters. They have also been discovered in bird feathers and the skin of chameleons.

It has already been mentioned that Professor Eli Yablonovitch of the University of California, Berkeley, proposed and created the crystals in the 1980s. Yablonovitch in 1987 visualized a photonic crystal by drawing the analogy of photons to electrons: a lattice of electromagnetic scatterers can tailor the properties of light in much the same way as crystalline solids do to electrons. In particular, when the lattice constants are on the order of the wavelength of light and the scattering strength of each scatterer is strong, the propagation of light waves inside such a lattice will be strongly modified by the photonic lattice structure. The basic problem is then to determine what the new photonic modes are inside such a lattice, usually specified by a position-dependent, periodic permittivity $\varepsilon(\vec{r})$.

Yablonovitch further admitted that the concept of photonic crystal was imbibed in the work of British physicist Lord Raleigh. As early as 1987, lord Raleigh suggested that a material with a repeating, regular structure—such as a crystal—could block light of particular wavelengths. This happens, Raleigh calculated, because if the light has a wavelength that is similar to the size of the repeated units in the structure, then the waves reflected off its internal surfaces will interfere and cancel each other out. That produces a 'stop-band' (later called a 'photonic band gap')—a range of light wavelengths that will be repelled by the crystal. Raleigh developed the idea of the 1D photonic crystal. Over an entire period of 100 years, no-one really thought of extending Raleigh's idea into two and three dimensions. That was where Eli Yablonovitch and Sajeev John stepped in, almost simultaneously—at a time when controlling light was a subject of great interest, for applications such as optical fibres and solar cells. But expanding the idea into more than one dimension, to block light in all directions, was not an easy task because the innovation demanded the right crystal structure to realize that, finally in 3D. A 1D photonic crystal is a simple stack of layers, but nobody knew till the mid-1980s what shape a 2D or 3D photonic crystal would be in reality. It took Yablonovitch's team four years, and a lot of failed experiments, to produce 'Yablonovite'—the first 3D photonic crystal. It consisted of a ceramic material, drilled with three intersecting series of cylindrical holes, 6 mm across [22], see figure 1.7. In such a structure, as Yablonovitch and his team understood, multiple scatterings work out such that no matter which way the light tries to go, it's blocked—in every imaginable direction.

Today, 2D photonic crystals are used in 'silicon photonics'—integrated circuits that use both light and current to transfer information. These are becoming common in large data centres. Other researchers have adapted the discovery to guide types of light that are useful in the surgery, including lasers. In the creation of Mother Nature we find the presence of the phenomenon of photonic crystals too. Butterfly wings

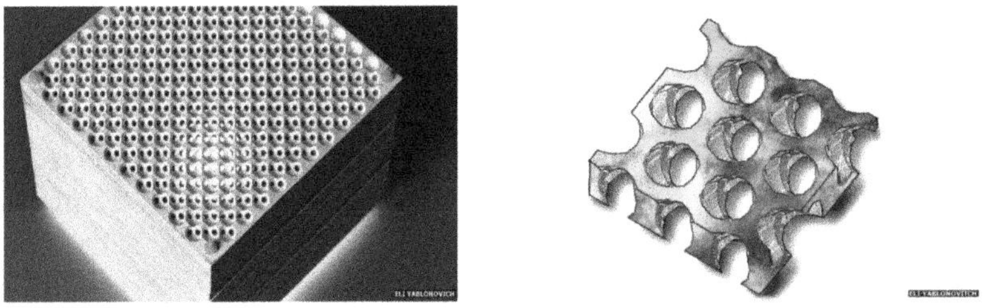

(a) (b)

Figure 1.7. (a) The photonic crystal designed by Eli Yablonovitch. (b) The engineered crystal consisted of an array of intersecting cylindrical holes. (Reprinted/adapted with permission from [22], copyright Eli Yablonovitch.)

and the colourful plumage of peacocks and some parrots all are due to the photonic crystal phenomena which were only understood after Yablonovitch and his fellow physicists fully described photonic crystals in the 1980s. Even the chameleon was recently shown to produce and control its colour using the shape of photonic crystals naturally present in their skin.

Photonic crystals have revolutionized the generation and propagation of light waves. Indeed, the existence of a photonic band gap in a range of frequencies prohibits light propagation in all possible directions. This has led not only to novel photon–atom bound states that alter the fundamental physics of light emission, but also to the concept of basic building blocks for optical materials that can be used to construct devices for practical applications. Deliberate defect structures inside the band gap add a new design dimension for versatile light control. Ultrahigh-Q cavities may be introduced as point defects in a perfect crystal to realize resonance channel add-drop filters and ultra-low threshold lasers. Channels for efficient light transmission are formed by line defects which can guide light through extremely sharp corners. Furthermore, by combining the index-guiding mechanism of slab waveguides with 2D photonic band-gap effects, control and manipulation of light in full three dimensions can be realized with present planar lithographic techniques. These developments echo with the past band-gap engineering in semiconductor electronics and suggest that photonic crystals may be used to tailor the properties of light in much the same way.

It would be worthwhile to discuss at the concluding part of this subsection the difference between metamaterial, natural dielectric, and PBG material or photonic crystal. The metamaterial is found to lie on the transition between ordinary structured dielectric and photonic crystals (see figure 1.5(b)).

Metamaterials are known to be effectively homogeneous, where $p \ll \lambda$ (p is the lattice period and λ is the wavelength of the operating signal), so that effective macroscopic ε and μ can be rigorously defined. They are thus said to behave like 'real' materials but with bulk constitutive properties. In fact, as long as $p < \lambda/4$, the difference between metamaterial and natural dielectric (such as glass and Teflon) is only *quantitative*. In the case of the natural dielectric material, the structural units inducing a given permittivity value are the constitutive molecules which are of the order of angstrom (10^{-10} m), while in the case of current metamaterial the structural units are of the order of centimeters in the microwave range; thus in the microwave L–X bands, the electrical size of the structural units in the former case is of the order of $p/\lambda \approx 10^{-9}$ while it is of the order of $p/\lambda \approx 10^{-1}$–$10^{-2}$ in the latter case. But qualitatively similar refractive phenomena occur in metamaterial as in conventional dielectrics. This is the point at which the fundamental originality of metamaterial lies: metamaterial represents artificial structures behaving in the same manner as conventional bulk dielectrics, hence the term 'material' in meta-materials, but having negative constitutive parameters. The subsequent and fundamental originality of metamaterials is that, due to their effective homogeneity, they can be extended to 2D and potentially 3D isotropic metamaterials.

Now, in a PBG structure the lattice period (p) is of the order of the operating wavelength (λ): i.e., $p \approx \lambda$ while for metamaterial, $p < \lambda$, may be 0.1λ. Therefore, interference effects cannot take place in metamaterial as the phase difference between unit cells is negligible and thus electromagnetic wave travels through the material probing only the average (i.e. effective) constitutive parameters. The effective negative refractive index in metamaterial is thus due to effective negative permittivity and effective negative permeability of the metamaterial structure. But PBG is diffraction/scattering limited and interference effects are dominant thus it operates in the Bragg regime resembling Bragg diffraction in x-ray. The negative refraction in PBG is not due to the constitutive properties in the photonic crystal but due to band-edge Bragg diffraction. Nevertheless, with photonic crystals many novel dispersion relationships can be realized, including ranges in which the frequency disperses negatively with wave vector as required for negative refraction. Using photonic crystals researchers have observed focussing, as predicted for NIM or metamaterial [23, 24].

However, a subtle difference of image formation with so-called superlensing by photonic crystal and metamaterial is worth mentioning at this point [25]. It is true that both are able to perform image formation with plane slab and both do not have an optical axis and thus need no strict alignment. So neither the superlens based on metamaterial nor that possible with photonic crystal, strictly speaking, is a lens in the conventional sense as they do not focus a parallel beam. Anyway, superlens of either type operates over distances on the order of the wavelengths of the electromagnetic signal and is an ideal candidate for small-scale integration. Further, there is essentially no physical limit on the aperture of this imaging system. Even with so many similarities between the two, the superlens with photonic crystal and that by metamaterial has an essential difference. In the metamaterial plane-slab lens the evanescent waves of the near-field exponentially grow within the metamaterial slab and are able to transfer the sub-wavelength details of the object to the image plane (see the discussion in section 1.5.2) but in photonic crystal lens this phenomenon is not present.

1.2.4 Artificial chiral materials

The term 'chiral' comes from the Greek word for the human hand—we all know that hands (left and right) are non-superimposeable. Due to the opposition of the fingers and thumbs, no matter how the two hands are oriented (except for the folded hands), it is impossible for both hands to exactly coincide. Chirality describes that something is different from its mirror image. As Lord Kelvin mentioned [26], chiral media and systems are ubiquitous. For example, our hands, screws, shells of snails and so on are clearly chiral. Additionally, at molecular scales, there are a variety of chiral molecules like amino acids and sugars. An achiral object is identical (superimposable) with its mirror image. Chiral objects have a 'handedness', for example, scissors, shoes and a corkscrew are the chiral objects we come across in our daily life. Achiral objects do not have a handedness, for example, a baseball bat (no writing or logos on it), a plain round ball, a pencil, a T-shirt and a nail.

Naturally available chiral materials like quartz, amino acids, sugar and so forth possess a small chirality parameter κ. But the emerging and varied applications at microwave/millimetre-wave and optical frequencies demand artificial chiral materials that have significantly high chirality parameter κ compared to any naturally available chiral material. In such artificial chiral material, right- and left-hand circularly polarized waves would propagate with different phase velocities and, in case the medium is lossy, absorption rates will also differ. The electromagnetic waves propagating through such chiral media would then show some interesting phenomena: (a) optical (electromagnetic) rotatory dispersion (ORD), causing a rotation of polarization; and (b) circular dichroism (CD), due to the different absorption coefficients of right- and left-handed circularly polarized waves—modifying the nature of field polarization and thus making linear polarization of a wave change into elliptical polarization. These properties have drawn considerable attention to chiral media and may open new potential applications in microwave and millimetre-wave technology: antennas and arrays, twist polarizers etc [27].

Traditionally, artificial chiral media at microwave frequencies can be fabricated by embedding conducting helices into a host medium. Nevertheless, chirality is a geometrical aspect; therefore helices are not the only possibility, so other type of inclusions, like metal cranks have been proposed also. The problems associated with the lack of homogeneity in chiral media based on random distribution of particles as helices or cranks can be reduced or even eliminated by designing periodical lattices. By an adequate distribution of metallic cranks it is possible to build chiral media with homogeneous, isotropic and reciprocal behaviour at microwave range [28], figure 1.8 shows an example of such medium.

When a metamaterial is made from chiral elements then it is considered to be a chiral metamaterial, and the chirality parameter κ will be non-zero. It may be mentioned herewith that there is a potential source of confusion in the metamaterial/chiral material literature as there are two conflicting uses of the terms left- and right-handed. Well, for chirality it refers to the two circularly polarized waves which are the propagating modes in chiral media. But for metamaterial it relates to the left-handed triplet of electric field, magnetic field and the Poynting vector and

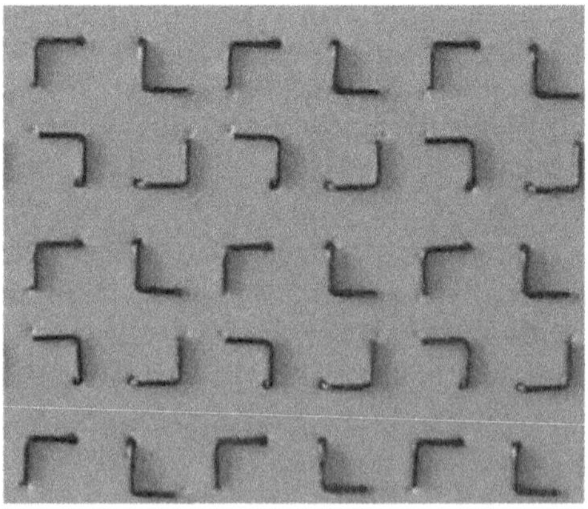

Figure 1.8. Typical chiral material with periodical lattice of cranks with segments introduced perpendicularly into the host medium, polyurethane foam.

characterized by negative refractive index arising out of negative permittivity and negative permeability of the medium.

Indeed, recently proposed designs of meta-atoms with metamaterials have enabled the realization of chiral metamaterials with unprecedented chiral optical properties e.g. strong optical activity, broadband optical activity, and nondispersive zero ellipticity. It has also been proposed as a way to achieve negative refraction [29] with chiral metamaterial which does not need negative permittivity and negative permeability of traditional metamterial to realize negative refractive index. Combining chiral metamaterials with nonlinear materials has opened up new possibilities in the field of nonlinear chirality as well as provided the foundation for switchable chiral devices.

Fabricating 3D chiral metamaterials is a challenging task, especially for optical wavelength applications, since chiral metamaterials are composed of more complex geometries than achiral metamaterials [30]. Many efforts have been made to fabricate chiral metamaterials using both top-down and bottom-up approaches. Indeed, many recent technologies like direct laser writing and block-copolymer self-assembly methods have brought progress in fabricating 3D chiral metamaterials in the visible frequency range. Nevertheless, it remains challenging to control optical activity and circular dichroism at visible wavelengths due to the limited chiral designs that are experimentally attainable and incomplete control of material properties such as losses and refractive index changes. It is believed that a combination of smart designs and material property control methods such as (quantum) gain and nonlinearities is expected to provide a new foundation for many practical applications of chirality in future.

1.3 Mathematical basis of the counter-intuitive properties of metamaterial

It has been mentioned in the introduction that metamaterial or LHM exhibits phase velocity (v_p) in the negative z-direction unlike the natural material (RHM), in which case v_p is in $+z$-direction; thus E, H, and k vectors form a left-handed triplet in the case of LHM, while a right-handed triplet in the case of RHM. However, the direction of group velocity (v_g) or the velocity of energy propagation (i.e., the direction of the Poynting vector $\vec{P}$) remains in the $+z$-direction in either case as demanded by causality. These facts of counter-intuitiveness of LHM with respect to RHM can be understood analytically as follows.

For plane TEM wave propagation in the z-direction in a source-free medium (see figure 1.9) we have electric and magnetic fields as:

$$\vec{E} = E_0 e^{j\left(\vec{k}_z\vec{r}-\omega t\right)}\hat{e} = |\vec{E}|\,\hat{e} \quad \text{and} \quad \vec{H} = H_0 e^{j\left(\vec{k}_z\vec{r}-\omega t\right)}\hat{h} = |\vec{H}|\,\hat{h} \tag{1.1a}$$

where $\hat{e}$ and $\hat{h}$ are unit vectors in the direction of the electric field and magnetic field while the unit vector in the direction of propagation is $\hat{k}$.

From constitutive relation: $\vec{B} = \mu\vec{H} = \mu\dfrac{\sqrt{\varepsilon}}{\sqrt{\mu}}\,|\vec{E}|(\hat{k}\times\hat{e}) = \sqrt{\mu\varepsilon}[\hat{k}\times|\vec{E}|\,\hat{e}]$,

since $\dfrac{|\vec{E}|}{|\vec{H}|} = \sqrt{\dfrac{\mu}{\varepsilon}}$.

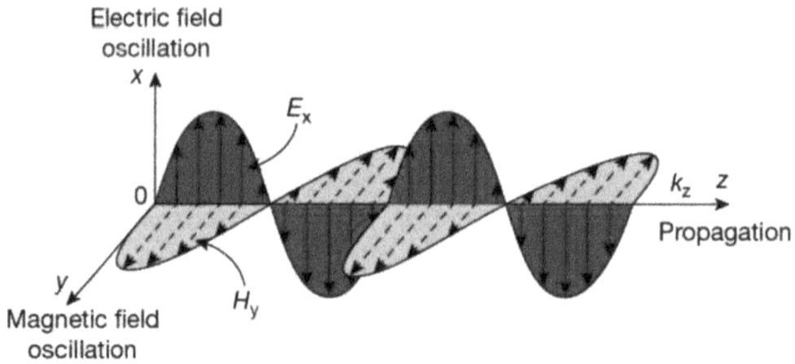

Figure 1.9. Transverse electromagnetic (TEM) wave propagating in the +*z*-direction in free space (full lines showing the *E*-field and dashed lines is the *H*-field).

Thus,

$$|\vec{k}|\vec{B} = \frac{1}{c}|\vec{k}|[\hat{k} \times \vec{E}] = \frac{|\vec{k}|}{\omega}[\vec{k} \times \vec{E}],$$

since $c = \frac{\omega}{|\vec{k}|}$ and $\vec{B} = \mu\vec{H}$

whence:

$$\vec{k} \times \vec{E} = \omega\mu\vec{H} \tag{1.1b}$$

Again from the constitutive relation:

$$\vec{D} = \varepsilon\vec{E} = \varepsilon\frac{\sqrt{\mu}}{\sqrt{\varepsilon}}|\vec{H}|\left[\hat{k} \times \left(-\hat{h}\right)\right] = -\sqrt{\mu\varepsilon}[\hat{k} \times |\vec{H}|\hat{h}]$$

thus:

$$|\vec{k}|\vec{D} = \frac{1}{c}|\vec{k}|[\hat{k} \times \vec{H}] = \frac{|\vec{k}|}{\omega}[\vec{k} \times \vec{H}]$$

whence:

$$\vec{k} \times \vec{H} = -\omega\varepsilon\vec{E} \tag{1.1c}$$

This represents the *familiar right-handed triplet* with $\vec{E}$, $\vec{H}$, $\vec{k}$, shown in figure 1.2(a). But in the case of LHM ε and μ are negative and since $|\varepsilon| = -\varepsilon > 0$ and $|\mu| = -\mu > 0$; we get from equations (1.1b) and (1.1c):

$$\vec{k} \times \vec{E} = -\omega|\mu|\vec{H} \text{ and } \vec{k} \times \vec{H} = +\omega|\varepsilon|\vec{E} \tag{1.1d}$$

Which forms the *unusual left-handed triplet* with $\vec{E}$, $\vec{H}$, $\vec{k}$ as shown in figure 1.2(b).

Thus the phase velocity in LHM can be written as:

$$v_p = \frac{\omega}{|\vec{k}|} = \frac{\omega}{\vec{k}}\hat{k}, \text{ with } \hat{k} = \vec{k}/|\vec{k}| \tag{1.1e}$$

As frequency is always positive while the wave number $\vec{k}$ in RHM is known to be in positive direction (outward propagating from the source), the phase velocity given by equation (1.1e) for LHM will be negative (inward propagation to the source, thus $\vec{k}$ is negative for LHM), giving rise to the so-called *backward wave*, and is thus opposite to that in RHM.

That phase velocity is negative ($v_p < 0$) in the case of LHM might appear to be disturbing at first glance. However, phase velocity relates to the propagation of the perturbations in the medium and not the energy. In contrast, a negative group velocity ($v_g < 0$) would violate causality as this is the velocity with which the electromagnetic energy moves; and being negative would mean transfer of energy towards the source (which is physically absurd!). This is justified because unlike the wave vector $\vec{k}$, the Poynting vector that gives the direction of energy flow is independent of the constitutive parameters of the medium (ε and μ) and depends only on $\vec{E}$ and $\vec{H}$.

1.4 Boundary conditions in metamaterial

For charge-free and source-free media the electromagnetic boundary conditions to be satisfied by the tangential and normal components of electric and magnetic field at the interface between RH–RH media are given by [31, 32]:

$$E_{1t} = E_{2t},\ H_{1t} = H_{2t};\ D_{1n} - D_{2n} = 0,\ B_{1n} = B_{2n} \tag{1.2a}$$

The indices t and n indicate the tangential and normal components, respectively. Looking at the relations of equation (1.2a) it may be guessed that the boundary conditions on the tangential component of E and H fields would be unaffected at the RH–LH interface, since the conditions on the tangential components do not depend on ε and μ. However, the normal components necessarily depend on constitutive parameters ε and μ (since $\vec{D} = \varepsilon\vec{E}$ and $B = \mu\vec{H}$) which are negative for LH medium. Thus the tangential components remain continuous in the RH–LH interface (as in equation 1.2(a)), but the normal components become antiparallel, as shown in figure 1.10, and are given by:

$$E_{\text{nRHM}} = -\frac{\varepsilon_{\text{LHM}}}{|\varepsilon_{\text{RHM}}|}E_{\text{nLHM}} \text{ and } H_{\text{nRHM}} = -\frac{\mu_{\text{LHM}}}{|\mu_{\text{RHM}}|}H_{\text{nLHM}} \tag{1.2b}$$

The consequence of antiparallel nature of the normal component of the wave vectors E and H at the interface between RH–LH media leads to negative refraction in

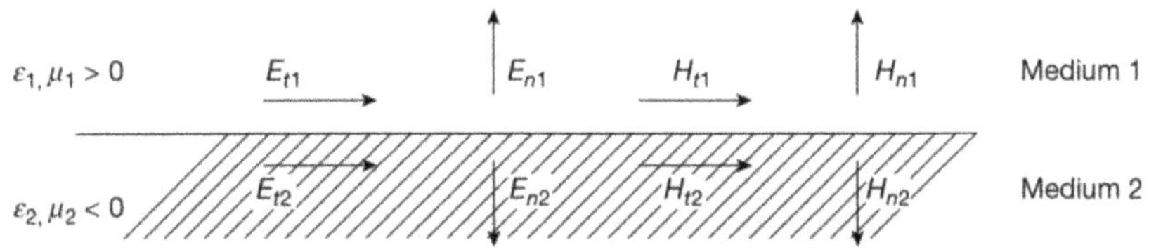

Figure 1.10. Boundary conditions at the interface between RH and LH media.

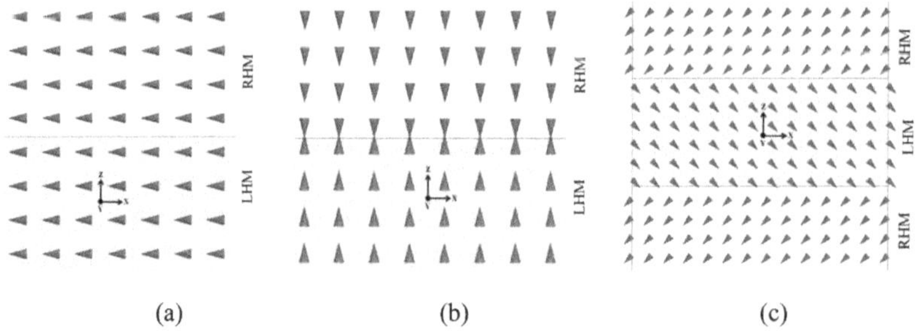

Figure 1.11. Simulation results in 3D electromagnetic field simulator, HFSS, for (a) tangential component of E field at the interface between RHM–LHM. (b) Normal component of E field at the interface between RHM–LHM. (c) Poynting vector in LHM–RHM–LHM interface (with 45° angle of incidence).

LHM. The near field focussing by a flat LH lens sandwiched between two RH media is again because of negative refraction taking place in RH–LH media. The detailed analysis for these and also that for reversal of Doppler effect, reversed Čerenkov radiation etc in LH media are available in suitable references [33].

It would be worthwhile at this step to model the field vector behaviour and power flow direction at LHM–RHM interface using 3D electromagnetic field simulator, Ansoft's High frequency structure simulator (HFSS) [34], see figure 1.11. This verifies the analytical results discussed so far in regard to the *counter-intuitive behaviour* of LHM yet not violating causality. It may be seen that, the tangential components of the electric field vector are in the same direction on each side of the boundary, whilst the normal components are in opposite directions (as demanded by the boundary conditions of RH–LH interface). But the time averaged Poynting vector, on the other hand, remains in the same direction in both the media (as demanded by causality and analytical result for the group velocity with which the electromagnetic energy flows).

1.5 Transmission properties of electromagnetic signal via metamaterial

To evaluate the properties of electromagnetic signal propagation via LH media we need to discuss a number of issues from different viewpoints.

1.5.1 Metamaterial plane-slab focussing

It is known from our knowledge of optics from school days that a plane slab of glass causes a light ray to get refracted through it and the emerging light ray suffers a lateral shift/displacement with respect to the incident light ray (see figure 1.12(a)). Also, it is known that a convex lens made of RHM, i.e., natural material, converges the incident parallel beam of light at the image plane while a concave lens acts as a diverging lens. But as a consequence of reversal of Snell's law by metamaterial a number of counter-intuitive phenomena take place with a metamaterial plane-slab and a metamaterial-based convex and concave lenses.

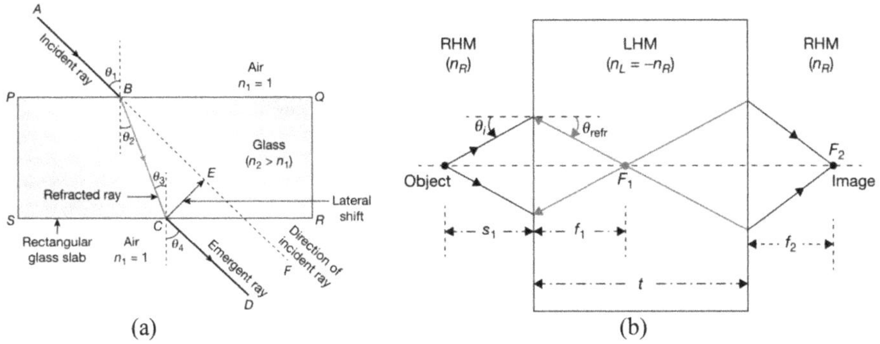

Figure 1.12. (a) Lateral displacement of light ray by an RHM plane slab. (b) Focussing by matamaterial (LHM) plane-slab.

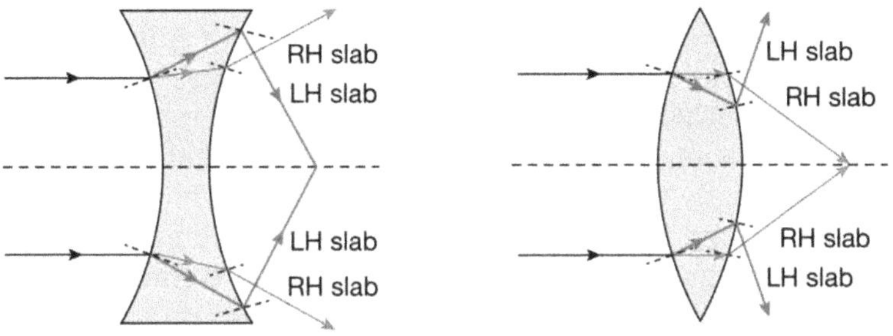

Figure 1.13. Convergence and divergence of incident rays by the concave and convex type negative index or metamaterial lenses.

Veselago [3], in 1967, noted that lens elements produced from NIM—if such could be ever found (not available in Nature and artificial ones were possible only after 1999 based on the works of J B Pendry)—would behave in a very different way from positive index lenses. For example, a concave negative index lens would act to focus an incident beam while a convex negative index lens would cause an incident beam to diverge rather than to converge, see figure 1.13.

In addition to changing the nature of convex and concave lenses, Veselago also noted that a *planar* slab with a refractive index equal to minus one could refocus the rays from a nearby source—something not possible with any positive index material. This property of a negative refractive index has proven to have significant consequences. A Fourier optic analysis of the planar negative index slab reveals that it can produce a focus with greater resolution than suggested by the 'diffraction limit' associated with all previously known passive optical elements. Because of this unusual property, the planar slab has been called a 'perfect lens' or 'superlens' although it has little in common with the lenses of conventional optics.

A question naturally arises how a negative index lens with $n = -1$ can focus an object to the image plane? This may be addressed as follows. For thin lenses, geometrical optics—valid for either positive or negative index—gives the result that the focal length is related to the radius of curvature of a lens by: $f = R/(n - 1)$. The denominator in the focal length formula implies an inherent distinction between positive and negative index lenses, based on the fact that an $n = +1$ material does not refract electromagnetic fields while $n = -1$ material does. The result is that negative index lens can be more compact with host of other benefits. Meta-lens made of NIM is found to be lighter than a positive index lens, a significant advantage for aerospace applications, and also possess much shorter focal distance, even though both lenses have the same radius of curvature.

Another point to note with meta-lens is the following. To make a conventional lens with the best possible resolution a wide aperture is sought. Each ray emanating from an object has wave vector component along the axis of the lens (z-axis), $k_z = k_0 \cos\theta$, and perpendicular to the axis (x-axis), $k_x = k_0 \sin\theta$. The former component is responsible for transporting the light from object to image and the latter represents a Fourier component of the image. For good resolution: the larger we can make k_x, the better. Naturally, the best that can be achieved is k_0 and hence the limit to resolution is: $\Delta \approx \pi/k_0 = \lambda/2$, where λ is the wavelength of radiation. This restriction of conventional optics is a huge problem in many areas of applications. Wavelength limits the feature size achieved in computer chips, and the storage capacity of DVDs. Even a modest relaxation of the wavelength limitation would be of great value. It will be seen subsequently that NIM can get over the problem of diffraction limit of conventional optics and hence the so-called wavelength limitation, and has the possibility to achieve great feats with its sub-wavelength resolution capability.

Anyway, let us now come back to our original discussion on metamaterial plane-slab focussing. The theoretical proposition of Veselago of the possibility of the reversal of Snell's law in a material having simultaneously negative permittivity and permeability led to the focussing by metamaterial plane-slab which was first established by Pendry [35] to be a remarkable phenomenon of metamaterial. Following this report of Pendry, a number of research works have been reported [36, 37] that showed that losses in the metamaterial significantly hampers the focussing by metamaterial plane-slab and hence the so-called metamaterial-based perfect lens; details of this loss issue of metamaterial on focussing will be discussed in a later section of this chapter.

A simple ray diagram shown in figure 1.12(b) indicates that the refracted wave within the slab makes a negative refracting angle (θ_{refr}) obeying Snell's law and instead of diverging the rays converge at a focal point F_1 within the metamaterial plane-slab. Again at the LHM–RHM interface during emergence from LHM slab the ray suffers negative refraction and meets at another focal point F_2 outside the slab [7]. Hence there exists a double focussing effect. In such counter-intuitive focussing by metamaterial plane-slab, depending on the angle of oblique incidence θ_i, a given f_1 (the internal focal length) will be set which will determine the external focal length f_2 and hence the final image point position at F_2. However, neither the superlens based on LHM nor that is possible with photonic crystal, as mentioned earlier, is a conventionally understood lens of optics as they do not focus a parallel beam. Anyway, such a superlens possesses several key advantages over conventional lenses which were discussed in section 1.2.3 (in the end part of PBG material).

The analytical expressions relating all the relevant parameters for such metamaterial plane-slab focussing (see figure 1.12(b)) can be derived assuming paraxial approximation as [7, 38]:

$$f_1 = s_1 \left| \frac{n_L}{n_R} \right| \frac{\cos \theta_{\text{refr}}}{\cos \theta_i}, \quad f_2 = s_1\left(\frac{t}{f_1} - 1\right), \quad t \geqslant -\frac{n_L}{n_R} s_1 \tag{1.3}$$

The Veselago–Pendry lens discussed above is true for *near-field focussing* with plane metamaterial slab under ideal conditions, i.e., $\varepsilon_{\text{RHM}} = \mu_{\text{RHM}} = 1$ and $\varepsilon_{\text{LHM}} = \mu_{\text{LHM}} = -1$, i.e., $n_{\text{R}} = +1$ and $n_{\text{L}} = -1$ and LHM has to be a loss-less medium. Such a plane-slab lens does not possess a focal length in the sense we understand in conventional optics. Further, it does not focus radiation from distant objects which we require in many practical applications like cameras, telescopes, antennas etc. Therefore, in these cases we would require negative index lenses with curved surface (see figure 1.13) that can focus far-field radiation in the same manner as traditional positive index lenses. Spherical profile lenses composed of negative index media can be more compact, they can be matched to free space, and they are found to have superior focussing performance.

1.5.2 Evanescent wave growth and sub-wavelength imaging with metamaterial plane-slab

When an electromagnetic wave emanates from an object, there exist both the propagating (or far-field) waves and evanescent (or near-field) waves; the latter contain the sub-wavelength (i.e. finer) details of the object. The term ‘evanescence’ has been derived from Latin which means ‘tending to vanish like vapour’. It is formed at the boundary between two media with different properties in respect of the wave motion, exhibiting an exponential decay of its amplitude; however, it is most intense within a quarter-wavelength or so of the irradiating signal from the surface of formation. Evanescent waves are found in the near-field region of a radio antenna too.

For an LHM/metamaterial plane-slab (see figure 1.12(b)), having $\varepsilon_{\text{LHM}} = -1$ and $\mu_{\text{LHM}} = -1$ such that: $n_{\text{LHM}} = -\sqrt{\varepsilon_{\text{LHM}}\mu_{\text{LHM}}}$, the impedance of the medium will equal that of the free-space or RHM. Thus propagating wave components will not suffer any reflection at the interface between RHM and LHM. However, a phase reversal of the electromagnetic wave occurs within the slab which enables it to refocus the beam by cancelling the phase shift that occurs as the light moves away from the source. This is similar to the Gouy phase shift that occurs in the case of focussing with conventional lenses made with RHM or natural material. Gouy shift is an additional phase shift occurring in the propagation of focussed Gaussian beams. Along the propagation, a Gaussian beam acquires a phase shift which differs from that for a plane wave with the same optical frequency. This difference is called the Gouy phase shift.

Sub-wavelength imaging, i.e. to obtain the finer details of the object (those of the order of fraction of the wavelength of the irradiating electromagnetic signal) at the image plane, is not possible in conventional optics with RHM due to the so-called ‘diffraction limit’. Physically, this means that the evanescent waves that contain the sub-wavelength details of the object get attenuated very fast (i.e. the amplitude decays exponentially as it propagates, but the spatial phase shift remains constant) and cannot reach the image plane to provide the sub-wavelength details of the object. However, in metamaterial i.e. in LHM, evanescent wave growth (i.e. replenishment of the amplitude decay) is understood to take place due to presence and coupling of the so-called surface plasmon polariton (SPP) waves in RHM–LHM interface making evanescent waves grow (a misnomer term used sometimes in literature is 'evanescent wave amplification'!) and hence sub-wavelength imaging is possible with LHM plane-slab overcoming the diffraction limit. This may be understood as follows.

It has already been mentioned that in LHM plane-slab focussing/sub-wavelength imaging a thin plane-slab of LH medium of refractive index −1 is embedded in an infinite RHM of refractive index +1, see figure 1.14, which has resemblance to figure 1.12(b) but redrawn here (with some more details) for explaining the phenomenon of evanescent wave growth. The intrinsic impedances of both the materials are equal to that of air so that there is no reflection due to impedance mismatch. This is the perfect lens of Vaselego–Pendry type. If the slab is in the near field of the source (object) resonant surface waves are excited along the interfaces that resemble the SPs observed on the surface of metals such as silver.

It was shown by Ruppin [39] that the RHM–LHM/LHM–RHM interface supports SPP waves excited by evanescent waves. To be more lucid, the situation inside the LHM plane-slab functions in the following way for the evanescent wave growth to take place. The electrons in metal undergo simple harmonic motion (SHM) caused by the time-varying E-field (tangential to the metal surface) of the incident evanescent component of the electromagnetic wave. The phenomenon occurs at the top and bottom surface of a practically finite sized LHM slab. The travelling wave of SHM-type from the top to the bottom of the LHM slab, however, is reflected from the side walls (left and right ones), i.e. the walls which allow

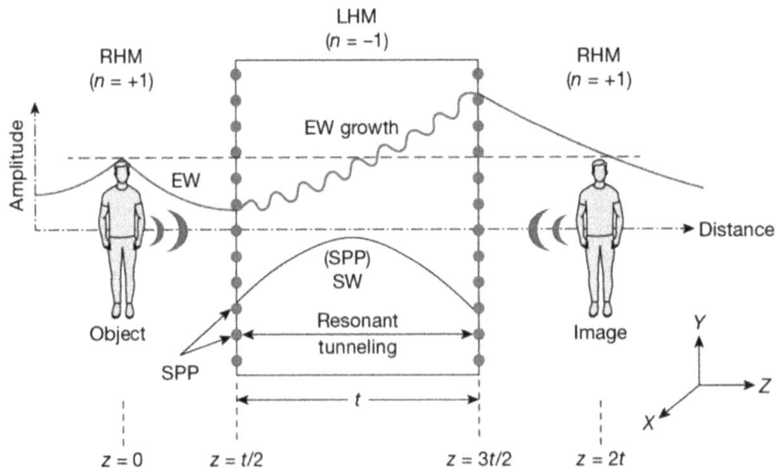

Figure 1.14. Presence and coupling of SPP waves in RHM–LHM interface of metamaterial causing evanescent wave (EW) growth.

propagating wave to flow under matched condition. Eventually, SPP standing wave (SW) is formed causing a cavity-like effect that mediates energy transfer from the evanescent waves at the object plane to the image plane. More specifically, a resonant-tunnelling effect [40] sets in, caused by coupling of SPP waves at the RHM–LHM interface (on the left of the LHM slab) and LHM–RHM interface (on the right of the LHM slab), see figure 1.14; those are separated by approximately a wavelength extent, which results in enhanced transmittance of the evanescent waves leading to sub-wavelength imaging of the object. This in essence is the concept of evanescent wave growth in LHM plane-slab resulting in the possibility of sub-wavelength imaging with LHM, i.e. metamaterial plane-slab. TLM (transmission-line modelling/transmission-line matrix method) full-wave solution for such a lens [33] indicates that it takes many periods of oscillations before the resonances build-up to a steady state, but, when it is reached, the near field of the source is reconstructed at the image location by the global interaction of the resonating 'plasmons'. It may be noted that the superlensing with plane metamaterial slab aided with evanescent wave growth caused by resonant tunnelling is a narrow band phenomenon: the validity of $\varepsilon = -1$ and $\mu = -1$ can be met only at one frequency because negative media are necessarily dispersive. Further, it may be stressed that evanescent wave growth (amplification!) does not imply a sustained input power (as in conventional amplifiers we understand in electronics engineering). Evanescent waves carry no power and hence in the absence of loss a large amplitude evanescent wave can be sustained indefinitely in a purely passive medium. It may be noted that Pendry analysed the system analytically in frequency domain [35], and Grbic and Eleftheriades [41] confirmed the super resolution capability of the lens by experiment (the details of which are given in section 5.3.3 in chapter 5).

It may be mentioned that, in the above discussion we have used two very important terms: 'diffraction limit' and 'SPP' and these two terms need a little bit of elaboration.

Classical electrodynamics imposes a fundamental resolution limit, called *diffraction limit* using conventional RHM lenses. More appropriately it is the *Abbe diffraction limit*, after the name of the scientist Ernst Abbe who in 1873 propounded that the spot size to which a light beam can be focussed by an optical microscope is of the order of the wavelength of the light signal used for the purpose. The issue of diffraction limit in terms of the propagating and evanescent wave coming out of a source and passing thorough a lens may be understood as follows [33, 35, 42]. Let us consider a luminous point source placed in front of a lens and radiation is emanating from the source at a frequency of ω. The electric field component of the radiation may be represented by 2D Fourier expression as:

$$\vec{E}(\vec{r}, t) = \sum_{\sigma, k_x, k_y} \vec{E}_\sigma(k_x, k_y) \times \exp(jk_z z + jk_x x + jk_y) \tag{1.4}$$

where we chose the axis of the lens to be the z-axis. Maxwell's equation tells us that

$$k_z = +\sqrt{\omega^2 c^{-2} - k_x^2 - k_y^2}, \quad \omega^2 c^{-2} > k_x^2 + k_y^2$$

The function of the lens is to apply a phase correction to each of these Fourier components, called the *propagating wave*, as a result of which at some distance beyond the lens the fields reassemble to a focus. And an image of the source appears.

But for the transverse component of the wave vector, k_z, known as the *evanescent wave*, has a different story about it! For these waves emanating from the source we have:

$$k_z = +j\sqrt{k_x^2 - k_y^2 - \omega^2 c^{-2}}, \quad \omega^2 c^{-2} < k_x^2 + k_y^2$$

These waves (the evanescent waves) are thus seen to decay exponentially with z and no phase correction can restore them to their proper amplitude. They are thus effectively removed from the image formation—the image is being formed only by the propagating waves.

A conventional lens is thus able to focus only the propagating waves, resulting in an imperfect image of the object, even if the lens diameter were infinite. The finer spatial details of the object (smaller than a wavelength), carried by the evanescent waves, are lost due to the strong attenuation that these waves experience when travelling from the object to the image through the lens. This loss of evanescent spectrum, which occurs even for an infinitely large lens, constitutes the origin of the so-called 'diffraction limit' in its ultimate form. The minimum resolvable feature Δx follows from the Fourier transform relationship: $k_{\max}\Delta x \sim 2\pi$, in other words $\Delta x \sim 2\pi/k_0 = \lambda_0$. The diffraction limit manifests as an image smeared over an area approximately one wavelength in diameter in conventional optics.

The surface plasmons (SPs) are waves that propagate along the surface of conductors, usually metals [43]. These are essentially light waves that are trapped on the surface of the conductor because of their interaction with the free electrons of the conductor. In this interaction, the free electrons respond collectively by oscillating in resonance with the light waves. The resonant interaction between the surface charge oscillation and the electromagnetic field of the light constitutes the SP, more specifically the SPP (surface palsmon polarioton). Polaritons are hybrid particles, e.g., in semiconductors an electron–hole pair forms an exciton polariton while the oscillating electrons at the surface of a metal create an SPP. SPs are of interest to a wide spectrum of scientists, ranging from physicists, chemists and material scientists to biologists. Once light has been converted into an SP mode on the flat metal surface it will propagate but will gradually attenuate (within an extent of quarter wavelength) owing to losses arising from absorption in the metal. SPPs on metal surface act to enhance the fields associated with the evanescent waves of the incident light waves, thus increasing the transmittance of the light signal. When the metal film is thin enough and has holes in it, a resonant tunneling effect sets in because the SP modes on the two surfaces of the metal can overlap and interact via the hole. The LHM plane-slab made of plasmonic metamaterial has metal inclusions (TW-SRR) in dielectric substrate and thus allows the setting-up of SP waves on the metal surface, see figure 1.14. The metallic inclusions on dielectric substrate can be thought of as simulating the metal with holes thus setting-up of the resonant tunneling effect becomes possible and hence increased transmittance of the evanescent waves, i.e. the growth of evanescent waves via the thin LHM plane-slab takes place.

SPs have been utilized almost exclusively at optical frequencies because they require negative permittivity which is readily available from metals like gold and silver at optical frequencies. In microwave range, such interface waves cannot be used in practical devices because their fields would be too loosely bound to the interface of available materials. However, with the advent of metamaterials (that exhibit tailorable constitutive parameters with negative permittivity/permeability) opened the way for novel SP solutions and structures at any frequency including that at microwave frequency too [44].

1.5.3 Mathematical analysis for characterization of transmission properties in LHM plane-slab

1.5.3.1 Introduction

The ray tracing of optics for refraction in RHM–RHM plane slab interfacing is well known to any school student, as shown in figure 1.15(a). But if we now consider that an LHM plane-slab is placed in the free-space (RHM medium) and an electromagnetic signal is incident obliquely on the LHM plane-slab from the left then the ray tracing will be as shown in figure 1.15(b).

In the following subsection we will derive at first the expressions for the Fresnel reflection and transmission coefficients (both for perpendicular and parallel polarized waves) from free-space to the LH medium (1–2 interface) and also from the LH medium to the free-space (2–3 interface), see figure 1.15(b). Finally the transmission characteristics of electromagnetic wave in the LH medium itself will be discussed.

1.5.3.2 Fresnel reflection and transmission coefficients from free-space to LH medium and from LH medium to free-space

Case I: For perpendicular polarization incident wave

Let a perpendicular polarization (E field vector being perpendicular to the plane of incidence, also called the s, H or TE_z) electromagnetic wave be incident on the x–y-

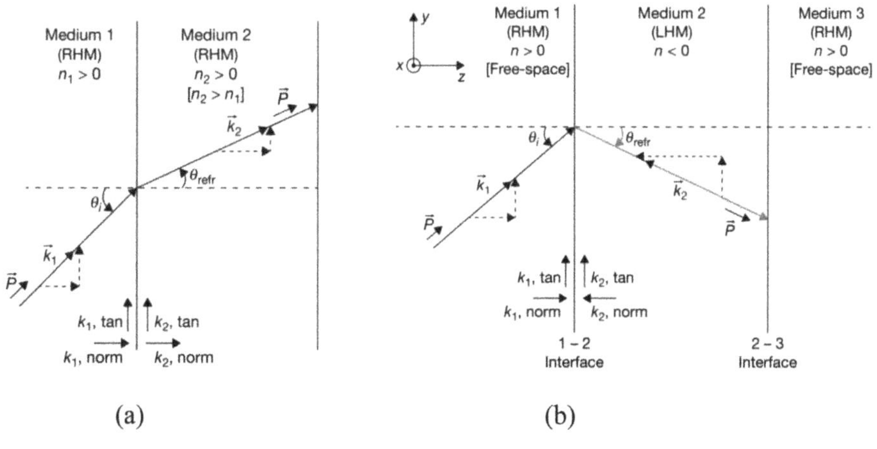

Figure 1.15. Ray tracing for (a) RHM–RHM interfacing and (b) RHM–LHM interfacing.

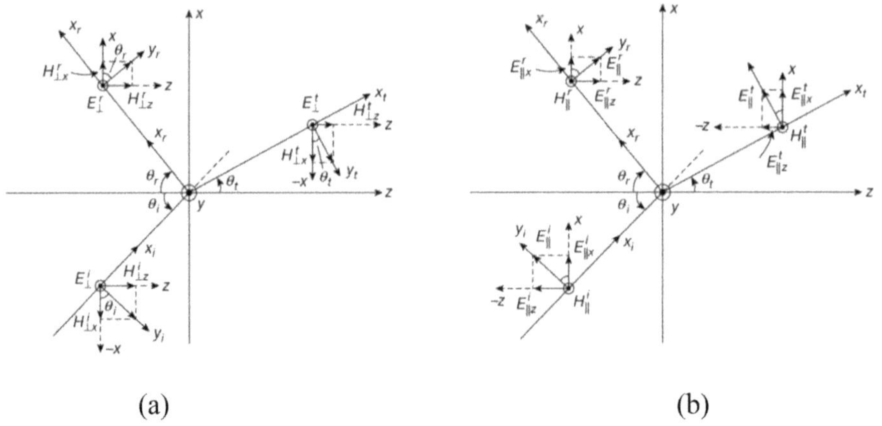

Figure 1.16. Oblique incidence of an electromagnetic wave on the interface between two media. (a) Perpendicular polarization (that is, the incident signal is an *s*-polarised wave). (b) Parallel polarization (that is, the incident signal is a *p*-polarised wave).

plane of the LHM plane-slab on the first interface 1–2 between the free-space and LHM slab, see figure 1.16(a). We can derive expressions for the Fresnel reflection and transmission coefficients for the reflection of signal in medium 1 and the transmission of signal in medium 2 as follows [31].

The incident and reflected electric and magnetic field intensity phasors in medium 1 may be written as:

Incident wave:

$$\vec{E}_{\perp}^{i}(x, z) = \vec{a}_y E_{\perp 0}^{i} e^{\gamma_1 x_i} = \vec{a}_y E_{\perp 0}^{i} e^{\gamma_1 (x \sin\theta_i + z \cos\theta_i)} \tag{1.5a}$$

$$\vec{H}_{\perp}^{i}(x, z) = \vec{a}_{y_i} \frac{E_{\perp 0}^{i}}{\eta_1} e^{\gamma_1 x_i} = \frac{E_{\perp 0}^{i}}{\eta_1}(-\vec{a}_x \cos\theta_i + \vec{a}_z \sin\theta_i) e^{\gamma_1 (x \sin\theta_i + z \cos\theta_i)} \tag{1.5b}$$

[Since $x_i = x \sin\theta_i + z \cos\theta_i$ and $\vec{a}_{y_i} = -\vec{a}_x \cos\theta_i + \vec{a}_z \sin\theta_i$].

Reflected wave:

$$\vec{E}_{\perp}^{r}(x, z) = \vec{a}_y E_{\perp 0}^{r} e^{\gamma_1 x_r} = \vec{a}_y E_{\perp 0}^{r} e^{\gamma_1 (x \sin\theta_r - z \cos\theta_r)} \tag{1.6a}$$

$$\vec{H}_{\perp}^{r}(x, z) = \vec{a}_{y_r} \frac{E_{\perp 0}^{r}}{\eta_1} e^{\gamma_1 x_r} = \frac{E_{\perp 0}^{r}}{\eta_1}(\vec{a}_x \cos\theta_r + \vec{a}_z \sin\theta_r) e^{\gamma_1 (x \sin\theta_r - z \cos\theta_r)} \tag{1.6b}$$

[Since $x_r = x \sin\theta_r - z \cos\theta_r$ and $\vec{a}_{y_r} = \vec{a}_x \cos\theta_r + \vec{a}_z \sin\theta_r$].

The transmitted electric and magnetic field intensity phasors in medium 2 are given by:

Transmitted wave:

$$\vec{E}_{\perp}^{t}(x, z) = \vec{a}_y E_{\perp 0}^{t} e^{\gamma_2 x_t} = \vec{a}_y E_{\perp 0}^{t} e^{\gamma_2 (x \sin\theta_t + z \cos\theta_t)} \tag{1.7a}$$

$$\vec{H}_{\perp}^{t}(x, z) = \vec{a}_{y_t} \frac{E_{\perp 0}^{t}}{\eta_2} e^{\gamma_2 x_t} = \frac{E_{\perp 0}^{t}}{\eta_2}(-\vec{a}_x \cos\theta_t + \vec{a}_z \sin\theta_t) e^{\gamma_2 (x \sin\theta_t + z \cos\theta_t)} \tag{1.7b}$$

[Since $x_t = x \sin\theta_t + z \cos\theta_t$ and $\vec{a}_{y_t} = -\vec{a}_x \cos\theta_t + \vec{a}_z \sin\theta_t$.]

The total electric field in medium 1 is the sum of the incident and reflected electric fields: $\vec{E}_{\perp}^{1} = \vec{E}_{\perp}^{i} + \vec{E}_{\perp}^{r}$ and the same is truc for the total magnetic field in medium 1: $\vec{H}_{\perp}^{1} = \vec{H}_{\perp}^{i} + \vec{H}_{\perp}^{r}$. Boundary conditions state that the tangential components of $\vec{E}$ and $\vec{H}$ each must be continuous across the boundary (x–y) plane between the two media. Since the electric fields in medium 1 and 2 have only the y-component, the boundary condition for $\vec{E}$ demands: $(\vec{E}_{\perp y}^{i} + \vec{E}_{\perp y}^{r})_{z=0} = (\vec{E}_{\perp y}^{t})_{z=0}$. Using equations (1.5a), (1.6a), and (1.7a) and setting $z = 0$ at the boundary, we have:

$$E_{\perp 0}^{i} e^{\gamma_1 x \sin\theta_i} + E_{\perp 0}^{r} e^{\gamma_1 x \sin\theta_r} = E_{\perp 0}^{t} e^{\gamma_2 x \sin\theta_t} \tag{1.8a}$$

Again, the magnetic fields in medium 1 and 2 have only x-component and hence the boundary conditions for $\vec{H}$ demands: $(\vec{H}_{\perp x}^{i} + \vec{H}_{\perp x}^{r})_{z=0} = (\vec{H}_{\perp x}^{t})_{z=0}$. On using equations (1.5b), (1.6b), and (1.7b) we have at the interface (1–2) boundary:

$$-\frac{E_{\perp 0}^{i}}{\eta_1}\cos\theta_i e^{\gamma_1 x \sin\theta_i} + \frac{E_{\perp 0}^{r}}{\eta_1}\cos\theta_r e^{\gamma_1 x \sin\theta_r} = -\frac{E_{\perp 0}^{t}}{\eta_2}\cos\theta_t e^{\gamma_2 x \sin\theta_t} \tag{1.8b}$$

The conditions represented by equations (1.8a) and (1.8b) must be satisfied for all values of x. the argument of all the three exponentials must be equal; leading to the phase matching condition:

$$\gamma_1 x \sin\theta_i = \gamma_1 x \sin\theta_r = \gamma_2 x \sin\theta_t \tag{1.9}$$

The first equality in equation (1.9) leads to Snell's law of reflection: ($\theta_r = \theta_i$), and the second equality leads to Snell's law of refraction: $(\sin\theta_t / \sin\theta_i) = (\gamma_1/\gamma_2)$.

Applying the equality given by equation (1.9) we get from equation (1.8) the following:

$$E_{\perp 0}^{i} + E_{\perp 0}^{r} = E_{\perp 0}^{t} \tag{1.10a}$$

$$(-E_{\perp 0}^{i} + E_{\perp 0}^{r})\frac{\cos\theta_i}{\eta_1} = -E_{\perp 0}^{t}\frac{\cos\theta_t}{\eta_2} \tag{1.10b}$$

These two equations can be solved simultaneously to yield the following expressions for the Fresnel reflection and transmission coefficients in the case of perpendicular polarization as:

$$\Gamma_{\perp} = \frac{E_{\perp 0}^{r}}{E_{\perp 0}^{i}} = \frac{\eta_2\cos\theta_i - \eta_1\cos\theta_t}{\eta_2\cos\theta_i + \eta_1\cos\theta_t} = \frac{\mu_{r2}k_{1z} - \mu_{r1}k_{2z}}{\mu_{r2}k_{1z} + \mu_{r1}k_{2z}} = r_{12} \tag{1.11a}$$

$$T_{\perp} = \frac{E_{\perp 0}^{t}}{E_{\perp 0}^{i}} = \frac{2\eta_2\cos\theta_i}{\eta_2\cos\theta_i + \eta_1\cos\theta_t} = \frac{2\mu_{r2}k_{1z}}{\mu_{r2}k_{1z} + \mu_{r1}k_{2z}} = t_{12} \tag{1.11b}$$

where μ_{r1} and μ_{r2} are the relative permeability of the medium 1 and medium 2 while k_{1z} and k_{2z} are the time phase of the travelling wave or the wavenumber in medium 1 and medium 2, respectively.

In a similar way, it can be derived that the Fresnel reflection and transmission coefficients for signal propagation from medium 2 (LHM) to medium 3 (free-space) are given by:

$$r_{23} = \frac{\mu_{r3}k_{2z} - \mu_{r2}k_{3z}}{\mu_{r2}k_{3z} + \mu_{r3}k_{2z}} = -r_{12} \tag{1.12a}$$

$$t_{23} = \frac{2\mu_{r3}k_{2z}}{\mu_{r2}k_{3z} + \mu_{r3}k_{2z}} \tag{1.12b}$$

It may be noted that in equations (1.11) and (1.12) $\mu_{r1} = \mu_{r3} = \mu_0$ is the permeability of the free-space while $k_{1z} = k_{3z} = k_0$ is the wavenumber in the free-space.

Case II: for parallel polarization incident wave

Similar to the above for perpendicular polarization, it may further be derived [31, 33] that the Fresnel reflection and transmission coefficients corresponding to the two boundaries on the left and right side of the LHM plane-slab for parallel polarization (E field vector being parallel to the plane of incidence, also called p, E or TM_z) wave, see figure 1.16(b), are given by:

$$\begin{aligned} \Gamma_{\|} &= \frac{E^r_{\|0}}{E^i_{\|0}} = \frac{\eta_2 \cos\theta_t - \eta_1 \cos\theta_i}{\eta_2 \cos\theta_t + \eta_1 \cos\theta_i} = \frac{\varepsilon_{r1}k_{2z} - \varepsilon_{r2}k_{1z}}{\varepsilon_{r1}k_{2z} + \varepsilon_{r2}k_{1z}} = r_{12} = -r_{23} \\ T_{\|} &= \frac{E^t_{\|0}}{E^i_{\|0}} = \frac{2\eta_2 \cos\theta_i}{\eta_2 \cos\theta_t + \eta_1 \cos\theta_i} = \frac{2(\varepsilon_{r1}\varepsilon_{r2}\mu_{r2}/\mu_{r1})k_{1z}}{\varepsilon_{r1}k_{2z} + \varepsilon_{r2}k_{1z}} = t_{12}, \end{aligned} \tag{1.13a}$$

while

$$t_{23} = \frac{2\varepsilon_{r3}k_{2z}}{\varepsilon_{r2}k_{3z} + \varepsilon_{r3}k_{2z}} \tag{1.13b}$$

Now, it may be noted that the medium 2 is LHM for which $k_{2z} = -|k_{2z}|$, $\varepsilon_{r2} = -|\varepsilon_{r2}|$, and $\mu_{r2} = -|\mu_{r2}|$. But the Fresnel coefficients calculated from above indicate that magnitude of these coefficients is the same whether it is RH–RH interface or RH–LH/LH–RH interface. This is because the Fresnel coefficients depend only on the tangential components of the fields which are not being affected whether the interface is of RH–RH or RH–LH (as was observed for boundary conditions discussed for RH–LH media in section 1.4).

1.5.3.3 Transmission characteristics of electromagnetic wave in the LHM plane-slab

There are various ways to evaluate the transmission characteristics of an electromagnetic wave via the LH medium. First, let us obtain the total transmission coefficient for the travelling electromagnetic wave in terms of multiple reflections taking place at the two interfaces (1–2 and 2–3) (for s-polarized incident wave), see figure 1.15, which can be written as [33, 35]:

$$
\begin{aligned}
T_{\text{TOTAL}(\perp)} &= t_{12}t_{23}e^{-jk_2d} + t_{12}t_{23}r_{12}r_{23}e^{-j3k_2d} + t_{12}t_{23}r_{12}^2r_{23}^2e^{-j5k_2d} + \ldots \\
&= \frac{t_{12}t_{23}e^{-jk_2d}}{1 - r_{12}r_{23}e^{-j2k_2d}} = |\, T(\omega)|\; e^{j\angle T(\omega)}
\end{aligned}
\tag{1.14}
$$

Here we have taken the distance between the faces (1–2) and (2–3) of the LHM slab, see figure 1.15, as d (instead of thickness t of the slab used earlier so that we do not confuse with the transmission coefficients t_{12}, t_{23} etc and the thickness of LHM slab).

The magnitude and phase expressions for the total transmission coefficient are thus:

$$
T_{\text{mag}} = \frac{t_{12}\; t_{23}}{[1 + (r_{12}r_{23})^2 - \{2r_{12}r_{23}\cos(2k_2d\}]^{1/2}}
\tag{1.15a}
$$

$$
T_{\text{pha}} = k_2\, d + \tan^{-1}\left\{\frac{r_{12}r_{23}\sin(2k_2d)}{1 - r_{12}r_{23}\cos(2k_2d)}\right\}
\tag{1.15b}
$$

In the above analysis we assumed that LHM is loss-less so that the imaginary parts of the constitutive properties are zero and hence that of the refractive index ($n_{2\text{reff}}(\omega) = \pm\sqrt{\varepsilon_{2\text{reff}}(\omega)\mu_{2\text{reff}}(\omega)}$). Also, for simplicity let us assume that the LHM is surrounded by vacuum (i.e. $n_1 = n_3 = 1$), which is always assumed to be loss-less for all practical purposes, and the incident signal is normal to the interface (i.e. $\theta_{\text{inc}} = 0$). Thus the wave numbers k_1 and k_2 are given by (ω/c) and $[(\omega/c)n_{2\text{reff}}\cos\theta_{\text{r}}]$, respectively.

The *evanescent wave growth* in an LH medium and possibility of *sub-wavelength imaging* is a revolutionary concept and an exotic property believed to be realizable with LHM/metamaterials plane-slab (see section 1.5.2). The evanescent wave is known to carry the sub-wavelength details of the object which normally decays in a natural (RH) material giving rise to 'diffraction limit' (resolving power being limited to the extent of a wavelength of the irradiating electromagnetic signal)—a well-known phenomenon in conventional optics. However, in a '*perfect lens*' or the so-called Veselago–Pendry type lens made of a plane, LHM slab is said to cause evanescent wave growth due to the presence of SPP at the RHM–LHM interface thereby making it possible to carry the sub-wavelength details of the object to the image plane with the possibility of '*sub-wavelength imaging*' with perfect lens; theoretically proposed by Pendry in 2000 [35]. Pendry's perfect lens demands $\varepsilon_{r1} = \mu_{r1} = -|\,\varepsilon_{r2}\,| = -|\,\mu_{r2}\,| = 1$ and hence $k_1 = k_2 = k$. For evanescent waves it is known that $k = -jq$ (say). Using the total transmission coefficient given by equation (1.14) and the Fresnel coefficients for s-polarized waves; we can now establish for an incident electromagnetic wave the conventional evanescent wave decay in RHM slab and the possibility of evanescent wave growth in LHM plane-slab.

For evanescent wave incidence on an RHM slab we can write equation (1.14) with the limit $\mu_{\text{r}1} \to +1$ and $\mu_{\text{r}2} \to +1$ (for perfect lens) as:

$$
T_{\perp}^{\text{evans(RH)}} = \frac{4k_1k_2}{(k_1 + k_2)^2}\,\frac{e^{-jk_2d}}{1 - \left(\dfrac{k_1 - k_2}{k_1 + k_2}\right)^2 e^{-j2k_2d}}
\tag{1.16a}
$$

And when we set $k_1 = k_2 = k = -jq$ (the evanescent wave approximation), we get:

$$T_{\perp}^{\text{evans(RH)}} = e^{-qd} \tag{1.16b}$$

Again for evanescent wave incidence on a LHM slab we can write equation (1.14) with the limit $\mu_{r1} \to -1$ and $\mu_{r2} \to -1$ as:

$$T_{\perp}^{\text{evans(LH)}} = \frac{-4k_1k_2}{(k_1 - k_2)^2} \frac{e^{-jk_2d}}{1 - \left(\frac{k_1 + k_2}{k_1 - k_2}\right)^2 e^{-j2k_2d}} \tag{1.17a}$$

On setting $k_1 = k_2 = k = -jq$ (the evanescent wave approximation) and noting that $\left(\frac{k_1 + k_2}{k_1 - k_2}\right)^2 e^{-jk_2d} >> 1$ we get with algebraic manipulation:

$$T_{\perp}^{\text{evans(LH)}} = e^{+qd} \tag{1.17b}$$

Referring to equation (1.16b) we observe that the evanescent wave will decay while transmitting through the RHM slab while equation (1.17b) indicates that the evanescent wave is enhanced or grows while passing through the LHM slab.

So far we have discussed electromagnetic signal propagation via LH medium considering the medium to be loss-less. However, practical LHM structures have finite losses which affect the propagation properties and thus a true 'perfect lens' realization is yet far from reality. The loss in the material has the effect of smearing the image being formed by the LHM plane-slab as evanescent wave growth is severely affected by the presence of loss in the LHM/metamaterial [37]. Extensive research is going on to mitigate these problems and to develop low-loss LHM in order to exploit the full potential of the so-called exotic properties of evanescent wave growth via LHM and better focussing with LHM plane-slab even at optical frequencies.

Another important property to characterize the transmission of propagating and evanescent waves via an LHM plane-slab is the thickness (d) of the LHM plane-slab [45]. The effects of the change of LHM plane-slab thickness on the propagating and evanescent waves are plotted in figures 1.17(a) and (b) for an s-polarized incident wave; in which operating frequency is 20 GHz and three slab thickness viz. $d = 10\lambda$, λ, and 0.1λ have been considered. It may be noted that for $d = 10\lambda$ and λ, the propagating wave is considerably attenuated near the two boundaries of the LH zone with RH zone. But with a sub-wavelength slab thickness ($d = 0.1\lambda$), the magnitude of propagating wave is less attenuated beyond 30 GHz, see figure 1.17(a). For evanescent waves, their growth through the LH slab is an important issue. It has been stated in much of the literature that a slab thickness of the order of λ is necessary for the growth of evanescent waves [46]. A further study [45] indicates that the growth of evanescent wave is directly related to the slab thickness (see figure 1.17(b)); the thinner the slab, the greater is the growth in terms of both magnitude and frequency band over which the evanescent wave growth process takes place in the LH zone. Thus it appears that a sub-wavelength thickness of LH slab will give better transmission properties both from the consideration of the propagating and evanescent waves. But for all practical purposes, the slab thickness

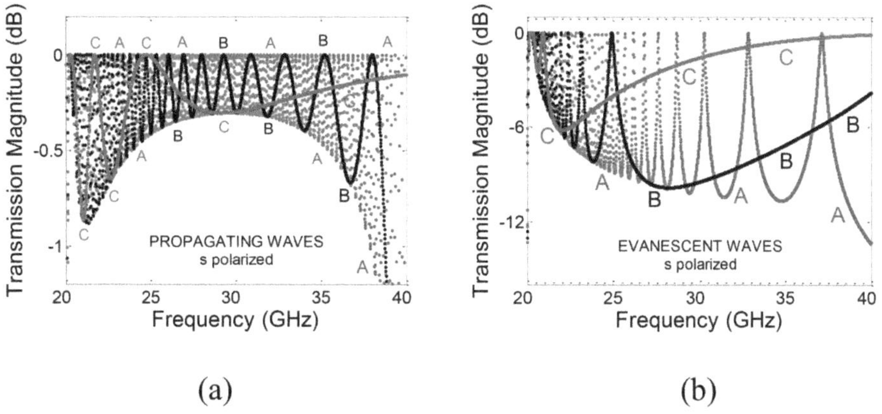

Figure 1.17. Magnitude of transmission function versus frequency with thickness (d) of an LHM plane-slab as the parameter. (a) For propagating waves. (b) For evanescent waves. In both cases: (A: d = 15 cm, B: d = 1.5 cm, C: d = 0.15 cm).

of the order of wavelength is preferable at higher frequency to accommodate with the tolerances of available fabrication technology for designing the LHM slab.

1.5.4 Further transmission characteristics of electromagnetic wave via metamaterial

An alternative approach to study the propagation or transmission characteristics of electromagnetic (EM) wave through an absorptive medium is the study of phase and group delay/velocity/index [45, 47]. The phase velocity v_p, group velocity v_g, phase delay τ_p, and group delay τ_g of a propagating EM signal through a medium are related by the following expressions:

$$\begin{aligned} v_p &= c_0/n = L/\tau_p \\ v_g &= c_0/[n + \omega(dn/d\omega)] = c_0/n_g = L/\tau_g \end{aligned} \tag{1.18}$$

where, n is the refractive index, also called the phase index of the medium and n_g is the group refractive index.

Phase delay τ_p is the measure of the time by which a particular phase of an EM wave propagating through an absorptive medium is retarded. A negative phase delay resulting from negative contribution of the phase index implies that the phase information is moving toward the source rather than moving away from it. Group delay τ_g measures the delay encountered by the envelope of the travelling EM wave. A negative group delay resulting from negative contribution of group index implies that the output peak response precedes the input peak response in a temporal axis. It may be mentioned that negative group delay and hence negative group velocity is a natural consequence of the Kramers–Krönig relations which themselves are based on linear and causal systems [47].

The plot in figure 1.18 shows the variation of group and phase delay of the EM wave with frequency. From this figure, we can identify two distinct frequency regions of operation corresponding to RHM and LHM. From 19.99 to 20 GHz, when τ_p is positive, we have the RHM region, while from 20 to 20.01 GHz is the

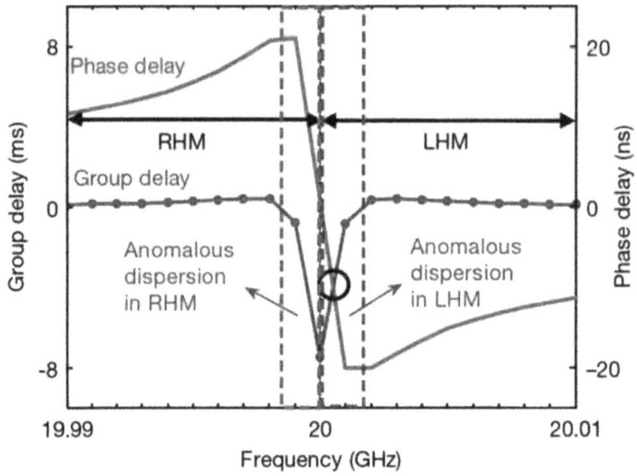

Figure 1.18. Group delay τ_g and phase delay τ_p versus frequency for electromagnetic signal propagating through LHM plane-slab (d = 15 mm), for perfectly overlapped $\varepsilon_{\text{reff}}$ and μ_{reff} profiles with both the resonant frequencies at 20 GHz.

LHM region when τ_p is negative. In RHM region itself, when τ_p and τ_g are both positive, we have the normal wave propagation in RHM. However, in the region when τ_p is positive but τ_g is negative, it is the anomalous wave propagation region for RHM. However, in LHM when τ_p is positive and τ_p is negative, we have BW propagation phenomena which characterize the LHM. But when τ_p is negative and simultaneously τ_g is also negative, we have anomalous wave propagation region for LHM. A further characteristics of this region is that here we observe a cross-over point where magnitude of τ_p is equal to magnitude of τ_g around $dn/dw = 0$ which is said to be a special characteristic exclusive to anomalous LHM region [48].

But for practical designs, skew-overlapped zones are to be studied [45]. From figure 1.19(a), it may be observed that RHM–LHM upon being buffered by electric

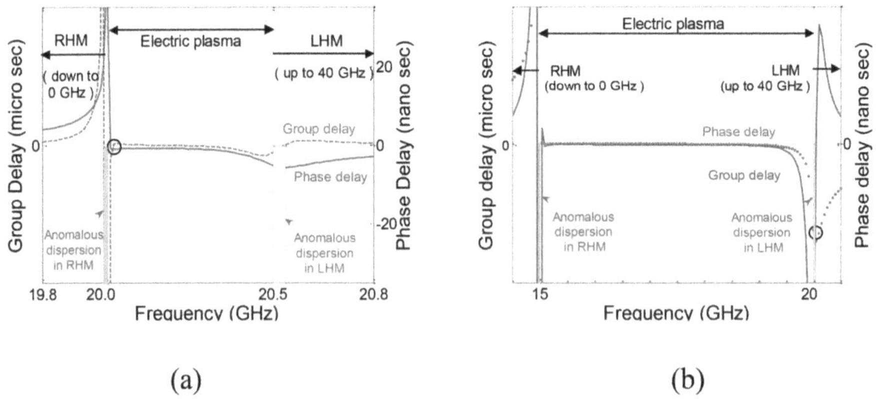

Figure 1.19. Group delay τ_g and phase delay τ_p versus frequency for electromagnetic signal propagating through LHM plane-slab (d = 15 mm) for skew-overlapped $\varepsilon_{\text{reff}}$ and μ_{reff} profiles with (a) f_{e0} = 20 GHz, f_{m0} = 20.5 GHz, $f_{ep} = f_{mp}$ = 40 GHz, (b) f_{e0} = 15 GHz, f_{m0} = 20 GHz, $f_{ep} = f_{mp}$ = 40 GHz; the circled point shows the cross-over.

plasma zone, the cross-over point in not seen to be within the LHM anomalous dispersion region, but it is in the electric plasma region. However, if the degree of skewness is altered, it may be observed from figure 1.19(b) that the cross-over point does fall in the anomalous dispersion zone of LHM.

Thus, it may be mentioned from this study that the cross-over point is not an exclusive characteristic of the LHM anomalous dispersion zone, and it may or may not be within the said region, depending on the degree of skewness of the overlapping zones of negative effective permittivity and permeability. This appears to be an important feature of skew-overlap LHM region.

1.6 Metamaterial can make things invisible—the metamaterial cloaking devices

1.6.1 Introduction

Among the many tropes found in science fiction and fantasy, few are more popular than the *cloaking device*. We are familiar with the invisibility cloak of Harry Potter (a character in J K Rowling's novel) or the 'Star Trek technology' that can make whole Romulan warships disappear. The invisibility cloak to the common person is a 'magic blanket' that has long fascinated moviegoers—from Indian classics like 'Maya Bazaar' to global bestsellers like 'Harry Potter'. But the science and technology of metamaterials is making such magic cloaks a reality. Since 2006, the development of a metamaterial-based cloaking device has been gaining pace with extreme enthusiasm. However, it must be noted that the science fiction movie type invisibility cloak is still a distant possibility, though not impossible.

Cloaking with metamaterial, that has generated so much media interest, is based on engineered anisotropic behaviour in metamaterials. The cloaking concept is to design a specific anisotropic behaviour into a metamaterial in such a way as to cause electromagnetic energy to propagate or bend around an object that is covered with it [49, 50]. In other words, the electromagnetic waves around an object are so deflected that the waves return to their original trajectories on the other side, as if they had passed through an empty space, which effectively renders the object invisible. If the anisotropic metamaterial is correctly designed, the cloaked object will neither scatter (reflect or transmit) nor absorb energy, and hence it appears to the electromagnetic fields as if the object is not present. One could in principle thus make an object invisible to the electromagnetic energy in any desired frequency band, from RF to optical. It may be said that the metamaterial cloaking recalls the principle of general relativity—mapping of free space into curved space—the anisotropic tailoring of ϵ and μ may be thought to take the place of mass in deflecting the light rays to avoid certain regions of space; in this case a region where the object might be hidden. However, there are still many technical hurdles to overcome to obtain the 'Holy Grail' of broadband cloaking with metamaterials.

The realization of cloaking of an object with a metasurface has also been experimented with. Here the scatterers composing a metasurface would be so chosen as to achieve a desired surface behaviour. For example, it is in principle possible to design a metasurface, so as to focus on electromagnetic plane wave to a desired region in space, much like a focussing antenna array. If the scatterers of the metasurface were designed in a manner such that they could be changed at will, one could as a result have a metasurface capable of changing the direction and frequency where the energy is focussed, i.e. a frequency- and space-agile surface. These concepts are currently being investigated and used to design metasurface-based cloaking devices too.

Acoustic metamaterials have also been researched to hide ships from SONAR (*SO*und *N*avigation *A*nd *R*anging), to render them sonically invisible. The sonar can spot ships by detecting echo of the transmitted sound wave scattered from the ship. To cloak ships from the sonar, we need to trick the sound waves to bend around the ship. This trick is not simple. In order to do that, we may take inspiration from the mirage formation by optical effects. Quite often we can notice a highway mirage on sunny days, where we see a displaced image of distant objects into the sky, while the real road with warm air on its surface seems immersed in water. This is because light travels with different speeds in the warm air and cool air; such a gradient in light speed causes light to bend away from the road surface. In acoustics, we can bend a sound wave in a similar fashion by changing the sound speed in different depths of the cloak. However, this is not a trivial job, as we have to make sure the sound waves from all angles are bent smoothly without scattering. When such acoustic cloaking is practically possible, not only can warships be cloaked, but high-resolution clinical ultrasound imaging will also be greatly benefitted.

1.6.2 Principle of cloaking with metamaterial and cloaking devices

We can see an object when light bouncing off it creates the reflections that enter our eye and forms an image on our retina. But if the light could be directed to flow smoothly around the object like water flowing past a smooth rock in a stream (with streamline motion of water and with no turbulence or wake whatsoever) there would be no reflection, no rays entering the eye, and nothing to see—not even a shadow. The principle of cloaking by metamaterials does just the same with judicious control, i.e., graded or anisotropic variation of negative refractive index around the object to be cloaked. With reference to figure 1.20(a), the object can be made invisible if there is no reflection from and also no transmission through or even no absorption of the incident electromagnetic signal in the object. The signal should just glide past the object. The trick is not a simple job as one has to make sure that waves from all angles are bending smoothly without scattering. In fact, when the electromagnetic signal is directed at the device, the wave would split, and it should be bent subtly around the device so that it is able to reform on the other side: resembling the river water flowing around a smooth rock, when no wakes are formed.

The first experimental cloaking device was developed in 2006 by D R Smith *et al* [50]. The physical concept of cloaking was, however, propounded by J B Pendry [49] of Imperial College, UK, and the design, fabrication and experimental acumen was that of D R Smith and his group of Duke University, US. The cloaking technique is based on transformation optics, in which a conformal coordinate transformation is

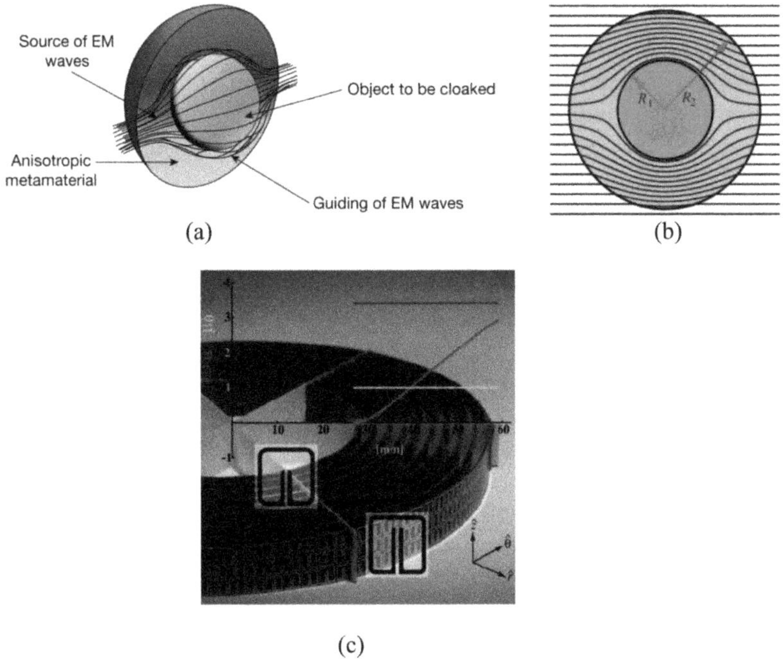

Figure 1.20. Principle of cloaking and cloaking device. (a) and (b) Schematics for explaining the principle of cloaking. (c) First microwave cloaking device (2006). (Reprinted/adapted from [51] copyright (2009), with permission from Elsevier.)

applied to Maxwell's equations to obtain a spatially distributed set of constitutive parameters that define the cloak. In other words, the technique relies on the transformation of coordinates, e.g., a point in the electromagnetic space is transformed into a sphere in the physical space, thus leading to the creation of a spherical volume where electromagnetic fields do not exist, but are instead guided around this volume, see figure 1.20(b). The paths of electromagnetic waves in the physical coordinate system can be controlled within a metamaterial with an appropriate spatial distribution of permittivity and permeability. For example, we consider the hidden object to be a sphere of radius R_1 and the cloaking region to be contained within the annulus region $R_1 < r < R_2$ for an incident plane wave, as in figure 1.20(b). The coordinate transformation gives permittivity $\vec{\varepsilon}$ and permeability $\vec{\mu}$ distributions in the region of $R_1 < r < R_2$ in the spherical coordinate system (r,θ,φ) as [49]:

$$\begin{aligned}\varepsilon_r' = \mu_r' &= \frac{R_2}{R_2 - R_1}\frac{(r' - R_1)^2}{r'^2}\\ \varepsilon_\theta' = \mu_\theta' &= \frac{R_2}{R_2 - R_1}\\ \varepsilon_\phi' = \mu_\phi' &= \frac{R_2}{R_2 - R_1}\end{aligned} \tag{1.19}$$

where $r' = R_1 + r(R_2 - R_1)/R_2$. Equation (1.19) indicates that both permittivity and permeability are identical and anisotropic, ensuring no reflection for arbitrary polarization. In practice, it is challenging to implement such extreme permittivity and permeability parameters.

The cloaking device that was designed by D R Smith's group was based on this theoretical background of Pendry at microwave frequency and was made up of 10 nested concentric fibre glass cylinders stamped with metallic (copper) split-ring resonator (SRR) of special design with a cylindrical gap in the middle where the object, a copper cylinder, (to be cloaked) was placed [51], see figure 1.20(c). By adjusting the length of each split and the curvature in the square corners of each unit cell of SRR the frequency of the electric and magnetic resonances could be shifted so that engineers were able to custom-design the value of the electric permittivity (ε) and magnetic permeability (μ) in each cell. The Duke team used metamaterial having gradually varying refractive indices with a value of 1 on the outside of the device and decreasing to zero in the centre. Their device could mask or make the object invisible from only one wavelength of the incident microwave signal corresponding to 8.5 GHz. The device was not perfect, causing shadowing of microwaves, i.e., distortions.

It may be noted that for 3 cm wavelength microwave signal the sub-wavelength SRR loops were to be about 3 mm. If they were to design the same at optical frequency range of 400 nm to 700 nm wavelength (see figure 4.7, the electromagnetic spectrum) then the SRR loop dimensions should have been 40 nm to 70 nm, a fantastically small dimension! However, such scaling methodology is not used in practical design for optical metamaterial—the reason for which has been discussed in chapter 4. Cai *et al* [52] presented a 2D optical metamaterial cloak consisting of metallic nanowires immersed in a cylindrical silica medium. The validity of cloaking was numerically confirmed at 632.8 nm, assuming a helium–neon laser.

The same group of Duke University, US, under the guidance of D R Smith, developed an improved cloaking device; see figure 1.21(a), again at microwave frequency in 2009 [53], however; this time it was broadband covering a frequency range of 13–16 GHz. In the new cloaking device, a beam of microwave aimed through the cloaking device at a 'bump' on a flat mirror surface bounced off the surface at the same angle as if the bump were not present. Additionally, the device prevented the formation of scattered beams that would normally be expected from such a perturbation. The cloak, which measures 20 inches by 4 inches and less than an inch high, is actually made from more than 10,000 individual pieces, non-resonant 'H'-shaped metallic elements, arranged in parallel rows. Each piece is made of the same fiber glass materials used in circuit boards and etched with copper. A specially developed algorithm (a series of complex mathematical commands) determines the shape and placement of each piece. Without the algorithm, proper designing and aligning of the pieces would have been extremely difficult and practically impossible. The breakthrough in this work is that it could handle a very wide range of wavelengths. However, the device was still a far cry from a wearable cloak—it could work only on a flat surface, and responds to microwaves, not light (but the realization of cloaking at optical frequencies is definitely a key step

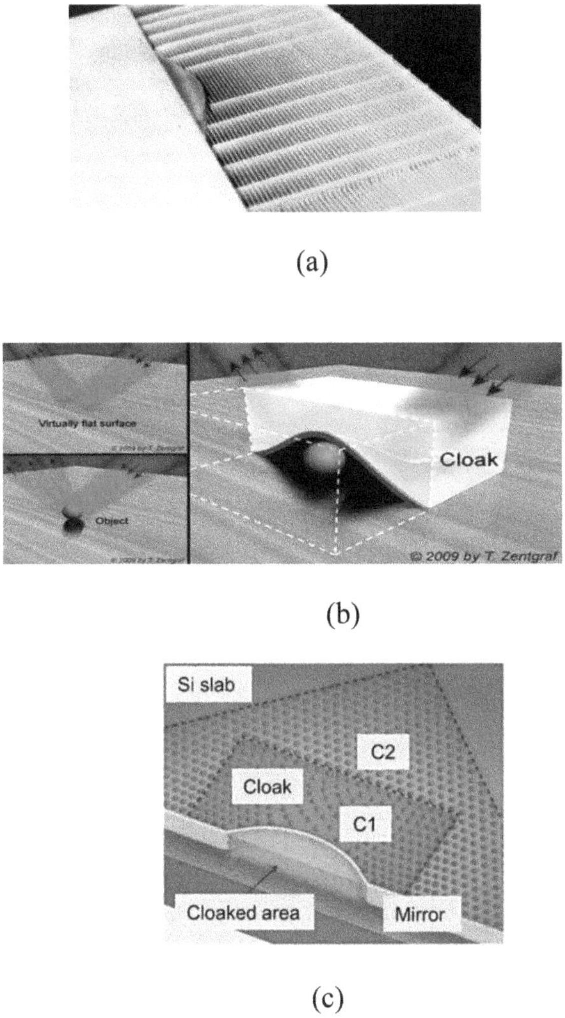

Figure 1.21. Carpet cloaking (a) Developed at microwave frequency by Duke University group (2009). (Reprinted/adapted with permission from [53] copyright The American Association for Advancement of Science.) (b) Images depicting the principle of carpet cloaking [Reprinted/adapted] with permission from [54]. (c) Developed at optical frequency by Berkeley Lab group (2009). (Reprinted/adapted from [55] with permission De Gruyter, copyright (2020) Jung *et al.*)

required towards achieving practical invisibility). Anyway, it is definitely a big step forward towards the development of the so-called *carpet cloaking*.

Carpet cloaking is capable of hiding microscopic objects. These may have potential use in optical computing; for example, such cloaks may be used to allow light to move more efficiently, by hiding the parts of a computer chip that get in the way of the beam. Also, expensive dielectric mirrors—special mirrors used to make printed circuits for electronics—can be ruined by tiny defects on their surfaces, which may be cloaked, making them look like perfect mirrors again. An object covered with a piece of cloth would normally be detectable based on its telltale bump, see figure 1.21(b), but using the metamaterial-based carpet cloaking device even the bump seems to vanish. In fact the carpet cloaking makes the reflection wave similar to that without the object by engineering the refractive index profile around the object. The carpet cloak relaxes the complexities of the requirements of permittivity and permeability to cloak an object, allowing experimental demonstrations in the microwave [53] and optical ranges [56, 57]. In one study [55, 56] an optical carpet cloak was implemented with a silicon-on-insulator wafer, wherein the silicon slab served as a 2D waveguide, see figure 1.21(c). The desired refractive index profile was realized by milling holes of varying densities in the silicon-on-insulator wafer. The cloaking phenomenon was observed in the range of 1400–1800 nm. In another study [57], the cloak was designed using quasi-conformal mapping and was fabricated in a silicon nitride waveguide on a specially developed nanoporous silicon oxide substrate with very low refractive index ($n < 1.25$). The spatial index variation was realized by etching holes of various sizes in the nitride layer at deep sub-wavelength scale creating a local effective medium index. The fabricated device demonstrated wideband invisibility throughout the visible spectrum with low loss.

Metasurface has also been used in the design of a very promising cloaking device at optical frequency. We have just now talked about carpet cloaking based on metamaterial, but such metamaterial-based optical cloaks use volumetric distribution of the material properties to gradually bend light and thereby obscure the cloaked region. Such a cloaking device is bulky and suffers from unnecessary phase shifts in the reflected light, making the cloaks detectable. However, metasurface-based skin cloaks [58, 59], see figure 1.22, operate at 730 nm wavelengths and are able to conceal a 3D arbitrarily shaped object by complete restoration of the phase of the reflected light. Compared to the metamaterial-based cloak, the skin cloak based on metasurface is miniaturized in size (only 80 nm, one-ninth of the wavelength) and scalable for hiding macroscopic objects, too.

Some useful review papers are available in the published literature [60, 61] which readers may refer to in order to get more details about the mathematical basis and applications of metamaterial and metasurface-based cloaking devices covering microwave-to-optical frequency range.

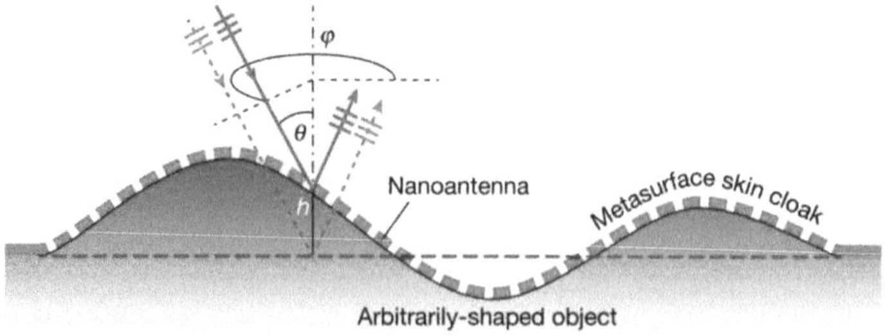

Figure 1.22. Schematic cross-section of metasurface skin cloak. (Reprinted/adapted with permission from [59] copyright (2021) Taylor and Francis Group LLC (Books) US.)

1.6.3 A comment on metamaterial/metasurface cloaking

The building of a practically useful invisibility cloak—the kind that could hide a person or military tank—requires crafting many little devices that pick up a ray of light on the far side of an object, away from the observer, and then relay that ray, row by row, around the object. When the ray arrives at the side facing the observer, it is re-emitted in the direction it would have taken had the object not been there at all. The issue is very weird—and appears to be total violation of the conventional laws of optics—but doable with metamaterial/metasurface and with the advanced technology of today, only patience and hard work with clear insight of the physics and engineering acumen is required.

Even if a 3D wideband cloaking device is successfully developed for practical use, there will be one problem which needs to be addressed. Anyone inside the invisibility cloak would not be able to see out, the person inside will be rendered apparently blind, for the same reason that an outside observer could not see in. '*If I cannot see you, you cannot see me—it would be like being inside a silvery bubble*,' as commented by J B Pendry. The would-be invisible person will have to figure out a way to cut out a visor, or perhaps de-cloak before accidentally walking into a wall!

1.7 Effect of losses in metamaterial performance

1.7.1 Introduction

Loss is a major hindrance to get desired performance from metamaterial (except for its use in absorber design where we desire to maximize loss in the material, as may be seen in section 4.2.3.1 in chapter 4). All the counter-intuitive phenomena like negative refractive index, sub-wavelength imaging with metamaterial plane-slab or the so-called 'superlens' are seriously affected by losses in the metamaterial. Even an enthusiastic report [62] to reverse the sequence of the well-known VIBGYOR with specially designed metamaterial prism, cloaking (that uses negative refraction by metamaterial in a subtle way) will be jeopardised if losses in the designed metamaterial cannot be reduced to a significant extent. In this section we discuss how losses in metamaterial affect the evanescent wave growth and subsequent focussing by LHM plane slab including the techniques to circumvent the problem of losses in metamaterials.

1.7.2 Effect of loss on evanescent wave growth and the focussing phenomenon by LHM plane-slab

Optical lenses have been used by scientists for centuries and their operation is well understood on the basis of classical optics: curved surfaces focus light by virtue of the refractive index contrast. Also their limitations are understood in terms of wave optics: no lens can focus light onto a circular area having diameter more than a wavelength—the phenomenon being known as the 'diffraction limit'.

Pendry [35] proposed the possibility of realising 'perfect lens' or 'superlens' with LHM, i.e. negative refractive index plane-slab within which the so-called 'evanescent wave growth' is understood to take place. But his analytical proposition was based on the assumption of purely ideal condition of loss-less metameterial medium having $\varepsilon = -1$, $\mu = -1$, that is $n = -1 + j0$, and the metamaterial medium is considered to be perfectly matched with the free-space impedance. However, Garcia *et al* [36] subsequently pointed out that no practical LHM will be perfectly loss-less and hence presence of loss will affect the growth of evanescent wave via the LHM plane-slab. As metamaterial or LHM is dispersive in nature, we need to include the effect of loss of the metamaterial medium when realizing 'evanescent wave growth' and hence the focussing phenomena by an LHM plane-slab.

Let us now discuss the issue of evanescent wave growth and focussing by metamaterial plane slab both for the loss-less and lossy cases based on the theoretical works reported in literature. Pendry's evanescent wave growth (for ideal loss-less case) takes place in an LHM plane-slab due to an exponentially growing factor: e^{+qd} (see equation 1.16(b)) for transmission of electromagnetic signal via LH media. But later reports [36, 37, 63] indicate that loss in LH media significantly affects this growth of evanescent wave within LH media and might severely jeopardize the focussing by LHM plane-slab. Garcia *et al* [36] showed with detailed mathematical analysis that the loss affects the growth of evanescent wave in LH media and another report by Ghosh and Kar [37] showed that in the extreme case the loss in the LH media might convert the evanescent wave 'growth' even into a 'decay' of the evanescent field in the LH media. Further, a transfer function analysis of an LHM plane-slab [63] by Shen *et al* has indicated that the focussing by metamaterial plane-slab becomes more and more smeared with increasing loss. The issue of loss in metamaterial plane-slab focussing has been reported in detail with computer-aided characterization by Ghosh and Kar [37] based on the analytical model equations provided by Garcia *et al* [36] and Shen *et al* [63] which will be discussed in the following.

Let us consider a plane slab of LHM, surrounded by RHM on both its left and right side, limited by the planes at $z = 0$ and $z = t$. We assume that the electric field of an evanescent component of an s-polarized electromagnetic wave is incident from the RHM to the LHM interface at $z = 0$. figure 1.23 [37] shows the electric field components of evanescent waves that exist at different regions [36] under unmatched

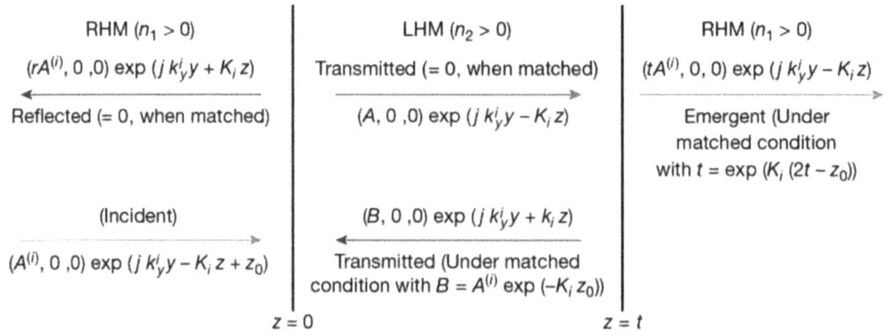

Figure 1.23. The evanescent wave propagation vectors in RHM and LHM regions for an RHM–LHM–RHM interfacing media.

and matched conditions (in fact, in practical situation there will be losses in the LH material and also the matching condition may not prevail in all situations).

As shown in figure 1.14, the evanescent wave is known to decrease in the RHM (on the left of LHM) with distance z; but with evanescent wave growth it is understood to get its decrease compensated with higher growth rate (as per Pendry's analysis [35] on the basis of ideal, perfectly matched situation). But for a practical lossy LHM the evanescent wave growth process in LHM will be affected with the increase of loss in the LH media, which will be seen shortly [36, 37]. For lossy LHM we may consider the refractive index of the LHM at some frequency ω as: $n = (-1 + jn_2)$, with $0 < n_2 \ll 1$. Here we have made use of: $n = -\sqrt{(-\varepsilon_r + j\varepsilon_i)(-\mu)}$. When $\mu = 1$, and $\varepsilon_r = 1$, one has: $n = \sqrt{1 - j\varepsilon_i} \approx -(1 - j\varepsilon_i/2)$, so that $n_2 = \varepsilon_i/2$. Then in LHM: $K = \sqrt{k_y^2 - k_0^2}$ becomes $K(1 - jn_2)$. Using all these conditions mentioned above and the necessary model equations [36] the plot of the normalized electric field amplitude with distance in the propagation direction is obtained with computer-aided characterization [37], as shown in figure 1.24.

The transfer function characterization of the focussing by LHM plane-slab, done with computer-aided technique [37], is based on the theoretical analysis reported by Shen *et al* [63]. Here the variation of the real part of the transfer function (i.e. the ratio of the electric field amplitude at the image plane to that at the object plane) with propagation factor; using energy dissipation factor δ (i.e. loss) as a parameter has been studied at first. This is followed by the characterization of the image-smearing with increasing value of δ (delta) in terms of the impulse function excitation of the LHM plane-slab with losses.

The mathematical model equations used for computer-aided characterization [37] has been summarized below [63].

It is assumed that the loss in LHM plane-slab, see figure 1.12(b), comes from the imaginary part of the complex permittivity of the metamaterial: $\varepsilon_L = -[\varepsilon_R(1 - j\delta)]$ with $\mu_L = -\mu_R$; and $0 < \delta \ll 1$.

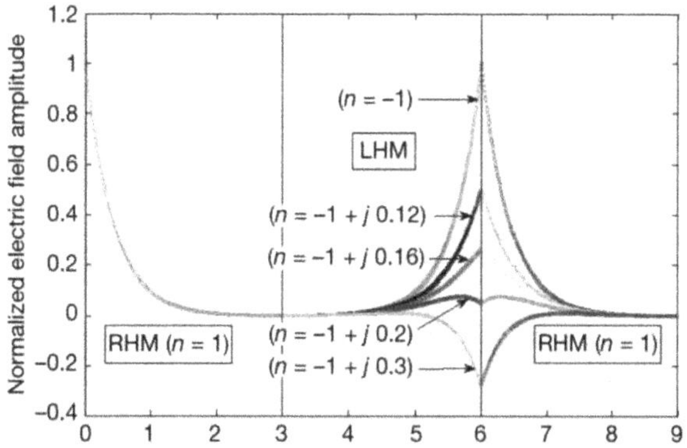

Figure 1.24. Plot showing the effect of loss on the growth of evanescent wave in the metamaterial (LHM) plane-slab. Increasing loss dampens the growth of evanescent wave via LHM.

The transfer function for LHM slab is given by [63]:

$$H(k_x) = T_1 P T_2 P' \tag{1.20}$$

where:

$$T_1 = 2/(1 - \zeta) + (1 + \zeta)P^2 T_2,$$

$$T_2 = 2/\left[1 - \left(\mu_L k_z / \mu_R k_z'\right)\right],$$

$$\zeta = 1/2\left[\left(\mu_R k_z' / \mu_L k_z\right) + \left(\mu_L k_z / \mu_R k_z'\right)\right],$$

$P = e^{-jk_z' t}$ and $P' = e^{jk_z(s_1 + f_2)}$

$$\begin{aligned} k_z &= [n_R^2 k_0^2 - k_x^2]^{1/2} \text{ (for propagating waves in RHM region)} \\ k_z &= j[k_x^2 - n_R^2 k_0^2]^{1/2} \text{ (for evanescent waves in RHM region)} \\ k_z' &= [n_L^2 k_0^2 - k_x^2]^{1/2} \text{ (for propagating waves in LHM region)} \\ k_z' &= j[k_x^2 - n_L^2 k_0^2]^{1/2} \text{ (for evanescent waves in LHM region)} \end{aligned} \tag{1.21}$$

In the above, T_1 is the effective transmission coefficient for plane wave at the RHM–LHM interface ($z < 0$ and $z > t$) and T_2 is the transmission coefficient inside the LHM slab. ς is related to the impedance matching of the LHM with the surrounding medium, i.e. $\varsigma = -1$, when the impedance is matched. P is the phase change factor or amplitude amplification factor as the wave passes the LHM slab and P' is the total phase change factor or amplitude decaying factor as the wave goes from the object plane to the left surface of the LHM slab and from the right surface of the LHM slab to the image plane.

The plot of transfer function of the metamaterial plane-slab with normalized propagation constant taking energy dissipation factor δ (Delta) as parameter is shown in figure 1.25.

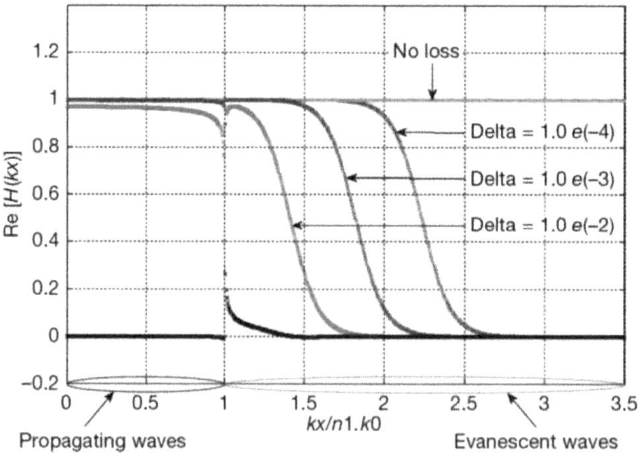

Figure 1.25. Plot of transfer function of the plane-slab versus normalized propagation constant taking energy dissipation factor δ (delta) as parameter.

It may be observed that the propagation is limited by two margins; one is the no-loss line that is parallel to the abscissa and the perfect-loss is the line parallel to the ordinate at unity marked in the abscissa axis. It is seen that the loss in the metamaterial slab seriously affects the propagation capability of the evanescent waves through the metamaterial plane-slab. The issue of sub-wavelength imaging with metamaterial which is understood to be possible with evanescent wave growth [35] is affected by loss in the metamaterial plane-slab [36, 37]. We have thus investigated how the loss affects the evanescent wave growth in LHM when the refractive index of the metamaterial medium has an imaginary part (i.e. loss in the metamaterial), see figure 1.24.

The quality of imaging may be studied by exciting the LHM slab by an impulse function and studying the impulse response of the LHM slab (which in all practical cases has finite loss). Figure 1.26 shows the normalized impulse response of LHM slab with loss or energy dissipation factor δ (delta) as a parameter. It may be observed that as the loss increases the sharpness of the sinc function decreases, and thus the image will be diffused or smeared or would be corrupted with aberrations.

It may thus finally be commented that the loss issue in metamaterial is one of the most important factors to impede its progress and large-scale acceptability for various applications. In order to overcome the troublesome losses in metamaterials, some researchers have tried to insert gain media like quantum dots (as in lasing spaser, discussed in chapter 4) and optical parametric amplification into the structure. In fact, loss compensation in metamaterials is a crucial step toward their practical utility in some specialized applications. Researchers even tried with using graphene, an allotrope of carbon, which is a very low-loss material, in designing some metasurface components (to be discussed in chapter 6). A very interesting review paper is available [64] that deals with various means of active gain control in metamaterials.

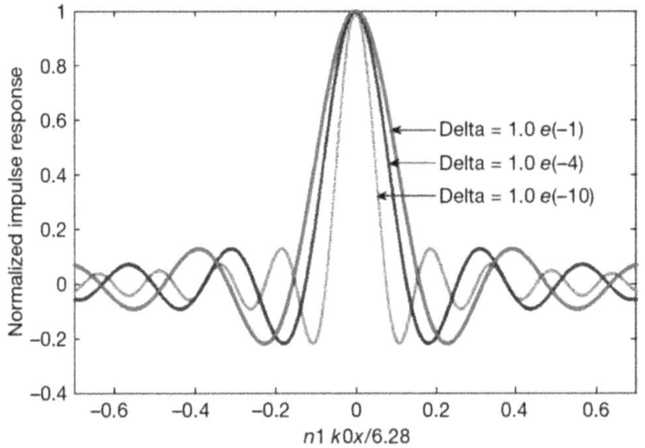

Figure 1.26. Response to an impulse excitation of the LHM slab (with loss). With the increase of energy dissipation factor, delta, the sharpness of sinc function decreases and the image is smeared.

References

[1] Shelby R, Smith D R and Schultz S 2001 Experimental verification of a negative index of refraction *Science* **292** 77–9

[2] Das B 2009 India joins the metamaterials club *Nat. India* **August** 273 https://www.nature.com/articles/nindia.2009.273

[3] Veselago V G 1968 The electrodynamics of substances with simultaneously negative values of ε and μ *Sov. Phys. Usp.* **10** 509–14

[4] Choi M, Lee S, Kim Y, Kang S B, Shin J, Kwak M H, Kang K Y, Lee Y H, Park N and Min B A 2011 A terahertz metamaterial with unnaturally high refractive index *Nature* **470** 369–73

[5] Pendry J B, Holden J, Stewart W J and Youngs I 1996 Extremely low frequency plasmons in metallic mesostructures *Phys. Rev. Lett.* **76** 4773–6

[6] Pendry J B, Holden J, Robbins D J and Stewart W J 1999 Magnetism from conductors and enhanced non linear phenomena *IEEE Trans. Microw. Theory Tech.* **47** 2075–84

[7] Eleftheriades G V, Iyer A K and Kremer P C 2002 Planar negative refractive index media using periodically L-C loaded transmission lines *IEEE Trans. Microw. Theory Tech.* **50** 2702–12

[8] Caloz C and Itoh T 2002 Application of the transmission line theory of left handed (LH) materials to the realization of a microstrip LH line *Proc. IEEE MTT-S Int. Symp.* **1** 412–5

[9] Duan Z, Wu B I and Chen M 2009 Review of Cherenkov radiation in double-negative metamaterials *Progress in Electromagnetic Symp. (Beijing, China, March 23–27 2009)* 65–7

[10] McGrgor I and Hock K M 2013 Metamaterial-based accelerating, bending and focussing structures *Proc. IPAC2013 (Shanghai, China)* 1286–8

[11] Chen J *et al* 2011 Observation of the inverse Doppler effect in negative-index materials at optical frequencies *Nat. Photon.* **5** 239–42

[12] Bose J C 1898 On the rotation of plane of polarisation of electric waves by a twisted structure *Proc. R. Soc.* **63** 146–52

[13] Lindman K F 1920 Über eine durch ein isotropes System von spiralförmigen Resonatoren erzeugte Rotationspolarisation der elektromagnetischen Wellen *Ann. Phys.* **368** 621–44

[14] Kock W E 1946 Metal-lens antennas *Proc. IRE* **34** 828–36

[15] Jones S S D and Brown J 1949 Metallic delay lenses *Nature* **163** 324–5

[16] Brown J 1953 Artificial dielectrics having refractive indices less than unity *Proc. IEE* **100** 51–62

[17] Sievenpiper D, Zhang L, Romulo F, Broas J, Alex´opolous N G and Yablonovitch E 1999 High-impedance electromagnetic surfaces with a forbidden frequency band *IEEE Trans. Microw. Theory Tech* **47** 2059–74

[18] Notomi M 2000 Theory of light propagation in strongly modulated photonic crystals: refraction-like behaviour in the vicinity of the photonic band gap *Phys. Rev.* B **62** 10696–705

[19] Luo C, Johnson S G, Johnnopoulos J D and Pendry J B 2003 Negative refraction without negative index in metallic photonic crystals *Opt. Express* **11** 746–54

[20] Yablonovitch E 1987 Inhibited spontaneous emission in solid-state physics and electronics *Phys. Rev. Lett.* **58** 2059–62

[21] John S 1987 Strong localization of photons in certain disordered dielectric superlattices *Phys. Rev. Lett.* **58** 2486–9

[22] Yablonovitch E 1993 Photonic band-gap structures *J. Opt. Soc. Am.* B **10** 283–95

[23] Parimi P V, Lu W T, Vodo P and Sridhar S 2003 Imaging by flat lens using negative refraction *Nature* **426** 404

[24] Cubukcu E, Aydin K, Ozby E, Foteinopolou S and Soukoulis C M 2003 Subwavelength resolution in a two-dimensional photonic-crystal-based superlens *Phys. Rev. Lett.* **91** 207401–4
[25] Qiu M, Xiao S, Berrier A, Anand S, Thylen L, Mulot M, Swillo M, Ruzan Z and He S 2005 Negative refraction in two-dimensional photonic crystals *Appl. Phys.* A **80** 1231–6
[26] Kelvin L 1894 *The Molecular Tactics of a Crystal* (Oxford: Clarendon)
[27] Barba I, Cabeceira A C L, García-Collado A J, Molina-Cuberos G J, Margineda J and Represa J 2011 Quasi-planar chiral materials for microwave frequencies *Electromagnetic Waves Propagation in Complex Matter* ed A Kishk (Wilimington, DE: Scitus Academics LLC) ch 7
[28] García-Collado A J, Molina-Cuberos G J, Margineda J, Núñez M J and Martín E 2010 Isotropic and homogeneous behaviour of chiral media based on periodical inclusions of cranks *IEEE Microw. Wirel. Compon. Lett.* **20** 176–7
[29] Pendry J B 2004 A chiral route to negative refraction *Science* **306** 1353–5
[30] Sang Soon O and Hess O 2015 Chiral metamaterials: enhancement and control of optical activity and circular dichroism *Nano Converg.* **2** 24
[31] Subal K 2022 *Microwave Engineering—Fundamentals, Design, and Applications* 2nd edn (Hyderabad: Universities Press)
[32] Brennan E, Fusco V and Schuchinsky A 2003 Investigation on properties of left-handed materials *High Frequency Postgraduate Student Colloqium (Belfast)* 65–8 (Cat. No. 03TH8707)
[33] Caloz C and Itoh T 2006 *Electronic Metamaterials: Transmission line Theory and Microwave Applications* (New York: Wiley Interscience)
[34] *Ansys High Frequency Structure Simulator* (Pittsburgh, PA: Ansoft Corporation) https://www.ansys.com/products/electronics/ansys-hfss
[35] Pendry J B 2000 Negative refraction makes a perfect lens *Phys. Rev. Lett.* **85** 3966–9
[36] Garcia N and Nieto-Vesperianas M 2002 Left handed materials do not make a perfect lens *Phys. Rev. Lett.* **88** 207403–6
[37] Ghosh M and Kar S 2013 Metamaterial plane-slab focussing and sub-wavelength imaging—the concept, analysis and characterization *Proc. Science and Information Conf. (London, October 7–9 2013)* 670–4 https://ieeexplore.ieee.org/xpl/conhome/6653326/proceeding
[38] Luo H, Hu W, Ren Z, Shu W and Li F 2006 Focussing and phase compensation of paraxial beams by a left-handed material slab *Opt. Commun.* **266** 327–31
[39] Ruppin R 2001 Surface polaritons of a left-handed material slab *J. Phys.: Condens. Matter* **13** 1811–9
[40] Barnes W L, Dereux A and Ebbesen T W 2003 Surface plasmon sub-wavelength optics *Nature* **424** 842–30
[41] Iyer A K and Elefteriades G V 2009 Free-space imaging beyond the diffraction limit using Veselago–Pendry transmission-line metamaterial superlens *IEEE Trans. Antennas Propag.* **57** 1720–7
[42] Grbic A and Eleftheriades G V 2004 Overcoming the diffraction limit with a planar left-handed transmission-line lens *Phys. Rev. Lett.* **92** 117403–6
[43] Barnes W L, Dereux A and Ebbesen T W 2003 Surface plasmon subwavelength optics *Nature* **424** 824–30
[44] Caloz C, Lee C J, Smith D R, Pendry J B and Itoh T 2004 Existence and properties of microwave surface plasmons at the interface between a right-handed and a left-handed media *IEEE AP-S USNC/URSI National Radio Science Meeting* 3 *(Monterey, CA, 20–25 June 2004)* 3151–4

[45] Roy T and Kar S 2012 Transmission properties of electromagnetic waves through left-handed material: a revisit *IETE J. Res.* **58** 77–82
[46] Alu A and Engheta N 2006 Physical insight into the 'growing' evanescent fields of doubly-negative metamaterial lenses using their circuit equivalence *IEEE Trans. Antennas Propag.* **54** 268–72
[47] Mojahedi M and Eleftheriades G V 2005 *Negative-RefractionMetamaterials Fundamental Principles and Applications* ed G V Eleftheriades and K G Balmain (Hoboken, NJ: IEEE Press and Wiley-Interscience) ch 10
[48] Mojahedi M, Mlloy K J, Eleftheriades G V, Woodley J and Chiao R Y 2003 Abnormal wave propagation in passive media *IEEE J. Sel. Top. Quantum Electron.* **9** 30–9
[49] Pendry J B, Schurig D and Smith D R 2006 Controlling electromagnetic fields *Science* **312** 1780–2
[50] Schurig D, Mock J J, Justice B J, Cummer S A, Pendry J B, Starr A F and Smith D R 2006 Metamaterial electromagnetic cloak at microwave frequencies *Science* **314** 977–80
[51] Alitalo P and Tretyakov 2009 Electromagnetic cloaking with metamaterials *Mater. Today* **12** 22–9
[52] Cai W, Chettiar U K, Kildishev A V and Shalaev V M 2007 Optical cloaking with metamaterials *Nat. Photon.* **1** 224–7
[53] Liu R, Ji C, Mock J J, Chin J Y and Smith D R 2009 Broadband ground-plane cloak *Science* **323** 366–9
[54] Yarris L C 2009 Blurring the line between magic and science: Berkeley researchers create an "invisibility cloak" https://newscenter.lbl.gov/2009/05/01/invisibility-cloak/
[55] Jung J, Park H, Park J, Chang T and Shin J 2020 Broadband metamaterials and metasurfaces: a review from the perspectives of materials and devices *Nanophotonics* **9** 3165–96
[56] Valentine L J, Zentgraf T, Bartal G and Zhang X 2009 An optical cloak made of dielectrics *Nat. Mater.* **8** 568–71
[57] Gharghi M, Gladden C, Zentgraf T, Liu Y, Yin X, Valentine J and Zhang X 2011 A carpet cloak for visible light *Nano Lett.* **11** 2825–8
[58] Ni X, Wong Z J, Mrejen M, Wang Y and Zhang X 2015 An ultrathin invisibility skin cloak for visible light *Science* **349** 1310–4
[59] Kar S 2021 Progress in metamaterial and metasurface technology and applications ed P K Choudhury *Metamaterials: Technology and Applications* (Boca Raton, FL: CRC Press of Taylor and Francis Group), ch 1 pp 1–29
[60] Vellucci S, Monti A, Barbuto M, Toscano A and Bilotti F 2021 Progress and perspective on advanced cloaking metasurfaces: from invisibility to intelligent antennas *EPJ Appl. Metamat.* **8** 19
[61] Lee K T, Ji C, Lizuka H and Banerjee D 2021 Optical cloaking and invisibility: from fiction toward a technological reality *J. Appl. Phys.* **129** 231101–18
[62] Silveirinha M G 2009 Anomalous refraction of light colours by a metamaterial prism *Phys. Rev. Lett.* **102** 193903–6
[63] Shen L and He S 2003 Studies of imaging characteristics for a slab of a lossy left-handed material *Phys. Lett.* A **309** 298–305
[64] Boardmann A D, Grimalsky V V, Kivshar Y S, Koshevaya S V, Lapine M, Litchinitser N M, Malnev V N, Noginov M, Rapoport Y G and Shalaev V M 2011 Active and tunable metamaterials *Laser Photonics Rev.* **5** 287–307

IOP Publishing

Metamaterials and Metasurfaces
Basics and trends
Subal Kar

Chapter 2

Plasmonic and transmission-line metamaterials

At the very early stage, the Universe was in the fourth state of matter—The Plasma

2.1 Introduction

The metamaterial that was successfully fabricated and tested at the University of California at San Diego (UCSD), US in 2001, to exhibit negative refractive index, was a plasmonic metamaterial at microwave frequency. It consisted of thin wires (TWs) acting as 'electric atom' and split-ring resonators (SRRs) acting as 'magnetic atom'. The combined structure of TW-SRR formed the first bi-dimensional plasmonic metamaterial in the world (see figure 1.5(a)(iii)). An array of TW is understood to behave as an electric plasma while a matrix of SRR behaves as a magnetic plasma, hence their combined structure forms the *plasmonic metamaterial*. Variants of TW and SRR were subsequently developed with improved characteristics in terms of frequency of operation, bandwidth and other issues and now we have different types of plasmonic metamaterial apart from the initially developed TW-SRR type. In this chapter we shall discuss all of them in detail, especially looking into their mutual advantages and disadvantages from an application point of view. As was mentioned in chapter 1, the first plasmonic metamaterial in India was made by our group at the University of Calcutta in collaboration with SAMEER (Society for Applied Microwave Electronics Engineering and Research), Kolkata centre and BARC (Bhabha Atomic Research Centre), Mumbai in 2009 and this was a combined structure of TW and LR (labyrinth resonator); the photograph of the same is shown in figure 2.1(a).

It may, however, be mentioned herewith that the very first metamaterial fabrication was announced in March 2000 [1] by the UCSD group, following Pendry's suggested TW and SRR. This structure combined parallel wires with split-ring couplet, see figure 2.2. Because this LHM (metamaterial) structure was mono-

doi:10.1088/978-0-7503-5532-2ch2

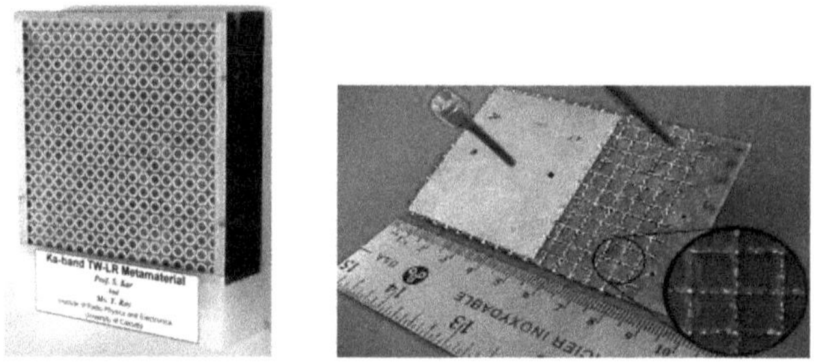

(a) (b)

Figure 2.1. (a) Plasmonic metamaterial designed at the University of Calcutta, India (2009). (b) Transmission-line metamaterial designed at Toronto University (2002).

Figure 2.2. First mono-dimensional (1D) experimental metamaterial structure, constituted of TWs and SRRs, designed and developed by the team of UCSD in 2000.

dimensional which responded to radiation coming from one plane only, their metamaterial thus exhibited negative refraction in one plane only, that is, it was strongly anisotropic. But in 2001, the same group of UCSD designed and fabricated and improved the structure with a 2D periodic version, a bi-directional structure, that functioned as an isotropic metamaterial of plasmonic type, see figure 1.5(a)(iii); thus the year 2001 can be reckoned as the successful year for the first metamaterial exhibiting negative refractive index. The metamaterial constructed by Shelby *et al* in 2001 (see figure 1.5(a)(iii)) was constructed of copper split-ring resonators and wires mounted on interlocking sheets of fibreglass circuit board. The total array was made up of 3 by 20 × 20 unit cells with overall dimensions of 10 × 100 × 100 mm.

Though the UCSD data of 2001 regarding the realization of negative refractive index with metamaterial was compelling, the concept of negative index of refraction proved counter-intuitive enough which prompted many other researchers to verify the same in their laboratories too. In 2003, the negative refraction experiment of the same sort of negative-index material was fabricated and tested at MIT, USA and confirmed the original findings [2]. Looking at the metamaterial wedges with different angles, the MIT group showed that the observed angle of refraction was consistent with Snell's law for metamaterial. In the same year, another group at Boeing Phantom Works also confirmed the negative refraction results in a separately designed sample [3]. In their measurements, the detector's distance from the sample was significantly larger than for the previous demonstrations.

A brief description of the experiment that was done on the first successful experimental demonstration of negative refraction properties of metamaterial in 2001 may be worthwhile in this perspective. Shelby *et al* used the bi-dimensional TW-SRR structure (figure 1.5(a)(iii)) that was cut into wedge-shaped piece of metamaterial and inserted into the experimental set-up shown in figure 2.3. The wave is incident on the flat side of the wedge and after transmission via the metamaterial sample the wave emerges towards the direction of the detector. The left-handedness of the TW-SRR structure was evidenced by the fact that a maximum of the transmission coefficient was measured in the negative angle (below the normal in the figure) with respect to the interface of the wedge, whereas a maximum in the positive angle (above the normal) was measured, as expected, when the wedge was replaced by a regular piece of Teflon (a right-handed material) with identical shape. The result reported was in qualitative and quantitative agreement with Snell's law, which reads: $k_1 \sin \theta_1 = k_2 \sin \theta_2$, or if the two media are isotropic (so that $k_1 = n_1 k_0$ and $k_2 = n_2 k_0$), $n_1 \sin \theta_1 = n_2 \sin \theta_2$, where k_i, n_i, and θ_i, respectively, represent the wavenumber, refractive index, and angle of the incident ray from the normal to the interface in each of the two media considered ($i = 1, 2, \ldots$).

In the few years after the first experimental demonstration of negative refractive index in 2001 by left-handed material of TW-SRR type, a large number of both theoretical and experimental reports confirmed the main properties of LH materials predicted by Veselago. Although in 2002 some controversies temporarily cast doubt on these novel concepts [4–6], these controversies were quickly shown to be based on physics misconceptions [7]. Now the negative refractive index concept of metamaterial

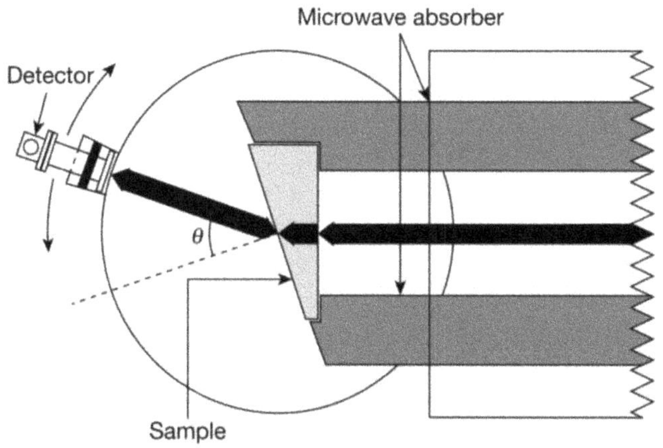

Figure 2.3. Experimental set-up for the determination of the negative refractive index of a wedge-shaped metamaterial sample.

is well established theoretically, numerically, and experimentally. The theoretical verifications are based on fundamental electromagnetic theory and transmission-line theory approaches. Experimental demonstrations are provided with TW-SRR (and its variants) bulk structures and planar transmission-line structures. Validation of metamaterial characterization has also been done successfully with various numerical techniques like finite-difference time domain (FDTD), finite elements method (FEM), transfer matrix algorithm (TMA), and transmission-line method (TLM).

Anyway, it was mentioned in chapter 1 the transmission-line metamaterial was first successfully developed at Toronto University, Canada in 2002; which is shown in figure 2.1(b). These were designed by periodically loading a host transmission line (conventional type) with dual of its equivalent circuit (i.e., replacing series inductance by series capacitance and shunt capacitance by shunt inductance) giving rise to the so-called periodically loaded transmission line (PLTL), exhibiting metamaterial property. Being non-resonant in nature, PLTL exhibits simultaneously low loss and broad bandwidth and they can be engineered in planar configuration thus inherently supporting 2D wave propagation and also compatible with modern microwave integrated circuits (MICs). This class of LH material is thus well suited for RF and microwave circuit design applications.

2.2 Plasmonic metamaterial

To form an artificial material (metamaterial), repeated (i.e. periodically arranged) elements are used to design so as to get a strong response to applied electromagnetic fields. As long as the size and spacing of the elements are much smaller than the wavelength of the incident electromagnetic signal, the incident radiation cannot distinguish the collection of elements from a homogeneous material. We can thus conceptually replace the inhomogeneous composite by an effectively continuous bulk material described by the material parameters ε and μ. At lower frequencies (much below the optical frequency) conductors are excellent candidates from which to form artificial materials, because their response to electromagnetic fields is large.

Metamaterials can be artificially constructed by combining sub-wavelength structures, one of which exhibits negative permittivity and the other that exhibits negative permeability below their plasma resonance frequency. In plasmonic metamaterial the negative permittivity can be realized with an array of metallic TWs or also with cut wires (CWs) [8, 9] below its electric plasma frequency while the negative permeability can be realized with a matrix of C-shaped metallic SRRs, below its magnetic plasma frequency. Each unit cell in such a periodic array of TWs and matrix of SRRs when irradiated with an electromagnetic signal acts, respectively, as an 'electric atom' and 'magnetic atom' mimicking the atomic arrangements as in the lattice of natural material. In practice, the TW array is fabricated by imprinting metallic (copper/gold) TWs on a dielectric substrate (FR4 or RT-duroid, depending on the frequency of operation), see figure 2.4(a), with printed circuit technology, while the SRR matrix is fabricated by imprinting metallic C-shaped rings on the dielectric substrate, as shown in figure 2.4(b). When the two together are combined in a single structure (as shown in figure 2.4(c)) we get the one dimensional

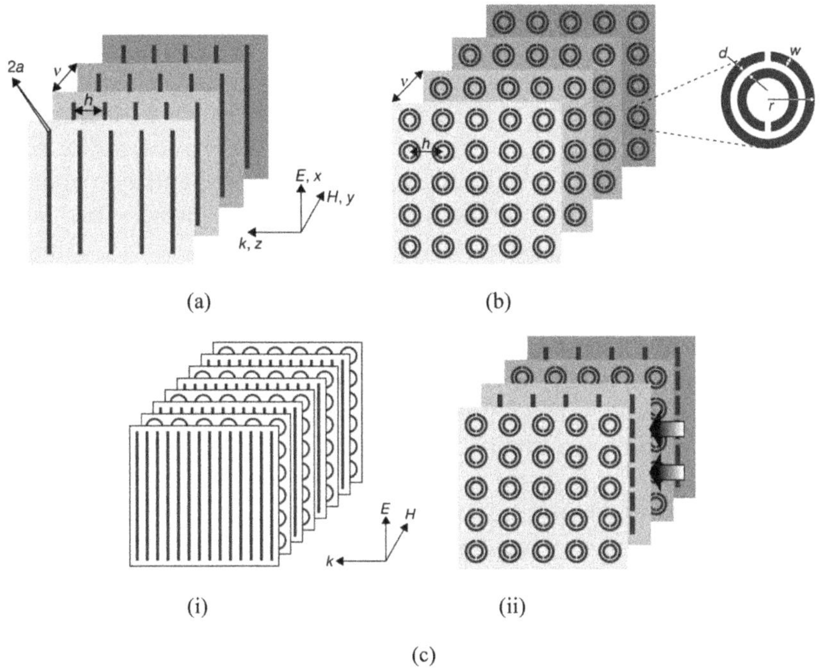

Figure 2.4. Realization of negative permittivity and negative permeability, respectively, with (a) an array of TWs and (b) a matrix of SRRs. (c) 1D plasmonic metamaterial with (i) TW-SRR and (ii) SRR–CW.

(1D) plasmonic metamaterial. It may be mentioned herewith that the term plasmonics is conventionally reserved in the domain of optical frequency; however, the TW array and SRR matrix behave as though they are plasmonic structures at microwave frequencies—thus the metamaterial constructed with TW-SRR is commonly termed as plasmonic metamaterial.

The realization of effective negative permittivity $\varepsilon_{\text{reff}}$ (with TW/CW array) and effective negative permeability μ_{reff} (with SRR matrix) is a bulk property and thus the effective negative refractive index n_{reff} of plasmonic metamaterial is given by:

$$n_{\text{reff}} = \sqrt{\varepsilon_{\text{reff}}\mu_{\text{reff}}} \tag{2.1}$$

It may be noted herewith that there arises an ambiguity in the sign of the square root if we simply use $\varepsilon_{\text{reff}} = -1$ and $\mu_{\text{reff}} = -1$, when from equation (2.1) we have still a positive value of n_{reff}. In general, ε and μ are complex functions of frequency and thus we need to write: $\varepsilon_{\text{reff}}(\omega) = |\varepsilon_{\text{reff}}(\omega)|e^{i\pi}$, for $\varepsilon_{\text{reff}}(\omega) < 0$ and $\mu_{\text{reff}}(\omega) = \left|\mu_{\text{reff}}(\omega)\right| e^{i\pi}$, for $\mu_{\text{reff}}(\omega) < 0$ (for those frequencies inside the left-handed band). Eventually we can write from equation (2.1): $n_{\text{reff}}(\omega) = \sqrt{\left|\varepsilon_{\text{reff}}(\omega)\mu_{\text{reff}}(\omega)\right|}\, e^{i\pi} = -\sqrt{\left|\varepsilon_{\text{reff}}(\omega)\mu_{\text{reff}}(\omega)\right|}$. However, it must be noted that the square root of either ε or μ alone must have a positive imaginary part—a necessity for a passive material. It may be noted on passing that a negative refractive index implies that the phase of a wave decreases rather than advances with passage through the negative refractive index medium.

Though somewhat less common than positive materials (having positive ε and positive μ), the materials having either negative ε or negative μ are nevertheless available in Nature. Materials with negative ε include metals like silver and gold, however, at optical frequencies; while materials with negative μ include resonant ferromagnetic or antiferromagnetic systems. However, a very basic question obviously comes to mind, why no materials with simultaneously negative ε and μ occur in Nature? Let us understand this as follows. In natural materials, the resonances that give rise to electric polarizations typically occur at very high frequencies—in optical regime for metals, and at least in the terahertz-to-infrared (IR) for semiconductors and insulators. On the other hand, resonances in magnetic materials typically occur at much lower frequencies and usually tail off towards the THz and microwave region (the reason for this is explained in chapter 4). In essence, the fundamental electronic and magnetic processes that give rise to resonance phenomena in natural materials do not occur at all with overlapping frequencies, although no physical law would preclude such overlaps from occurring. It is found from the Lorentz/Drude model of a material that negative electric or magnetic response occurs near to the resonance. Hence because of the seeming separation in frequency response over which the electric and magnetic resonant phenomena occurs in natural material, Veselago's theoretical analysis of materials (1967) with ε and μ simultaneously negative might have remained a curious and un-attempted exercise in electromagnetic theory until the 1990s. However, in the mid-1990s, researchers began looking into the possibility of engineering artificial materials to have tailored electromagnetic response. Although the field of artificial materials, especially artificial dielectrics, dates back to 1940; subsequent advances in fabrication technology and computational electromagnetics—coupled with the emerging awareness of the importance of materials with negative constitutive properties—led to a resurgence of efforts in developing new structures/artificial materials with novel material properties like photonic crystals/electromagnetic band-gap (EBG) materials and metamaterials.

It may be noted herewith that as the negative material parameters occur near a resonance, the negative refractive index materials will have to face two important consequences. Firstly, negative material parameters will exhibit frequency dispersion: that is to say, they will vary as a function of frequency. Secondly, the usable bandwidth of such artificial materials having negative constitutive properties will be relatively narrow compared with the natural materials having positive constitutive properties.

Now let us discuss the issue of the realization of negative permittivity with TW array and also negative permeability with SRR matrix.

2.2.1 Negative permittivity with TW and CW array

Essentially, metals are plasmas; since they consist of an ionized 'gas' of free electrons. Below plasma frequency (the frequency at which it becomes transparent), the real component of the permittivity (ε) of bulk metals is negative. This $\varepsilon < 0$ response results from the free electrons in the metals that screen external electromagnetic radiation. However, the natural plasma frequencies of metals normally occur in the ultra-violet region of the electromagnetic spectrum, in which wavelengths are extremely short and

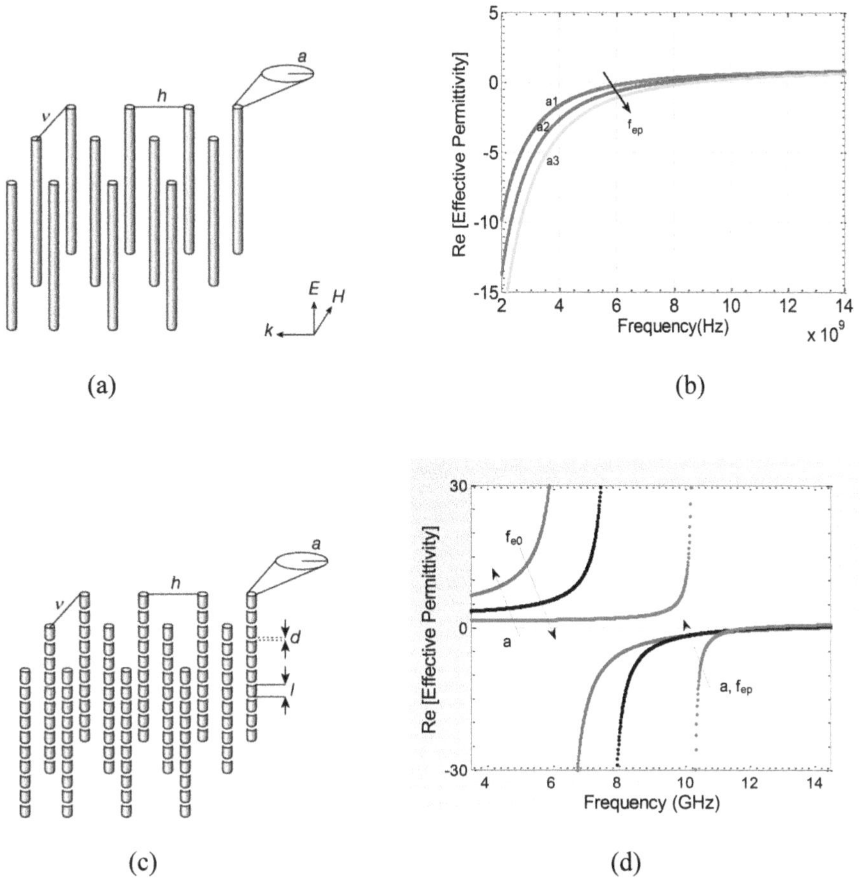

Figure 2.5. (a) An array of metallic TWs, (b) variation of real part of effective permittivity with frequency for different wire radius ($a1 = 0.15$ mm, $a2 = 0.45$ mm, $a3 = 0.9$ mm) with fixed lattice constants h and ν. (c) An array of CW and (d) frequency-dependent effective permittivity of CW for different values of a.

hence fabrication complexity and cost is exorbitant. Although the permittivity is negative at frequencies below the plasma frequency, the approach towards absorptive resonances at lower frequencies increases the dissipation leading to the complex nature of the permittivity. To realize the plasma frequency in the microwave region, Pendry *et al* [10] solved the problem from a novel standpoint with a detailed physics-oriented understanding. They used an array of thin metallic wires (see figure 2.5(a)) which could exhibit negative permittivity below plasma frequency in the microwave frequency range. The plasma frequency of TW structure is given by:

$$\omega_p^2 = n_{\text{eff}} e^2 / \varepsilon_0 m_{\text{eff}} \tag{2.2}$$

In naturally occurring materials, n refers to the actual density of the charge carriers (usually electrons) and m_{eff} is the effective mass of electrons. But in TW structure n (i.e. n_{eff}) and m_{eff} are related to the geometry of the lattice rather than the fundamental charge carriers giving TW structure to have a much greater flexibility in design and realization of its characteristics with respect to frequency of operation compared to conventional materials. Spatial confinement of the electrons to TWs decreases the effective electron concentration ($n_{\text{eff}} = n\pi r^2/a^2$) in the volume of the structure. Also, the self-inductance of the wire-array manifests itself as a greatly enhanced effective mass of the electrons confined to the wires. All these effects in TW structure eventually help to decrease the plasma frequency of it by many orders of magnitude according to equation (2.2). Because the plasma frequency can be tuned by the geometry of TW structure, the region of moderately negative values of permittivity can be made to occur at nearly any frequency range from low RF to the optical. Thus by using an array of thin metallic wires, having the virtue of its macroscopic plasma-like behaviour, Pendry *et al* [10] could realize an effective negative permittivity at microwave frequencies. In practice, as mentioned above, thin metallic wires made of copper or gold are printed in the form of an array on low-loss dielectric to realize negative permittivity with TW at microwave frequency.

The analytical basis of realizing negative permittivity for designing plasmonic metamaterial may be understood as follows. Figure 2.4(a) shows an array of metallic TWs irradiated with an electromagnetic field with the E-field parallel to the length of the wire. The exciting electric field induces a current along the wires and generates equivalent dipole moments resulting in a frequency-dependent effective relative permittivity given by [10]:

$$\varepsilon_{\text{reff}}(\omega) = 1 - \frac{\omega_{\text{ep}}^2}{\omega^2 + j\omega\xi_e} \tag{2.3}$$

This recalls the frequency-dependent permittivity obtained with plasmonic model for metals—known as Drude model (see appendix A).

The frequency-dependent effective permittivity represented by equation (2.3) exhibits non-resonant high-pass characteristics and the real part of the permittivity is negative for $\omega < \omega_{\text{ep}}$ (see figure 2.5(b)). However, sometimes from a designer's point of view a resonant characteristic for the frequency-dependent effective relative permittivity is desirable. The negative permittivity with resonant characteristics (see figure 2.5(d)) can be realized with discontinuous TW or the CW array shown in figure 2.5(c) [8, 9]. For CW structures, the negative permittivity range does not extend to zero frequency, and a stop-band appears around the resonance frequency. This may be looked upon as a variant of the continuous TW originally proposed by Pendry *et al* [10]. Here the resonant characteristic arises due to the inductance represented by each length of the CW and capacitance due to the gap between the CWs. The effective relative permittivity of CW is given by [9]:

$$\varepsilon_{\text{reff}} = 1 - \frac{\omega_{\text{ep}}^2 - \omega_{e0}^2}{\omega^2 - \omega_{e0}^2 + j\omega\xi_e} \tag{2.4}$$

This recalls the frequency-dependent permittivity given by the resonant model for materials known as Lorentz model (see appendix A).

In equations (2.3) and (2.4) $\omega_{\text{ep}} = 2\pi f_{\text{ep}}$; f_{ep} is the electric plasma frequency, the upper bound for negative permittivity, $\omega_{e0} = 2\pi f_{e0}$; f_{e0} is the electric resonant frequency, the lower bound for negative permittivity, and ξ_e is the damping factor due to metal loss imparted by the TW (or CW) metal imprints given by $\xi_e = \varepsilon_0(h\omega_{ep}/a)^2/\pi\sigma$, where, h is the horizontal lattice constant, i.e., the centre-to-centre spacing between two adjacent wire sections, a is the radius of each wire, and σ is the conductivity of the metal. It may be noted that in equation (2.3) or in equation (2.4) the plasma frequency ω_{ep}, and resonance frequency ω_{e0} are determined by the geometry of the lattice of the periodic structure rather than by the charge, effective mass, and density of electrons, as in the case in naturally occurring materials.

The generic expressions for f_{e0} and f_{ep} are given by:

$$f_{e0} = \frac{c_0}{\pi a}\sqrt{\frac{(N_y - 1)d}{N_y(l + d)\ln(h/a)}} \tag{2.5a}$$

and

$$f_{ep} = \frac{c_0}{2\pi h}\sqrt{\frac{2X\pi}{\ln(h/a)}} \tag{2.5b}$$

For TW: $d = 0$, $X = 1$, and for CW: $d \neq 0$, $X = 2$.

For derivations of equations (2.5a) and (2.5b) see appendix B.

2.2.2 Negative permeability with SRR

The path to achieving magnetic response from conductors or non-magnetic materials was a challenge and different from obtaining the electric response mentioned above. This follows from the basic definition of the magnetic moment: $\vec{m} = 0.5\int_V \vec{r} \times \vec{j}\,\mathrm{dV}$ for current density $\vec{j}$. It may be seen that a magnetic response may be obtained if a local current can be induced to circulate in closed loops. Introducing a resonance characteristic in the element would enable a strong magnetic response, potentially one that can lead to a negative μ. In naturally available magnetic materials, it is known that magnetism results from the force created by circular motion of electrons in the atoms. Thus in 1999, J B Pendry and his colleagues [11] proposed a variety of structures (tubes or loops of conductors with a gap/split inserted) that, they predicated, is capable of producing artificial and tailorable magnetism from non-magnetic material. They proposed, in addition to other structures, a coupled split-ring structure, see figure 2.6(a)(ii) (the single split-ring is shown in figure 2.6(a)(i)), which is basically an L–C tuner. It can tune to the incident magnetic field of the electromagnetic wave similar to a receiving antenna of radio and acts as a magnetic resonator. To be more lucid, a time-varying magnetic field of incident electromagnetic signal would induce an electromotive force, as per Faraday's law of electromagnetic induction, in the plane of the element, the SRR, driving currents within the conductor. Because of the split/gap in the SRR, the circulating current in the split-ring will result in a build-up of charge across the gap with the energy stored as capacitance. The small gap/split in the plane of SRR thus introduces a capacitance (C) into the circuit and gives rise to a resonance at a frequency: $\omega_0 = 1/\sqrt{LC}$ (where L is the inductance of the metallic ring) set by the geometry of the

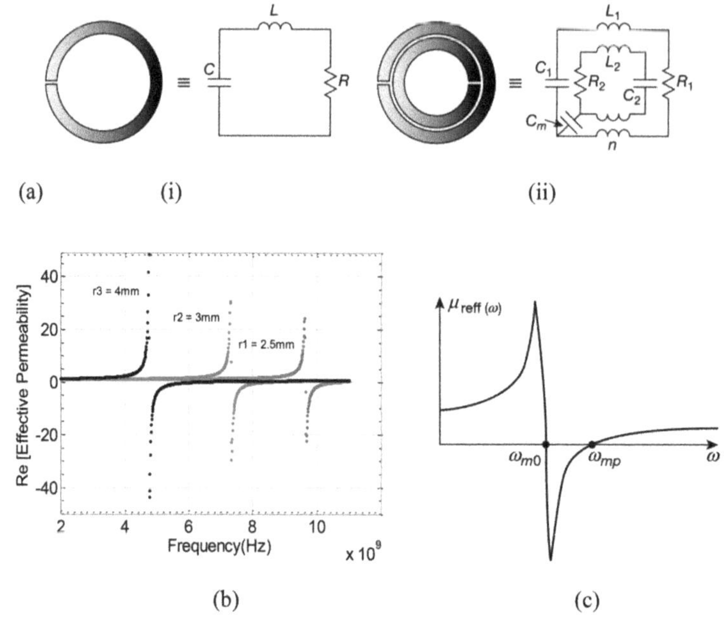

Figure 2.6. (a) SRRs with electrical equivalent circuit. (i) Single structure and (ii) coupled structure. (b) Resonant characteristics of SRR with different radius of the structure: $r3 > r2 > r1$. (c) SRR characteristics showing ω_{m0} and ω_{mp}.

element. For frequencies below ω_0, currents in the SRR can keep up with the driving force produced by the externally varying magnetic field of the incident electromagnetic signal and a positive response is achieved. However, as the rate of change (frequency) of the external magnetic field is increased, the currents can no longer keep up and eventually begin to lag, resulting in an out-of-phase or negative response. Such an SRR in its various forms can be viewed as the so-called 'magnetic atom', i.e. magnetic inclusion structure for designing the metamaterial. Thus Pendry *et al* [11] envisioned that a controllable artificial magnetic property could be realized with a periodic array of minuscule metallic SRRs printed on a dielectric substrate, see figure 2.4(b), which collectively behaves as a magnetic plasma (analogous to electric plasma of metallic TW/CW array printed on dielectric substrate, as is shown in figure 2.4(a)), and exhibit a negative permeability below the magnetic plasma frequency (see figures 2.6(b) and (c)).

It may be observed from the lumped circuit equivalent of single-ring and double-ring (couplet) SRR, shown in figure 2.6(a)(i) and (ii) that in single-ring configuration, the circuit model is that of the simple RLC resonator (where R, L, and C, respectively, represent the resistance, inductance and capacitance of the single-ring SRR, see figure 2.6(a)(i)) with resonant frequency: $\omega_0 = 1/\sqrt{LC}$. But in the double-ring configuration, capacitive coupling and inductive coupling between the larger and smaller rings are modelled by a coupling capacitance (C_m), caused by the distributed capacitance between the two rings, and by a transformer (with transforming ratio n), respectively. However, the double SRR is essentially equivalent to the single SRR if mutual coupling is weak, because the dimensions of the two rings are very close to each other in practical design (being sub-wavelength) so that $L_1 \approx L_2 \approx L$ and $C_1 \approx C_2 \approx C$, resulting in a combined resonant frequency close to that of the single SRR. A question naturally arises, if a single-ring structure already acts as a magnetic resonator, providing a negative μ regime, what may be the benefits from the addition of a second ring? Also, why in almost all experimental efforts is the double-ring SRR the preferable building block? The answer is very straightforward—the double-ring SRR possesses larger magnetic moment (and hence a larger amount of negative permeability) due to higher current density in the couplet. For even more magnetic moment and hence negative permeability multiple split-ring resonators (MSRRs) are used. Instead of the circular ring proposed by Pendry *et al* as discussed above, squared rings [12] represent a valid alternative when miniaturization is desired. With the same linear dimensions, in fact, the square ring has a longer strip, leading to a lower resonant frequency compared to the one obtained with a circular counterpart.

The analytical basis of realizing negative permeability with SRR may be understood as follows. Figure 2.4(b) shows sheets of SRR matrix arranged in an 1D structure in which each of the SRRs is being excited by the H-field of the electromagnetic wave applied at right angles to the plane of the SRR matrix sheets so as to induce resonating currents in the loop and generate magnetic dipole moments resulting in a frequency-dependent effective relative permeability given by [11].

$$\mu_{\text{reff}} = 1 - \frac{F\omega^2}{\omega^2 - {\omega_{m0}}^2 + j\omega\xi_m} \tag{2.6}$$

This reminds the frequency-dependent permeability of Lorentz resonant model for materials (see appendix A).

In equation (2.6), $F\,(= \pi r^2/h^2)$ is the fractional area occupied by the rings in the unit cell, i.e. fill-factor with h as the horizontal lattice constant; $\omega_{m0} = c\sqrt{\dfrac{3\nu}{\pi\,\ln(2wr^3/d)}}$ is the magnetic resonant frequency with ν vertical lattice spacing, w is width of the rings, r radius of the outer ring, d radial spacing between the rings; $\zeta_m = 2\nu R_c/r\mu_0$ is the damping factor due to metal losses with R_c the resistance of the metallic strips per unit circumferential length of the ring. It is interesting to note that the SRR structure has a magnetic response caused by the artificial magnetic dipole moments provided by the ring resonators even though the SRRs are made of non-magnetic material. It may be observed that over the frequency range $\omega_{m0} < \omega < \omega_{\text{mp}}$, where $\omega_{\text{mp}} = \omega_{m0}/\sqrt{1 - F^2}$ is the magnetic plasma frequency; $\text{Re}(\mu_{\text{reff}}) < 0$, (see figures 2.6(b) and (c)). It may be noted that the magnetic resonant frequency (ω_{m0}) is the frequency beyond which μ_{reff} diverges and the separation between ω_{m0} and ω_{mp} is considered to be the bandwidth of operation for negative permeability exhibited by the structure, resulting in a stop-band to the propagating wave.

2.2.3 A note—artificial magnetism with non-magnetic material

The magnetic inclusion structure SRR (and its variants to be discussed shortly) used for metamaterial design is basically artificial magnetism realized with non-magnetic material like copper, silver, gold etc. J B Pendry in his seminal research paper in 1999 [11] first proposed SR (spiral resonator) and split-ring resonator (SRR) as

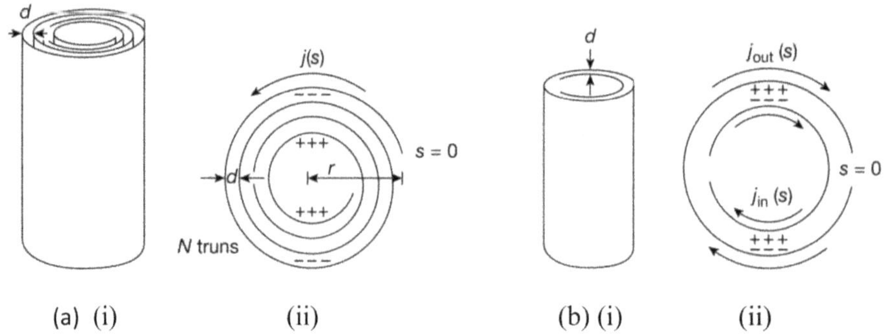

Figure 2.7. (a) 'Swiss roll' magnet-less magnetic medium. (b) Coupled cylinders with cuts in diametrically opposite positions. In both the cases (i) the structure (ii) plan-view showing the electrical situation.

tailorable magnetic inclusion structure realizable with non-magnetic material. The idea of such artificial magnetism has been borrowed from the lesson learnt from nature that magnetic polarization in materials follows indirectly from the flow of orbital currents or from unpaired electron spins. The basic structures that Pendry discussed in his paper for realizing artificially an effective magnetism from non-magnetic material is a 'Swiss roll' structure; see figure 2.7(a), and a concentric cylinder structure with appropriate cut in diametrically opposite locations in the two cylinders, see figure 2.7(b).

A Swiss roll is a cylinder constituted by closely packed concentric layers of metal obtained by winding a conducting sheet in spiral around a central mandrel. The overall structure is made by a bundle of parallel rolls arranged in closely packed array, see figure 2.7(a). When it is excited by a magnetic field parallel to the axis of the rolls, an electric current is induced in the spiral of the rolls, and the distributed capacitances between the turns enables transverse current flow. Due to the absence of the metal connectivity in its transverse plane the structure cuts-off direct current (DC) and allows only AC current flow in RF operation from the self-capacitance of the rolls, which induces a negative real part of the permeability just above the resonance frequency. The SR is realized with a cross-section of Swiss roll.

The coupled SRR, as shown in figure 2.7(b), was obtained by taking a cross-section of two layers of metal sheets rolled into cylinders and placed concentrically having longitudinal slits cut on the surface (with the slits rotated by 180° with each other). SR, SRR are the magnetic inclusion structures that Pendry proposed for the first time to realize artificial magnetism with non-magnetic material—that has revolutionized the practical research with left-handed material (metamaterial) which was theoretically predicted by Vaselego in 1967.

It may be mentioned here that the artificial magnetism realizable with Swiss roll structure was used in practical experimental demonstration by utilizing it in magnetic resonance imaging (MRI) [13] used for imaging soft tissues in sensitive portions of the body like the brain. Comparable image quality was obtained with Swiss roll structure as with 'body coils' built into the structure of the magnet. The benefit with Swiss roll artificial magnetism is that the DC magnetism is absent here and thus the static and low frequency magnetic fields are not perturbed, thus better spectral data and image quality is possible. Hence this magnet-less magnetic structure may be considered as a new paradigm for the manipulation of RF flux in MRI and spectroscopy systems. A related Swiss roll-based endoscope for near-field imaging, utilising a Swiss roll medium of much greater permeability and lower loss with highly uniaxial anisotropy, has also been reported [14].

The background story of Pendry's invention of artificial and tailorable magnetism with non-magnetic materials is fascinating. Back in the 1990s, J B Pendry, Professor of Imperial College, London and an expert in condensed-matter physics with a special interest in electromagnetism, was consulting with Marconi Materials Technology UK. The British company manufactured a radiation-absorbing carbon material that could hide battleships from radar detection but didn't know the physics of how it worked. Pendry helped them to get the insight of the physics behind it and discovered that the electrical properties allowing the material to absorb radiation came not from the carbon per se but from the shape of its long, fine fibers. The Marconi company was so pleased with Pendry's scientific insight into carbon fibers that the company wanted to know whether he might have any new tricks up his sleeve. Pendry proposed trying to change the magnetic properties of a material. He wondered: could he take a material that was not intrinsically magnetic and try to realize magnetism from it by altering its internal physical structure alone?

Ordinarily either a material has an innate ability to be magnetized—generating forces from electrons moving inside them—or it does not. To invest this quality where it does not naturally occur, Pendry envisioned a theoretical metamaterial: a precisely crafted composite material that could selectively mimic properties of a conventional magnetic substance like iron. '*Magnetism involves charges going around in a circle*,' Pendry says. '*If electrons in atoms could do this, then we could do it on a larger scale*.' This kind of material, he hypothesized, could be manufactured from minuscule loops of copper wire (copper is not naturally magnetic) embedded in a dielectric material like fibreglass. Pendry predicted that when current flowed through those loops, a magnetic response would occur. There were many nuances to this scheme. If one cuts the loop one could make a magnetic resonator (this is the SRR), which would act like a controllable switch to control the magnetic properties of this artificial magnetic structure on command by the designer.

This concept of realizing tailorable artificial magnetism by Pendry might appear to be quite simple, yet it is very elegant. In fact, most of the elegant concepts are apparently simple and this inherent simplicity gives the glory of their elegance.

Pendry knew he was in uncharted territory whose hues, tints and shades may turn out to be remarkable one day, but at first he could not comprehend the magnitude of his utterly new idea of realizing artificial and tailorable magnetism (from non-magnetic material) with magnetic inclusion structure, SRR, of the so-called metamaterial. Anyway, the rest is history that has gifted us with exotic application possibilities of metamaterial like the cloaking device, superlens/supermicroscope making sub-wavelength imaging possible and many more useful and practically viable applications covering the electromagnetic spectrum from RF through THz to optical frequency domain in a span of just three decades!

2.2.4 Variants of SRR

With the first realization of negative permeability by Pendry *et al* in 1999 [11] using an SRR, it has received enthusiastic acceptance for various metamaterial-inspired designs, many researchers have modified it for better performance in terms of higher or lower frequency application, bandwidth, losses, bi-anisotropy and so forth. The SRR can be either a couplet as mentioned above, see figure 2.6(a), or multiple inclusion SRR, i.e. MSRR, as shown in figure 2.8(a) for increased magnetization and hence increased magnitude of effective permeability [15, 16]. However, SRR and MSRR both suffer from the problem of bi-anisotropy. This is caused due to asymmetric placement of slits in the two rings which results in non-zero electric dipole moment as the currents flowing across one slit do not cancel each other.

A brief mention of the difference between bi-isotropic and bi-anisotropic media would be worthwhile at this point. If in a medium electric field causes magnetic polarization, and the magnetic field induces electric polarization, i.e. magneto-electric coupling takes place—such a medium is termed as bi-isotropic. The media which exhibit magneto-electric coupling and which are also anisotropic, we term as bi-anisotropic; SRR and MSRR are such media.

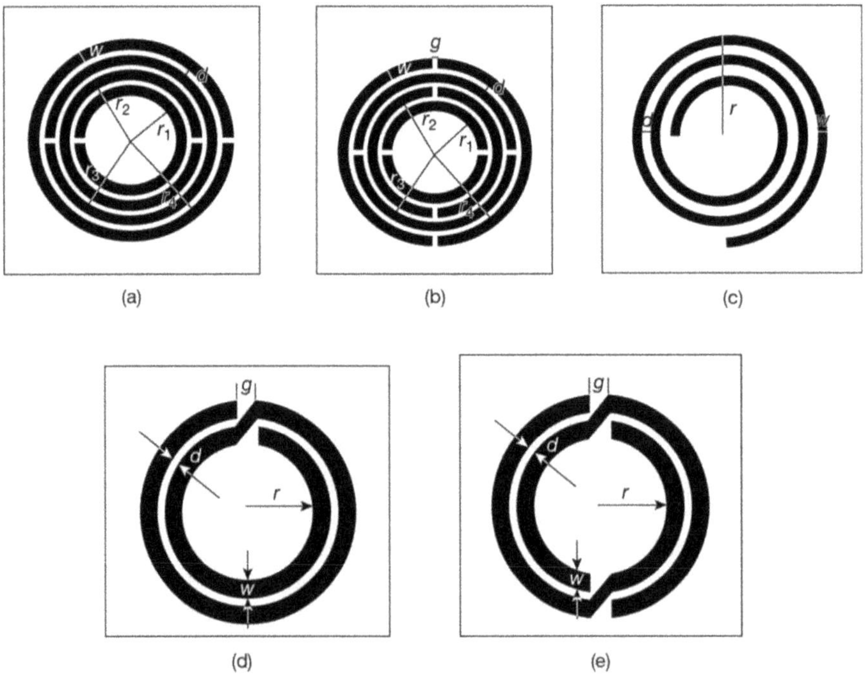

Figure 2.8. Variants of SRR: (a) MSRR, (b) LR, (c) SR, (d) TTSR, and (e) NBSR.

Anyway, the imbalance in the currents on both rings can be remedied by using a symmetric resonator structure, which can be done by using two slits in both rings and rotating one of them with respect to the other by 90°. Such a placement of rings results in the cancellation of the currents flowing across each slit by the current flowing on the slit that is located on the same ring. Thus we get the double-split-ring resonator (DSRR) or more aesthetically can be called a labyrinth resonator (LR) [15, 16], see figure 2.8(b). Another magnetic inclusion structure, which is a planar version of Pendry's Swiss-role structure [11], is the spiral resonator (SR) [17], see figure 2.8(c). Further, two variants of SRR/SR are two-turn spiral resonator (TTSR) and non-bi-anisotropic spiral resonator (NBSR) [18–20], see respectively figure 2.8(d), and figure 2.8(e), having advantages in some applications. In TTSR, two rings of SRR have cut on the same side and two ends are cross-joined for connectivity; this can also be considered as another form of SR of figure 2.8(c). NBSR is the symmetric version of TTSR where bi-anisotropy is reduced. SRR and all its variants are magnetic inclusion structures exhibiting resonant characteristics as depicted in figure 2.6(c).

A question naturally arises why we need to go for MSRR, SR, and LR instead of using the basic SRR structure for the design of various microwave and millimetre-wave components? Apart from realising higher magnetization and other issues discussed above regarding MSRR, SR, and LR, there are other practical reasons which need to be addressed in engineering design and we will discuss them in the following.

At microwave frequency the structures useful for miniaturization are MSRR and SR. But, when moving towards higher frequencies, the issue of miniaturization is not important and in fact miniaturization is avoided due to the technological limitations to print sufficiently small inclusions. In such cases, the design strategy is different and LR becomes the magnetic inclusion structure of best choice. The SRRs are designed with a physical dimension of the order of $\lambda/20$ at microwave frequency for a given assigned space occupancy of the inclusion when both L and C are realizable with it reaching their maximum value [21]. For further miniaturization, by lowering its resonant frequency, one may load an SRR in such a way as to increase its overall capacitance or inductance or both. Increasing the inductance while keeping the area of the SRR unaltered is rather challenging. Therefore miniaturization is usually realized by increasing only the capacitance.

One possibility is to add lumped capacitors in the split-gaps and between the gaps of the two rings of the SRR structure. Addition of such capacitance in parallel with both the split-gap's self-capacitance and the distributed capacitance between the two rings of SRR will increase the overall capacitance and lower the resonant frequency. But this solution is not a practical one when a large number of magnetic inclusions are needed to design a microwave component. Another practically viable possibility is to increase the capacitance of a resonant magnetic inclusion by increasing the distributed capacitance between the rings with innovative physical design of the structure. Typical designs capable of increasing the distributed capacitance are the MSRR, which consists of a number of split rings, and the SR, made of N turns.

The MSRR magnetic inclusion, see figure 2.8(a), which is a generalization of the regular SRR for $N > 2$, is characterized by an increased distributed capacitance. Considering the voltage distribution along the rings, it may easily be observed that the distributed capacitances between any pair of adjacent rings are all connected in parallel and, thus, they sum up to give an increased capacitive effect. It may further be noticed that while increasing the number of rings in the MSRR structure (for a given assigned space occupancy) by packing more and more inner structures, after a few rings, both the capacitance and inductance value saturate and the resonant frequency cannot be decreased after that limit any further. Thus MSRR may typically lead to miniaturization, with just a few numbers of rings, in the linear dimensions of the inclusion to the order of $\lambda/40$ to $\lambda/50$ [21]. Further miniaturization, if desired, may be obtained by using the so-called SR, see figure 2.8(c). The achievable miniaturization with SR is of the order of $\lambda/170$, which is useful for a lot of interesting applications (low-profile antennas and ultra-thin microwave absorbers) [21].

When magnetic inclusion structure has to be designed for millimetre-wave frequency, the miniaturization of structure is not followed but it is so designed that its resonant frequency falls in the millimetre-wave frequency range of operation. This means that, starting from the regular SRR, we need to decrease either the capacitance or inductance, or both, to make the inclusion resonate in the desired millimetre-wave frequency range. In order to decrease the inductance of the SRR one may think of the possibility of reducing the area occupied by the resonator. With this view in mind, we may continue to squeeze the SRR dimensions until it does not clash with the available technology to produce ever smaller sub-wavelength inclusions. When this limit is reached, the other possibility is to decrease the capacitance by increasing the split lengths and adding further splits in the SRR design and this is how we get to the so-called LR structure. Such magnetic inclusion structure comes from SRR just by adding a pair of two more splits. This structure divides the individual rings of the whole structure, see figure 2.8(b), into four sections. The distributed capacitance associated with each section is then 1/4 of the total distributed capacitance between the external and internal rings. Analysing the voltage distribution along the rings indicates that all the four capacitances are in series, leading to a significant reduction of the capacitive effects, and thus, helping to increase the resonant frequency of the LR structure. Further it may be noted that, since the distributed capacitance of the LR is smaller than that in the SRR, MSRR, and SR, the split capacitance cannot be ignored anymore, especially for small values of the split (g) of the LR structure.

A generalized expression for the effective negative permeability of magnetic inclusion structure (like MSRR, SR, and LR) can be given by [15]:

$$\mu_{\text{reff}} = 1 - \frac{F}{1 + j\dfrac{2R_c v}{\omega r \mu_0 (N-1)} - \dfrac{v}{\pi r^2 \mu_0 \omega^2 \left(C_1 \dfrac{2\pi r}{M} + \dfrac{N}{2} C_2 \right) (N-1)^2}} \tag{2.7}$$

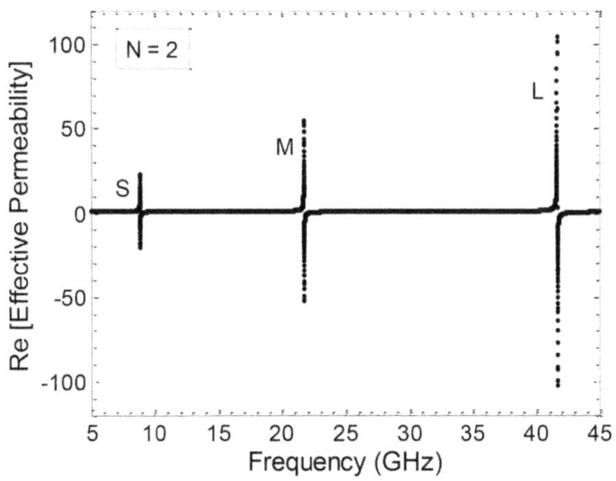

Figure 2.9. A comparative plot of the real part of effective permeability with frequency for SR (S), MSRR (M) and LR (L).

where F (= $\pi r^2/h^2$) is the fill-factor with h as the horizontal lattice constant; R_c is the resistance of the metallic strips per unit circumferential length of the ring; N is the number of metallic inclusions. The constant $M = 6$ (for MSRR), 1 (for SR), and 24 for LR. $C_1 = (\varepsilon_0/\pi)\ln[wN/d(N-1)]$ is the distributed capacitance between the strips of the adjacent rings per unit circumferential length of the ring with d being the separation between individual rings; and $C_2 = (\varepsilon_0 w t_M/g)$ is the gap capacitance between the cuts along a ring with t_M the thickness of the metal strip of the ring and g is the cut of the gap, for MSRR and SR this capacitance can be neglected. The derivation of equation (2.7) is given in appendix C.

In all the magnetic inclusion structures SRR, SR, LR the magnetic resonant frequency and hence the magnetic plasma frequency are controlled by the distributed capacitance caused by simultaneous currents flowing between adjacent rings induced by the H-field of the incident electromagnetic wave. The three structures in fact differ in the contribution of this effective synthesized capacitance. For example, SR has no splits like MSRR and thus will posses more capacitance while LR will have less capacitance than MSRR. Thus with more capacitive loading, the resonant frequency of SR will be less than that of MSRR while the resonant frequency of LR will be more than that of MSRR because of less capacitive loading. A typical comparative study of the three structures has been shown in figure 2.9 and table 2.1 [15]. It may be observed that LR will be a better magnetic inclusion structure at higher microwave frequency in terms of larger bandwidth of operation and design tolerances together with not being bi-anisotropic like SRR/MSRR.

Other variants of SRR structures have also been discussed in literature, see figure 2.10, providing size miniaturization at higher frequency of operation while designing metamaterial based couplers and filters. The U-shaped split-ring resonator (USRR) is a variant of SRR where two U-shaped strips are placed inverted on each other, which makes it easy to fabricate at higher frequencies. The USRR structure was initially designed for infrared frequency applications [22]. In the broadside

Table 2.1. Comparison of the characteristics of LR, MSRR, SR.

Structure type	r (mm)	Δf (GHz)	f_{m0} (GHz)
LR	1.000	0.670	41.8
MSRR	0.648	0.278	21.5
SR	0.357	0.081	8.8

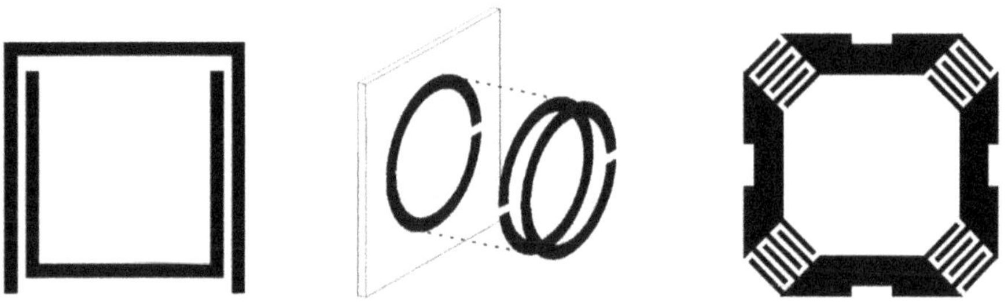

(a) (b) (c)

Figure 2.10. Other variants of SRR and magnetic inclusion structure (a) USRR, (b) BC-SRR and (c) ICRR.

coupled split-ring resonator (BC-SRR), two split-rings of an SRR are on two sides of a substrate having a broadside coupling [23]; it is an alternative of an SRR to avoid bi-anisotropic effects. In inter-digital capacitor loaded ring resonator (ICRR), a square ring is loaded with inter-digital capacitors at four corners [24]. ICRR is the only negative permeability structure that exhibits totally polarization-independent behaviour with respect to the applied electric field due to its unique shape. A comparative study on the characteristics of different magnetic inclusion structures is available in the literature [25].

2.2.4.1 An equivalent circuit characterization methodology for MSRR, SR, and LR

A very useful and practical technique has been developed by Bilotti *et al* [21] to model and design artificial magnetic inclusion structures like MSRR, SR and LR. We will discuss here in brief this quasi-static equivalent circuit approach for characterization of these important magnetic inclusion structures. The quasi-static approach is used as the inclusions are much smaller than the operating wavelength, which is, indeed, the case when artificial magnetic inclusions are used to fabricate metamaterial samples.

In their work, Bilotti *et al* [21] gave a complete circuit model that has taken into account the presence of a dielectric substrate where the magnetic inclusions are printed on, see figure 2.11, and considered the losses both in the dielectrics and the metallic conductors. They validated their analytical results obtained from the equivalent circuit model with full-wave simulation and experimental measurements too.

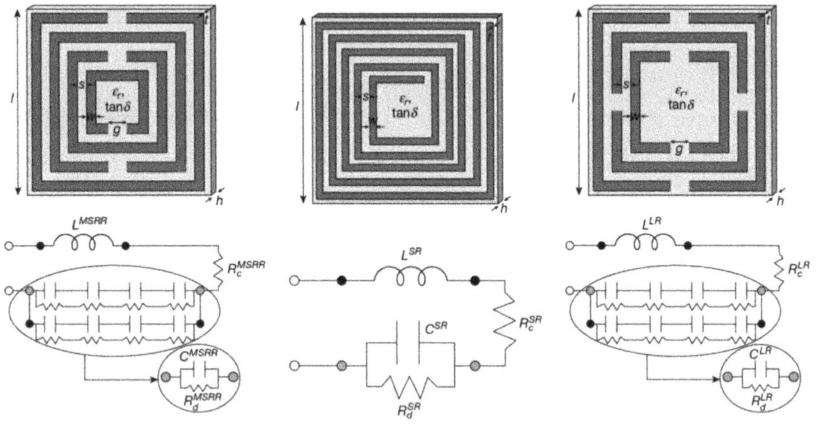

(a) (b) (c)

Figure 2.11. Sketch with geometrical dimensions and the corresponding quasi-static equivalent circuit model of (a) MSRR, (b) SR, and (c) LR.

The expressions for the circuit components in the equivalent circuit model are given below for the three magnetic inclusion structures.

Multiple split-ring resonators (MSRRs)

$$L^{\mathrm{MSRR}} = \frac{\mu_0}{2}\frac{l_{\mathrm{avg}}}{4}4.86\left[\ln\left(\frac{0.98}{\rho}\right) + 1.84\rho\right] \tag{2.8a}$$

where μ_0 is the permeability of vacuum, l is the side length of the external ring, w is the width of the strips, s is the separation between two adjacent strips, $l_{\mathrm{avg}} = 4[1 - (N - 1)(w + s)]$ is the average strip length calculated over all the N rings, and $\rho = (N - 1)(w + s)/[l - (N - 1)(w + s)]$ is the so-called filling ratio.

It may be observed that the presence of dielectric substrate, upon which the MSRRs is printed, does not affect the inductance (this is true for SR and LR too). However, it will be seen that the presence of the dielectric affects the distributed capacitance between the rings, as expected.

The capacitance is given by:

$$C^{\mathrm{MSRR}} = \frac{N - 1}{2}[2l - (2N - 1)(w + s)]C_0 \tag{2.8b}$$

with

$$C_0 = \varepsilon_0\varepsilon_r^{\mathrm{sub}}(\varepsilon_r, h, w, s)\frac{K(\sqrt{1 - k^2})}{K(k)} \tag{2.8c}$$

where ε_0 is the permittivity of vacuum, K is the complete elliptic integral of first kind, $k = s/(s + 2w)$, and the effective relative permittivity $\epsilon_r^{\mathrm{sub}}$ related to the dielectric filling the substrate is given by:

$$\varepsilon_{\mathrm{r}}^{\mathrm{sub}}(\varepsilon_r, h, w, s) = 1 + \frac{2}{\pi}\mathrm{arctg}\left[\frac{h}{2\pi(w+s)}\right](\varepsilon_r - 1) \tag{2.8d}$$

Referring to figure 2.11(a), the series resistance $R_{\mathrm{c}}^{\mathrm{MSRR}}$ (representing the losses in the conductor) and shunt resistance $R_{\mathrm{d}}^{\mathrm{MSRR}}$ (that takes care of the losses in the dielectric substrate) are given by:

$$R_{\mathrm{c}}^{\mathrm{MSRR}} = \frac{\rho_c}{wt}\frac{L^{\mathrm{MSRR}}}{\mu_0} \tag{2.8e}$$

and

$$R_{\mathrm{d}}^{\mathrm{MSRR}} = \frac{1}{\sigma_d}\frac{s}{h[l-(2w+s)]}\frac{l_{\mathrm{avg}}}{4l} \tag{2.8f}$$

where ρ_{c} is the electrical resistivity of the metal and σ_{d} is the conductivity of the dielectric substrate. Using equations (2.8a)–(2.8f) and the circuit representation depicted in figure 2.11(a) it is possible to calculate analytically the resonant frequency of the MSRR inclusion.

Spiral resonator (SR)

$$L^{\mathrm{SR}} = \frac{\mu_0}{2\pi}l_{\mathrm{avg}}^{\mathrm{SR}}\left[\ln\left(\frac{l_{\mathrm{avg}}^{\mathrm{SR}}}{2w}\right) + \frac{1}{2}\right] \tag{2.9a}$$

$$C^{\mathrm{SR}} = \frac{l}{4(w+s)}\frac{N^2}{N^2+1} \times \left[l(N-1) - \frac{N^2-1}{2}(w+s)\right]C_0 \tag{2.9b}$$

where N is the number of turns of the spiral, l is the side length of the external turn, w is the width of the strips, s is the separation between two adjacent turns, C_0 is as defined for the MSRR, and $l_{\mathrm{avg}}^{\mathrm{SR}} = 4l - [2(N+1) - 3/N](s+w)$ is the average length of the spiral turn.

It is worth noting that, as in the case of the MSRR, here also squared version of the SR has been considered. Anyway, the same formulation also applies to the circular counterparts, simply by changing the length of the straight segments with the length of the circular ones.

The series resistance due to the losses in the conductor is given by:

$$R_c^{\mathrm{SR}} = \frac{\rho_c}{\mathrm{wt}}\frac{L^{\mathrm{SR}}}{\mu_0} \tag{2.9c}$$

while the shunt resistance arising out of the dissipation in the dielectric substrate is given by:

$$R_d^{\mathrm{SR}} = \frac{1}{\sigma_d}\frac{s}{4h[l-(2w+s)]}\frac{l_{\mathrm{avg}}^{\mathrm{SR}}}{4l} \tag{2.9d}$$

Also, in this case, from the equivalent circuit depicted in figure 2.11(b) and equations (2.9a)–(2.9d), it is possible to analytically determine the resonant frequency of the spiral resonator.

Labyrinth resonator (LR)

For higher microwave frequencies, miniaturization is not always desired, as mentioned earlier, and it is found that an LR is the most suitable magnetic inclusion structure. As for the other structures like MSRR and SR the inductance of LR is not affected by the presence of the dielectric substrate and is given by:

$$L^{\mathrm{LR}} = \frac{\mu_0}{2\pi}\frac{l_{\mathrm{avg}}^{\mathrm{LR}}}{4}\left[\ln\left(\frac{l_{\mathrm{avg}}^{\mathrm{LR}}}{w}\right) - 2\right] \tag{2.10a}$$

In LR, the capacitance is constituted of two contributions. One is due to the distributed capacitance (as was the case for MSRR and SR) and the other is due to the gap capacitance. In fact, in LR it is no longer advisable to neglect the gap capacitance. Since the effect of the distributed capacitance is significantly reduced in LR structure, the length of the gap g (see figure 2.11(c)) starts playing an important role. The gap capacitance, in fact, may be of the order of magnitude of the total distributed capacitance.

Considering the series and parallel connections of the capacitances in figure 2.11(c) (shown below the LR structure) the distributed capacitance between any pair of adjacent rings is given by:

$$C_{\mathrm{Dsitributed}}^{\mathrm{LR}} = \frac{C_0}{16}\left\{(N-1)\left[4(l-g) - \frac{N}{2}(s+2w)\right]\right\} \tag{2.10b}$$

While the gap capacitance is given by the sum of the capacitances of the $2N$ gaps:

$$C_{\mathrm{Gap}}^{\mathrm{LR}} = 2N\varepsilon_0\varepsilon_{\mathrm{r}}^{\mathrm{Sub}}(\varepsilon_r, h, w, s)\frac{2w+\sqrt{2}g}{\pi}\mathrm{arccos}h\left[\frac{2w+g}{g}\right] \tag{2.10c}$$

Since the gap capacitance appears in parallel with the distributed capacitance in LR, the total capacitance is given by:

$$C^{\mathrm{LR}} = C_{\mathrm{Distributed}}^{\mathrm{LR}} + C_{\mathrm{Gap}}^{\mathrm{LR}} \tag{2.10d}$$

The series resistance due to losses in the conductor of the LR resonator is obtained in the same way as for MSRR and SR as:

$$R_c^{\mathrm{LR}} = \frac{\rho_c}{\mathrm{wt}}\frac{L^{\mathrm{LR}}}{\mu_0} \tag{2.10e}$$

But the shunt resistance due to losses in the dielectric medium is given by the two following contributions:

$$R_{d1}^{\mathrm{LR}} = \frac{1}{\sigma_d} \frac{s}{h[4l - 4(2w + s) - 2g]} \frac{l_{\mathrm{avg}}^{\mathrm{lR}}}{4l} \tag{2.10f}$$

$$R_{d2}^{\mathrm{LR}} = \frac{1}{\sigma_d} \frac{g}{2\mathrm{hwN}} \tag{2.10g}$$

where the average length of the rings is given by:

$$l_{\mathrm{avg}}^{\mathrm{LR}} = 4l - 2g - (N - 1)(s + w) \tag{2.10h}$$

It may be noted that R_{d1}^{LR}, which is the resistance associated with dielectric losses between the strips, has been calculated as for the MSRR and SR, while R_{d2}^{LR}, which represents the dielectric losses in the cuts of the gap, is derived straightforwardly from the definition of resistance and considering that all the $2N$ contributions are connected in parallel. The final expression of the equivalent shunt resistance is thus given by:

$$R_{\mathrm{deq}}^{\mathrm{LR}} = \frac{R_{d1}^{\mathrm{LR}} R_{d2}^{\mathrm{LR}}}{R_{d1}^{\mathrm{LR}} + R_{d2}^{\mathrm{LR}}} \tag{2.10i}$$

Here also, using equations (2.10a)–(2.10i), it is possible to calculate analytically the resonant frequency of individual LRs.

2.2.5 A few more negative permittivity and negative permeability structures

Apart from the wire media and the SRR and its variants discussed so far, there are other negative permittivity and negative permeability structures that have been used for various applications demanded by the designers. One such structure is the so-called complementary split-ring resonator (CSRR), see figure 2.12(a), proposed by Falcone [26] which is complementary, or the dual or electric analogue of the magnetic inclusion structure SRR. It is known that SRR responses to the magnetic field of the incident electromagnetic radiation while CSRR responds to the electric field of the electromagnetic radiation, as per Babinet's principle. CSRR as an electric inclusion structure of metmaterial is very important from an application point of view, for example we will see in chapter 3 that a CSRR loaded microstrip antenna is a very useful improvisation of conventional microstrip patch antenna. Machac *et al* [27] proposed the single electric dipole and spider dipole shown in figures 2.12(b) and (c). Single dipole is an anisotropic medium due to its dependence on the direction of electric field applied while spider dipole is polarization independent.

Another negative permittivity structure is shown in figure 2.12(d) introduced by Ruopeng *et al* [28] where the resonance of the medium was set by the internal inductance and capacitance of the unit cell rather than the cell-to-cell capacitive coupling. These electric LC (ELC) resonators are formed by the combination of two

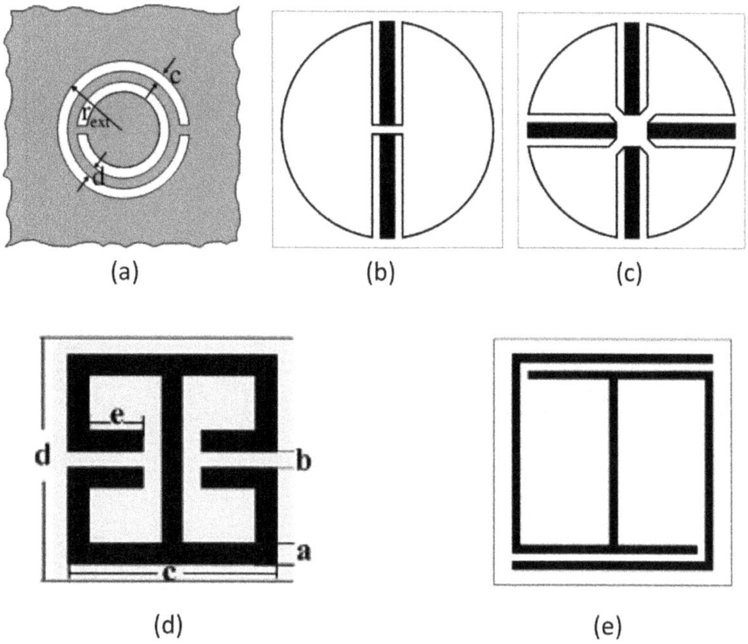

Figure 2.12. Various other negative permittivity structures: (a) CSRR, (b) single electric dipole, (c) spider dipole, (d) ELC resonator, and (e) a double H-shaped resonator.

identical SRRs placed back to back. When excited by an electric field, currents flow in such a manner as to cancel out any resonant magnetic response, leaving only a resonant electric response. ELC resonators are convenient to control and can be made relatively insensitive to the cell-to-cell coupling.

Blaha *et al* [29] introduced a new kind of planar structure denoted as a double H-shaped resonator (DHR), as shown in figure 2.12(e). This structure is of a resonant nature, and consequently its operation is frequency selective. The interaction of the exciting field with the particle results in overall negative effective permittivity in a narrow frequency band above its resonant frequency.

We have discussed SRR in section 2.2.2 and its variants in section 2.3 including other magnetic inclusion structures, but there are also some other magnetic inclusion structures developed from time to time (which are basically variants of SRR) to realize negative permeability, together with negative permittivity demanded by some applications. One such structure is axially symmetric SRR [30], shown in figure 2.13(a), especially designed for infrared frequencies. This structure provides good transmission properties and has good field symmetry. However, for waveguide mounting the contact issue of the rod structure in it sometimes becomes a problem. Omega particle [30, 31], see figure 2.13(b), also known as omega SRR, is a single structure that combines split-rings and a rod. The two rings being back to back helps to cancel the problem of bi-anisotropy present

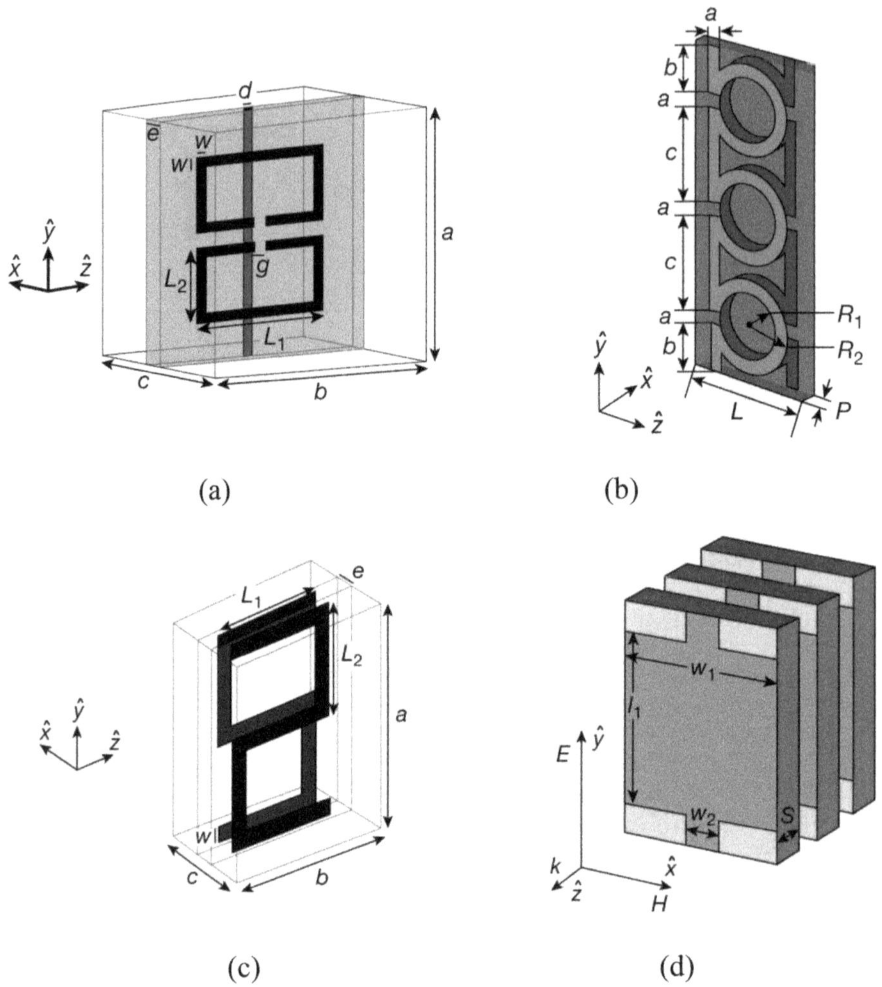

Figure 2.13. (a) Axially symmetric SRR, (b) omega particle, (c) S-shaped SRR, and (d) fishnet structure.

with conventional SRRs. It is suitable for 3D metamaterials and it can also be fabricated using hot press technique. Here also the contact issue of the rod remains a problem for waveguide mounting. The S-shaped split-ring or simply S-ring (its popular name) [30, 32], shown in figure 2.13(c), is another structure having wide bandwidth and low losses. It is capable of giving negative refractive index from single structure and can have multiple bands if slightly modified. The foremost importance of S-ring is that it does not require the addition of a rod to exhibit a negative permittivity at similar frequencies to where it exhibits a negative permeability. There is no rod issue with S-ring metamaterial structure, thus it is a significant advantage of this structure for waveguide mounting. The fishnet structure [33], shown in figure 2.13(d), is another magnetic inclusion structure which have many unique variants used up to optical frequencies.

2.3 Transmission-line metamaterial

2.3.1 Introduction

The successful realization of negative refractive index with plasmonic metamaterial realized with TW-SRR combination in 2001 was definitely a great stepping stone for taking forward metamaterial research and for making V G Veselago's visionary speculation of LHM (i.e. substances with simultaneously negative values of ε and μ) a reality. But the negative permittivity and negative permeability realisable with TW and SRR, respectively, near their electric and magnetic plasma frequency makes the plasmonic metamaterial of TW-SRR type highly lossy and having narrow bandwidth of operation. But from a practical engineering application point of view the need is actually for low-loss and broadband metamaterial/LH media. The limitations/weaknesses of resonant-type LH media prompted the LHM researchers to look for an alternative architecture. In fact, recognizing the analogy between the LH waves possible with the dual of the normal transmission line, three groups [34–36] introduced almost simultaneously in June 2002, a transmission line approach of metamaterials. In December 2002 even the realization of negative refractive index with transmission-line metamaterial was also reported [37] for the first time.

It may be mentioned herewith that LH structures are characterized by the so-called backward waves (i.e. the phase velocity is in the negative z-direction and antiparallel to the group velocity) in addition to its negative refractive index and many other related properties. But the backward-wave propagation was known for decades in periodic structures [38, 39] since the 1940s. But the novelty of LH materials is in the fact that they are effectively homogeneous fundamental-mode structures, fully characterized in terms of their constitutive parameters ε and μ (hence the name metamaterial or in short MTM). But the previously known backward-wave structures were scattering media based on the propagation of (negative) space harmonics.

2.3.2 LH/CRLH transmission line and their characteristics

As mentioned above, an LH transmission line is the *dual* of the conventional transmission line, i.e. the unit cell of an LH transmission line would consist of a capacitance in the series arm and an inductance in the shunt arm. The practical implementation of transmission line metamaterial, i.e. LH transmission line is done by periodically loading a host conventional transmission line (keeping periodicity p much smaller than the wavelength of operation) with series capacitance (C_L) and shunt inductance (L_L), as shown in figure 2.14(a), which is thus basically a periodically loaded transmission line (PLTL) synthesizing an LH media or LHM transmission line. This is a 1D structure while 2D and 3D structures have different configurations [40]. Each unit cell of LH transmission line in figure 2.14(a) is seen to be a so-called *composite right/left-handed* (CRLH) transmission line (see figure 2.14(b))—this is because in a practical LH transmission line mixed contributions of both the LH and RH structures are to be considered. In fact, in the physically realizable planar LH structures, see figure 2.14(c), in addition to the LH series-C/shunt-L the contributions

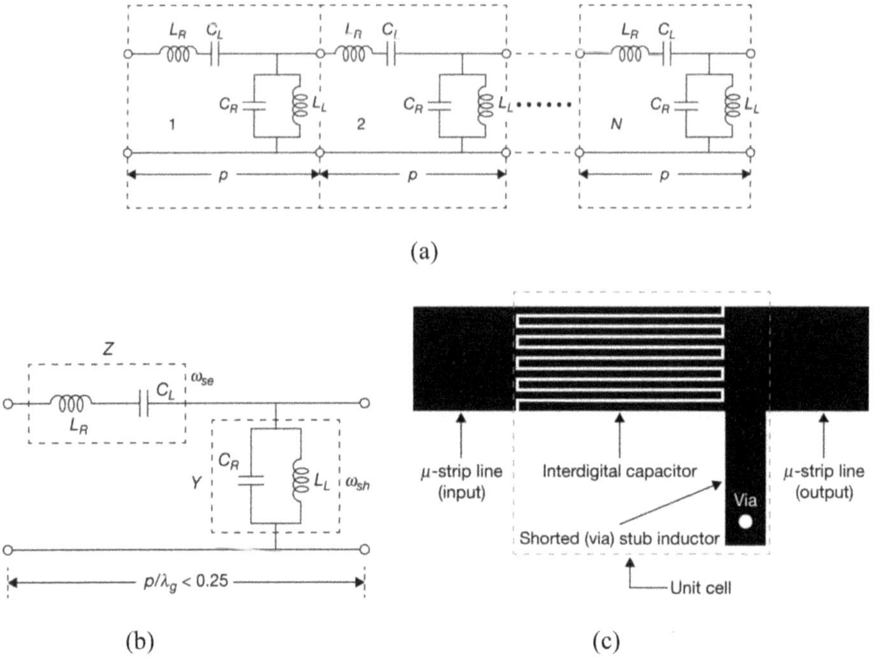

Figure 2.14. (a) Periodically loaded N-section host (RH; L_R, C_R) transmission line with LH elements (C_L, L_L) in which $p < \lambda_g/4$ is the periodicity (unit cell dimension). (b) CRLH transmission line representing each unit cell of PLTL. (c) Physically realizable planer, i.e. microstrip (μ-strip) LH structure with interdigital capacitor and stub inductor.

of RH series-L/shunt-C comes, respectively, due to the magnetic flux induced by the currents in the fingers of the inter-digital capacitors (IDCs) and due to the voltage gradient that exists between the fingers of the IDC with the ground plane of the planer (commonly microstrip) structure. The characteristics of CRLH have been accurately described and applied to a number of guided-, radiated-, and refracted wave applications [40]. Anyway, it may be noted that the basic difference between the periodic structures made with normal material or RHM have a periodicity (p) of the order of half guide-wavelength (λ_g) (i.e. $p \approx \lambda_g/2$) while for metamaterial or LHM, $p < \lambda_g$ (typically $p/\lambda_g < 0.25$, say 0.1); when effective (average) macroscopic constitutive parameters ε_{eff} and μ_{eff} comes into play.

Let us now derive the characteristics of basic LH and CRLH transmission line [40] starting from the telegrapher's equation as is normally done for conventional transmission lines too [41]. Referring to figure 2.14(b), we can write the telegrapher's equation for CRLH transmission line for cosine reference on assuming time harmonic variation for voltage and current as:

$$\frac{dV}{dz} = -\text{ZI} \text{ and } \frac{dI}{dz} = -\text{YV} \tag{2.11}$$

where

$$Z = j\left(\omega L_R - \frac{1}{\omega C_L}\right) \text{ and } Y = j\left(\omega C_R - \frac{1}{\omega L_L}\right) \tag{2.12}$$

In equation (2.11) V and I are the position dependent voltage and currents [$V = V(z)$ and $I = I(z)$] along the line, respectively. By solving simultaneously the two equations in (2.11) we get the wave equations for V and I as:

$$\frac{d^2V}{dz^2} - \gamma^2 V = 0 \text{ and } \frac{d^2I}{dz^2} = -\gamma^2 I \tag{2.13}$$

where γ is the complex propagation constant expressed in the per unit immittances as:

$$\gamma = \alpha + j\beta = \sqrt{ZY} \tag{2.14}$$

and is associated with $+z/-z$-propagating ($e^{-\gamma z}/e^{+\gamma z}$) travelling wave solutions:

$$V(z) = V_0^+ e^{-\gamma z} + V_0^- e^{+\gamma z} \tag{2.15a}$$

$$I(z) = I_0^+ e^{-\gamma z} + I_0^- e_0^{+\gamma z} = \frac{\gamma}{Z}[V_0^+ e^{-\gamma z} - V_0^- e^{+\gamma z}] \tag{2.15b}$$

The second equality in equation (2.15b) was obtained by taking the derivative of equation (2.15a) and equating the resulting expression with the first equation of equation (2.11).

The characteristic impedance Z_c is then given by:

$$Z_c = \frac{V_0^+}{I_0^+} = -\frac{V_0^-}{I_0^-} = \frac{Z}{\gamma} = \sqrt{\frac{Z}{Y}} \tag{2.16}$$

Now let us summarize the characteristics of both the basic LH transmission line and CRLH transmission line. In both the cases we shall consider the simple loss-less case, however, the detailed analysis of lossy case is available in advanced texts [40]. Further, detailed analysis of 2D MTM transmission line is available in suitable references [37, 40].

2.3.2.1 Basic LH transmission line characteristics

Refer to figure 2.14(a), which is an incremental section of a loss-less LH transmission line. For this we can now obtain expressions for complex propagation constant γ, propagation constant β, the characteristic impedance Z_c, the phase velocity v_p, and the group velocity v_g as follows:

$$\gamma = \alpha + j\beta = \sqrt{ZY} = \frac{1}{j\omega\sqrt{L_L C_L}} = -j\frac{1}{\omega\sqrt{L_L C_L}}$$

$$\beta = -\frac{1}{\omega\sqrt{L_L C_L}} < 0$$

Table 2.2. Equivalent circuit with different parameters of RH and LH transmission lines.

Parameters	β	Z_c	V_p	V_g	n
RHM 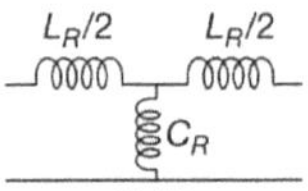	$\omega\sqrt{L_R C_R}$	$\sqrt{\frac{L_R}{C_R}}$	$\frac{1}{\sqrt{L_R C_R}}$	$\frac{1}{\sqrt{L_R C_R}}$	$\frac{\sqrt{L_R C_R}}{\sqrt{\mu_0 \varepsilon_0}}$
LHM 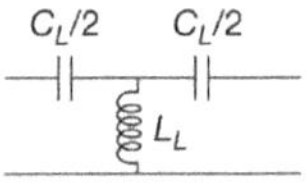	$-\frac{1}{\omega\sqrt{L_L C_L}}$	$\sqrt{\frac{L_L}{C_L}}$	$-\omega^2\sqrt{L_L C_L}$	$+\omega^2\sqrt{L_L C_L}$	$-\frac{1}{\omega^2\sqrt{L_L C_L \mu_0 \varepsilon_0}}$

$$Z_c = \sqrt{\frac{Z}{Y}} = +\sqrt{\frac{L_L}{C_L}}$$

$$v_p = \frac{\omega}{\beta} = -\omega^2\sqrt{L_L C_L} < 0$$

$$v_g = \left(\frac{\partial \beta}{\partial \omega}\right)^{-1} = +\omega^2\sqrt{L_L C_L} > 0$$

$$n = \frac{c}{v_p} = -\frac{1}{\omega^2\sqrt{L_L C_L}\sqrt{\mu_0 \varepsilon_0}} < 0 \tag{2.17}$$

A comparison of the characteristics of RH and LH transmission line has been shown in table 2.2.

It may be observed from equation (2.17) and table 2.2 that for LH transmission line the phase velocity, v_p (i.e. the velocity with which the wavefront moves) is antiparallel with the group velocity, v_g (i.e. the velocity with which the energy of the electromagnetic wave moves or the Poynting vector moves); unlike RH transmission line for which both move in the direction of propagation. Thus we may say that the LH transmission line exhibits the well-known phenomenon of *backward wave* as the wavefront moves in the backward ($-z$-direction) while the Poynting vector remains directed in the $+z$-direction.

Further, it may be noted that even though the LH media is dispersive and ε and μ are negative—it does not violate causality. It is known that for frequency dispersive ε and μ, from Poynting's theorem, the expression for the mean value of the internal energy per unit volume of the medium is [40]:

$$\overline{W} = \frac{\partial(\varepsilon\omega)}{\partial\omega} \mid E \mid^2 + \frac{\partial(\mu\omega)}{\partial\omega} \mid H \mid^2 \tag{2.18}$$

This is known as entropy condition for dispersive media. By the law of entropy, specifying that the entropy of a system is an ever increasing quantity, we must have:

$$\overline{W} > 0$$

Hence for the dispersive case the following two inequalities need to be satisfied:

$$\frac{d(\varepsilon\omega)}{d\omega} > 0$$

$$\frac{d(\mu\omega)}{d\omega} > 0$$

Thus, even when ε, $\mu < 0$, their spectral derivatives remain positive. Hence, causality is not violated.

So far we have discussed the LH transmission line as dual of conventional transmission line in analogous form. But this does not give physical insight of realizing negative permittivity with series C and negative permeability with shunt L by periodically loading a conventional transmission line. However, physical understanding and rigorous mathematical analysis based on electromagnetic treatment of the whole problem is available in an advanced textbook [42]. Further, it may be noted that we have discussed so far 1D LH transmission line. Detailed analysis of 2D or planer LH transmission line is also available in advanced texts [40, 42]. But we have not repeated those discussions and analysis here for brevity of presentation in this book. Interested readers may consult the referred texts for those details.

2.3.2.2 CRLH transmission line characteristics and ZOR

Referring to figure 2.14(b) it may be observed that at low frequencies, the host transmission line (RHM) circuit elements L_R and C_R behave practically as short ($\omega L_R \to 0$) and open ($1/\omega C_R \to \infty$), respectively, and hence the CRLH behaviour is then determined with series C_L and shunt L_L. This exhibits pure left-handed (PLH) characteristics with a *high-pass* nature (with negative phase constant) and having phase velocity antiparallel with the group velocity (see equation 2.17). There is thus an LH stop-band below a certain cut-off frequency (ω_{cL}); see the dispersion diagram, ω versus β depicted in figure 2.15.

At high frequencies, C_L and L_L behaves, respectively, as short ($1/\omega C_L \to 0$) and open ($\omega L_L \to \infty$) so that the equivalent CRLH is then expected to exhibit a *low-pass* nature (pure right-handed, PRH, with positive phase constant and having parallel phase and group velocity, see table 2.2) with series L_R and shunt C_R; giving a RH stop-band above a certain cut-off frequency (ω_{cR}). In general, the series resonance frequency ($\omega_{se} = \sqrt{L_R C_L}$) and shunt resonance frequency ($\omega_{sh} = \sqrt{L_L C_R}$) are different; and then the line is called an *un-balanced* CRLH line. These CRLH curves are seen to differ significantly from the PLH and PRH curves as those are determined by the combined effects of LH and RH contributions at all frequencies. However, if these two frequencies are made equal, i.e. $\omega_{se} = \omega_{sh} = \omega_0$, we call it *balanced* CRLH line, when $L_R C_L = L_L C_R$ equivalently meaning $Z_L = Z_R$ (where Z_L and Z_R are, respectively, the characteristic impedance of PLH and PRH line; as the characteristic impedance Z_c of high frequency loss-less line is $\sqrt{L/C}$). Under this condition, the gap disappears and at the *transition frequency* ($\omega_0 = \sqrt{\omega_{se}\omega_{sh}}$) an *infinite wavelength* ($\lambda_g = 2\pi/|\beta|$) is achieved (as $\beta = 0$ at this frequency) implying that the balanced condition allows matching over an infinite bandwidth. Thus from a practical point of view, the balanced CRLH transmission line is preferred as it

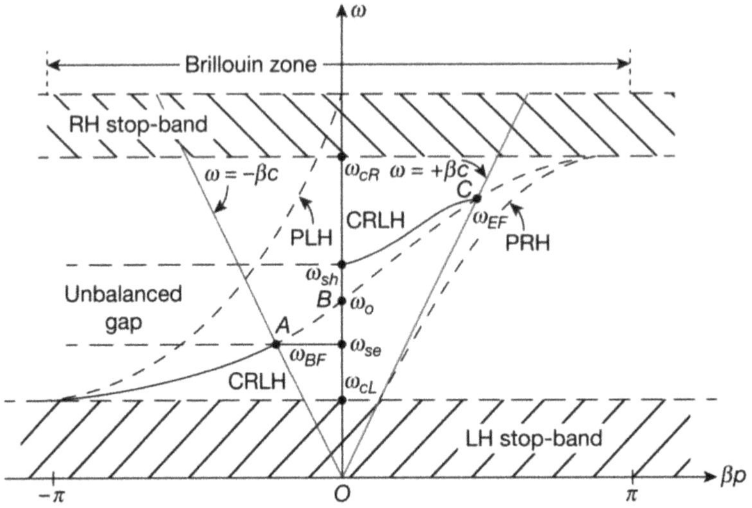

Figure 2.15. Dispersion diagram (ω versus β) indicating PLH, PRH and CRLH characteristics.

can be matched over a broad frequency band, whereas the un-balanced one can be matched over a restricted bandwidth (strictly at one single frequency).

It is interesting to note from figure 2.15 that at ω_0 phase shift along a line of length l is zero [$\varphi = -(\beta)l = 0$, as $\beta = 0$]. When frequency is decreased below ω_0 towards zero, β goes on taking higher and higher negative value making phase shift along the line to increase with positive value [$\varphi = -(-\beta)l = +\beta l$] tending to infinity as $\omega \rightarrow 0$. But when frequency is increased above ω_0 towards infinity, β goes on taking higher and higher positive value making phase shift along the line to increase with negative value [$\varphi = -(\beta)l$] tending to infinity as $\omega \rightarrow \infty$.

It may further be noted that the CRLH structure is never operated at the edges of Brillouin zone where $p \approx \lambda_g/2$ (though RH and LH stop bands exists there); instead, it is used only in the vicinity of the transition frequency where $p < \lambda_g/4$, ensuring effective material property, where p is the average unit cells size and λ_g is the wavelength of the propagating electromagnetic signal in the composite CRLH medium.

Depending upon the nature of termination, the CRLH will support travelling wave or standing wave. If it is terminated with matched load it will naturally support travelling wave but if it is terminated with either open or short it will support standing wave, when it can be used as a *resonator*. Although PRH transmission line (i.e. the conventional transmission line and waveguide) also exhibit resonator characteristics with similar terminations, we would observe shortly that CRLH transmission line resonator has some unusual characteristics. It is said to allow negative and zeroth-order resonances, which has the potential to design dual-band and size-independent resonator (in contrast, normal resonator length should be at least half-wavelength or any integral multiple of it). This makes transmission line type planar LH material very attractive for the design of miniaturized microwave circuit components and antennas.

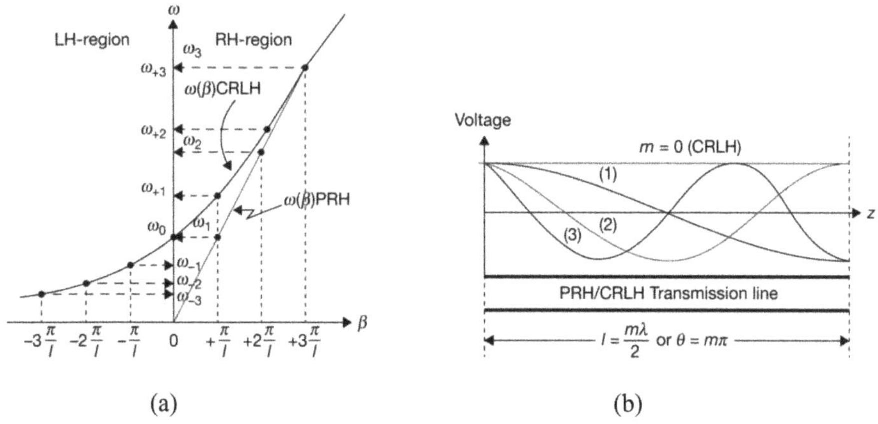

Figure 2.16. (a) Dispersion curves of conventional (PRH) and CRLH transmission line resonators. (b) Field distribution of the resonance modes: (1) m = 1 (PRH), ±1 (CRLH); (2) m = 2 (PRH), ±2 (CRLH); (3) m = 3 (PRH), ±3 (CRLH).

In a conventional distributed circuit resonator (realizable with transmission line or waveguide), the resonance frequencies ω_m correspond to the frequencies where the physical length l of the structure is a multiple of half-wavelength (which means that the electrical length $\theta = \beta l$ is a multiple of π), i.e.:

$$l_m = m\frac{\lambda}{2} \text{ or } \theta_m = \beta l_m = m\pi, \text{ with } m = +1, +2, ..., +\infty \tag{2.19a}$$

It may be observed from figure 2.16(a) that the dispersion curve $\omega(\beta)_{\text{PRH}}$ is linear and the resonance frequencies are obtained by sampling the dispersion curve with sampling rate of π/l in the β variable. An infinite number of resonances are possible with conventional (PRH) transmission line since such a line ideally has infinite bandwidth, extending from $\omega = 0$ to $\omega = \infty$. Since the electrical length of PRH transmission line can be only positive, we can have only non-zero and positive resonances in the case of PRH. The field distribution of the resonance modes, i.e. the voltage distribution (for the case of an open circuited transmission line) with conventional (PRH) transmission line resonator is shown in figure 2.16(b). As the dispersion curve is linear, the higher order resonances (ω_m) are just the harmonics of the fundamental resonance (ω_1), i.e. $\omega_m = m{\cdot}\omega_1$; where m is the harmonic number.

The dispersion curve $\omega(\beta)_{\text{CRLH}}$ for a CRLH resonator based on balanced CRLH transmission is also shown in figure 2.16(a) with typical field distributions of the resonance modes in figure 2.16(b). The propagation constant (β) of a balanced CRLH line can be expressed as the sum of the propagation constant of a PRH transmission line (with linear and positive ω versus β) and of a PLH transmission line (having negative and hyperbolic ω versus β characteristics). It may be observed from figure 2.16(a) that CRLH transmission line can have $\beta = 0$ (at transition frequency ω_0) as well as $\beta < 1$ (in the LH region), which is not the case with PRH or conventional transmission line. Thus the electrical length $\theta = \beta l$ can be zero and negative, which implies that the resonance index m takes values: 0, ±1, ±2, ...±∞ with:

$$l_m = | m | \frac{\lambda}{2} \text{ or } \theta_m = \beta l_m = m\pi \tag{2.19b}$$

Here also, resonance frequencies are obtained by sampling the dispersion curve with sampling rate of π/l in the β variable and an infinite number of resonances are possible with ideal CRLH resonator. But the CRLH resonator has a number of significant differences when compared with the conventional (PRH) transmission line resonator. Firstly, *negative resonances* (at $m < 0$) and *zero resonance* (at $m = 0$) exist in CRLH resonator in addition to the usual positive resonances (at $m > 0$) observed for PRH resonator. The mode with $m = 0$ is particularly interesting as this mode corresponds to a flat-field distribution (with no voltage gradient along the length of the CRLH line, see figure 2.16(b)) and therefore is expected to be independent of the physical length of the transmission line; which implies on theoretical ground that it can be used to realize arbitrarily small resonator, popularly known as *zeroth-order resonator* (ZOR). Secondly, the field distribution of all modes with $m = \pm 1, \pm 2, \ldots$ would have identical impedance (as they are having same voltage distribution, as shown in figure 2.16(b)); this may be useful for designing dual-band CRLH filters and antennas. Thirdly, the non-linear nature of the dispersion curve of CRLH transmission line indicates that, especially in the LH region, the resonant frequencies are not in the harmonic ratios and the resonant spectrum gets more and more compressed as we move towards lower frequencies.

References

[1] Smith D R, Padilla W J, Vier D C, Nemat-Nasser S C and Schultz S 2000 Composite medium with simultaneously negative permeability and permittivity *Phys. Rev. Lett.* **84** 4184–7

[2] Houck A A, Brock J B and Chuang I L 2003 Experimental observations of a left-handed material that obeys Snell's law *Phys. Rev. Lett.* **90** 137401–4

[3] Parazzoli C J, Greegor R B, Li K, Koltenbah B E C and Tanielian M 2003 Experimental verification and simulation of negative index of refraction using Snell's law *Phys. Rev. Lett.* **90** 107401–4

[4] Garcia N and Nieto-Vesperianas M 2002 Left handed materials do not make a perfect lens *Phys. Rev. Lett.* **88** 207403–6

[5] Garcia N and Nieto-Vesperians M 2002 Is there an experimental verification of negative index of refraction yet? *Opt. Lett.* **27** 885–7

[6] Valanju P M, Walser R M and Valanju A P 2002 Wave refraction in negative-index media: always positive and very inhomogeneous *Phys. Rev. Lett.* **88** 187401–4

[7] Smith D R, Schurig D and Pendry J B 2002 Negative refraction of modulated electromagnetic waves *Appl. Phys. Lett.* **81** 2713–5

[8] Ozbay E, Aydin K, Cubukcu E E and Bayindir M 2003 Transmission and reflection properties of composite double negative metamaterials in the free space *IEEE Trans. Antennas Propag.* **51** 2592–5

[9] Kar S, Roy T, Gangooly P and Pal S 2013 Analytical characterization of cut-wire and thin-wire structures for metamaterial applications *2013 Science and Information Conf. (London, October 7–9, 2013)* 665–9 https://ieeexplore.ieee.org/xpl/conhome/6653326/proceeding

[10] Pendry J B, Holden J, Stewart W J and Youngs I 1996 Extremely low frequency plasmons in metallic mesostructures *Phys. Rev. Lett.* **76** 4773–6

[11] Pendry J B, Holden J, Robbins D J and Stewart W J 1999 Magnetism from conductors and enhanced non linear phenomena *IEEE Trans. Microw. Theory Tech.* **47** 2075–84

[12] Katsarakis N, Koschny T, Kafesaki M, Economou E N and Soukoulis C M 2004 Electric coupling to the magnetic resonance of split-ring resonators *Appl. Phys. Lett.* **84** 2943–5

[13] Whiltshire M C K, Pendry J B, Young I R, Larkman D J, Gilderdale D J and Hajnal J V 2001 Microstructured magnetic materials for RF flux guides in magnetic resonance imaging *Science* **291** 849–51

[14] Whiltshire M C K, Hajnal J V, Pendry J B, Edwards D J and Stevens C J 2003 Metamaterial endoscope for magnetic field transfer: near field imaging with magnetic wires *Opt. Express* **11** 709–15

[15] Roy T, Banerjee D and Kar S 2009 Studies on multiple inclusion magnetic structures useful for millimeter-wave left-handed metamaterial applications *IETE J. Res.* **55** 83–9

[16] Bulu I, Caglayan H and Ozbay E 2005 Experimental demonstration of labyrinth-based left-handed materials *Opt. Express* **13** 10238–47

[17] Bilotti F, Toscano A and Vegni L 2007 Design of spiral and multiple split-ring resonators for the realization of miniaturized metamaterial samples *IEEE Trans. Antennas Propag.* **55** 2258–67

[18] Chatterjee S, Majumder A, Kumar A, Das S and Kar S 2012 Analytical and simulation studies on spiral resonator (SR) and its variants-TTSR and NBSR *Presented in 99th Indian Science Congress (KIIT University, Bubaneswar, 3–7 January 2012)*

[19] Baena J D, Bonache J, Martin F, Sillero R M, Lopetegi T, Laso A G and Portillo M F 2005 Equivalent-circuit model for split-ring resonators and complementary split-ring resonators coupled to planar transmission lines *IEEE Trans. Microw. Theory Tech* **53** 1451–61

[20] Baena J D, Marque´s R and Medina F 2004 Artificial magnetic metamaterial design by using spiral resonators *Phys. Rev.* B **69** 014402

[21] Bilotti F, Toscano A, Vegni L, Aydin K, Boratay K and Ozbay E 2007 Equivalent circuit model for design of metamaterials based on artificial magnetic inclusions *IEEE Trans., Microw. Theory Tech.* **55** 2865–73

[22] Zhou J, Koschny T and Soukoulis C M 2007 Magnetic and electric excitations in split ring resonators *Opt. Express* **15** 17881–90

[23] Marques R, Mesa F, Martel J and Medina F 2003 Comparative analysis of edge- and broadside-coupled split ring resonators for metamaterial design—theory and experiments *IEEE Trans. Antennas Propag.* **51** 2572–81

[24] Iyer A K and Eleftheriades G V 2006 Volumetric layered transmission-line metamaterial exhibiting a negative refractive index *J. Opt. Soc. Am.* B **23** 553–70

[25] Kumar A, Majumder A, Das S and Kar S 2013 Simulation based characterization of negative permeability plasmonic structures at X band *Science and Information Conf. (London, October 7–9, 2013)* 675–79 https://ieeexplore.ieee.org/xpl/conhome/6653326/proceeding

[26] Falcone F, Lopetegi T, Laso M A, Baena J D, Bonache J, Beruete M, Marqués R, Martín F and Sorolla M 2004 Babinet principle applied to the design of metasurfaces and metamaterials *Phys. Rev. Lett.* **93** 197401–4

[27] Machac J, Protiva P and Zehentner J 2007 Isotropic epsilon-negative particles *2007 IEEE/MTT-S Int. Microwave Symp. (Honolulu, HI, 3–8 June 2007)* 1831–4

[28] Liu R, Degiron A, Mock J J and Smith D R 2007 Negative index material composed of electric and magnetic resonators *Appl. Phys. Lett.* **90** 263504

[29] Blaha M, Machac J and Rytir M 2009 A double H-shaped resonator and its use as an isotropic ENG metamaterial *Int. J. Microw. Wirel. Technol.* **1** 315–21
[30] Engheta N and Zilokowski R W (ed) 2006 *Metamaterials— Physics and Engineering Explorations* (Piscataway, NJ: IEEE Press, Wiley Interscience)
[31] Huangfu J, Ran L, Chen H, Zhang X, Chen K, Grzegorczyk T M and Kong J A 2004 Experimental confirmation of negative refractive index of a metamaterial composed of Ω-like metallic patterns *Appl. Phys. Lett.* **84** 1537–9
[32] Chen H, Ran L, Huangfu J, Zhang X, Chen K, Grzegorczyk T M and Kong J A 2004 Left-handed metamaterials composed of only S-shaped resonators *Phys. Rev.* E **70** 057605
[33] Kafesaki M, Tsiapa I, Katsarakis N, Soukoulis C M and Economou E N 2007 Left-handed metamaterials: the fishnet structure and its variations *Phys. Rev.* B **75** 235114
[34] Iyer A K and Eleftheriades G V 2002 Negative refractive index metamaterials supporting 2D waves *IEEE-MTT Int. Symp* vol 2 *(Seattle, WA, 2–7 June 2002)* 412–5
[35] Olnier A A 2002 A periodic-structure negative-refractive-index medium without resonant elements *Proc. IEEE-AP-S USNC/URSI Nat. Radio Science Meeting (San Antonio, TX, 16–21 June 2002)* 41–4
[36] Caloz C and Itoh T 2002 Application of the transmission line theory of left-handed (LH) materials to the realization of a microstrip LH transmission line *Proc. IEEE-AP-S USNC/URSI Nat. Radio Science Meeting (San Antonio, TX, 16–21 June 2002)* 412–5
[37] Eleftheriades G V, Iyer A K and Kremer P C 2002 Planar negative refractive index media using periodically L-C Loaded transmission Lines *IEEE Trans. Microw. Theory Tech.* **50** 2702–12
[38] Brillouin L 1946 *Wave Propagation in Periodic Structures* (New York: McGraw-Hill)
[39] Pierce J R 1950 *Traveling-Wave Tubes* (New York: Van Nostrand)
[40] Caloz C and Itoh T 2006 *Electronic Metamaterials: Transmission Line Theory and Microwave Applications* (New York: Wiley Interscience)
[41] Subal K 2022 *Microwave Engineering—Fundamentals, Design, and Applications* 2nd edn (Hyderabad: Universities Press)
[42] Eleftheriades G V and Balmain K G (ed) 2005 *Negative-Refraction Metamaterials—Fundamental Principles and Applications* (Piscataway, NJ: IEEE Press, Wiley Interscience)

Chapter 3

Matamaterial-inspired passive components, antennas and active devices

Inspiration gives one the strength to move forward

3.1 Introduction

After Veselago's seminal paper on LHM, more than 30 years elapsed until the first LH material (metamaterial) was conceived and experimentally demonstrated. But with the advent of successful metamaterial design, fabrication and experimentation since the beginning of the 21st century, the development of metamaterial-inspired passive components, antennas and active devices followed soon. With the incorporation of some of the elementary metamaterial inclusions like split-ring resonator (SRR) (or its complimentary version, CSRR), composite right/left-handed (CRLH) etc, has led to the development of more efficient and size-miniaturized microwave passive components and antennas including other active devices. Figure 3.1 shows an assortment of different matamaterial-inspired microwave passive components and antennas designed and developed by different researchers [1].

In this chapter, we will first discuss the design of CRLH-based filter and CSRR-loaded microstrip antenna design that has been done by our group of metamaterial research at the University of Calcutta. Then we will briefly review the development of various metamaterial-inspired passive components like couplers, phase-shifters and antennas like leaky-wave antenna, zero-order and dual-band ring antennas and so forth done by other research groups around the world. This will be followed by the recent scenario of the developments of various microwave and THz active devices (both solid-state and vacuum tube type) with metamaterial assistance. Be it the passive components, antennas, and active devices—metamaterial-inspiration/assistance will be seen to be capable of size miniaturization, performance improvement and new application potential.

doi:10.1088/978-0-7503-5532-2ch3

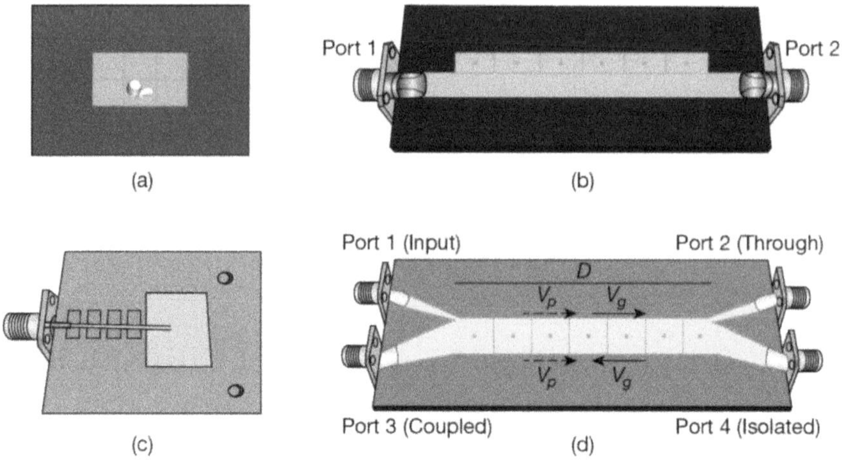

Figure 3.1. Metamaterial-inspired microwave passive components and antennas. (a) Low-profile omnidirectional antenna. (b) Braodband band-stop filter. (c) Patch antenna using SRRs. (d) Backward-wave directional coupler.

Apart from that, many innovative ideas have been proposed since the first design of artificial materials in the 1940s, many of which have opened up new dimensions in setting-up various trends in the development of metamaterial passive components and antennas including active devices. One section will be fully devoted for a concise discussion on this particular topic at a suitable part of this chapter.

3.2 CRLH-based microwave filter design

3.2.1 Design of unit cell of CRLH

Microwave passive components like filter, phase shifter, coupler etc and antennas can be designed with CRLH using multiple unit cells of CRLH to realize improved performance characteristics and size reduction of the component. For all these applications, the design methodology of unit cell of CRLH using interdigital capacitor and stub inductor is to be understood first. The methodology is discussed in the following steps [2].

Step 1: First we select an appropriate transition frequency, ω_0 (which normally represents the centre of the operational bandwidth) given by:

$$\omega_0 = \sqrt{\omega_{se}\omega_{sh}} = \frac{1}{\sqrt[4]{L_R C_R L_L C_L}} = \sqrt{\omega_R \omega_L} \tag{3.1a}$$

where $\omega_R = \sqrt{L_R C_R}$ and $\omega_L = \sqrt{L_L C_L}$

Step 2: Matching condition is applied to the ports of the CRLH structure making:

$$Z_c = Z_R = \sqrt{\frac{L_R}{C_R}} \tag{3.1b}$$

and

$$Z_c = Z_L = \sqrt{\frac{L_L}{C_L}} \tag{3.1c}$$

where Z_c is the characteristic impedance of the CRLH transmission-line (TL) network.

It may be observed that we have so far three equations given in equations (3.1a), (3.1b) and (3.1c) to determine four unknowns L_R, C_R, L_L, C_L.

Step 3: The fourth equation required to determine all the four unknowns mentioned above may be the bandwidth (a designer's constraint) which extends from high-pass LH cut-off ω_{cL} to low-pass RH cut-off ω_{cR}, see figure 2.15, and in the balanced case is given by:

$$\omega_{cL} = \omega_R \left| 1 - \sqrt{1 + \frac{\omega_L}{\omega_R}} \right| \text{ and } \omega_{cR} = \omega_R \left(1 + \sqrt{1 + \frac{\omega_L}{\omega_R}} \right) \tag{3.1di}$$

It may be noted that the cut-offs do not depend on the number of cells but only on the value of *LC* parameters. The fractional bandwidth may be written as:

$$\text{FBW} = 2\frac{(\omega_{cR} - \omega_{cL})}{(\omega_{cL} + \omega_{cR})} \tag{3.1dii}$$

An appropriate number of cells *N* need to be specified to realize a desired performance from the component to be designed.

Step 4: The network can now be implemented by practically realizing the inductances L_R, L_L, and capacitances C_R, C_L with a suitable technology. For instance in microstrip technology, inductors can be implemented in the form of spiral inductor strips or simply stub strips; while capacitors may be implemented by metal-insulator-metal (MIM) structures or with interdigital structure (see figure 2.14(c)).

Step 5: The characteristic parameters of the effectively homogeneous CRLH TL network are finally determined by the relations:

$$L'_R = \frac{L_R}{p},\ C'_R = \frac{C_R}{p},\ L'_L = L_L.\,p,\ C'_L = C_L.\,p \tag{3.1e}$$

where *p* represents the physical length of the unit cell (this is required as L_R, C_R, L_L, C_L all were defined in terms of the quantities designated in the incremental model of CRLH, see figure 2.14(b)).

A typical microstrip-based CRLH TL of nine unit cells is shown in figure 3.2 in which each unit cell consists of *interdigital capacitors* and *stub inductors* shorted to ground plane by via (see figure 2.14(c)).

The unit cells may be symmetric or asymmetric type as shown in figure 3.3; where W_{IDC}, l_{IDC}, g_{IDC}, l_{stub}, and W_{stub}, are the finger width of interdigital lines, finger length of interdigital lines, gap between the fingers, length of the grounded stub, and the width of the grounded stub, respectively.

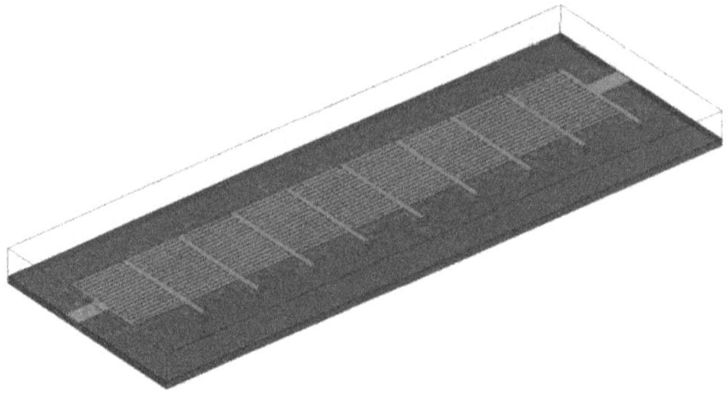

Figure 3.2. A microstrip-based nine-cell CRLH TL using interdigital capacitors and shorted stub inductors.

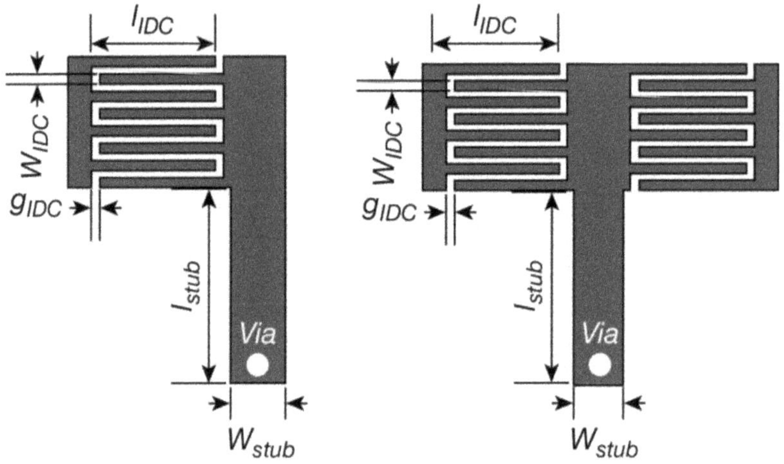

(a) (b)

Figure 3.3. Unit cells of CRLH TL: (a) asymmetric type and (b) symmetric type.

The equivalent circuit of a symmetric type unit cell of CRLH TL is shown in figure 3.4. This is a T-network having two impedance branches with capacitance $2C_L$ and inductance $L_R/2$ and an admittance branch with inductance L_L and capacitance C_R. The contributions L_L and C_L are due to the stub inductor and interdigital capacitor and L_R and C_R arises due to parasitic reactance caused by the magnetic flux generated by the currents flowing along the fingers of the interdigital capacitor and the voltage gradient that exist between the fingers of the said capacitor with the ground plane. The parasitic reactance has an increasing effect with increasing frequency.

Approximate formulas for calculating the inductance L_L and capacitance C_L may be obtained as follows. The shorted stub can be realized with a shorted TL section whose input impedance is an inductive reactance [3] when:

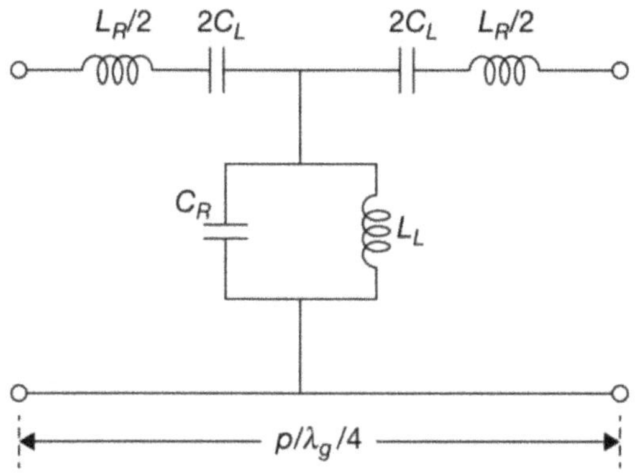

Figure 3.4. T-equivalent circuit of symmetric type unit cell of CRLH TL.

$$L_L \approx \frac{R_0}{\omega} \tan(\beta l) \tag{3.1f}$$

where R_0 is the characteristic impedance of the host TL, β the propagation constant and l is the length of the stub.

The capacitance of the interdigital capacitor may be calculated from [3]

$$C = (\varepsilon_r + 1)l[(N - 3)A_1 + A_2]\, pF \tag{3.1g}$$

with

$$\begin{aligned} A_1 &= 4.41 \tanh\left\{0.55\left(\frac{h}{w}\right)^{0.45}\right\}10^{-6}\ \text{pF}/\mu\text{m}, \\ A_2 &= 9.92 \tanh\left\{0.52\left(\frac{h}{w}\right)^{0.5}\right\}10^{-6}\ \text{pF}/\mu\text{m} \end{aligned} \tag{3.1h}$$

where l, w and h represent the finger length (usually $\leqslant \lambda/4$), finger width and the height of the substrate, respectively, and N is the number of fingers of the interdigital capacitor.

The approximate formulas to calculate L_L and C_L are just the starting guess for the design, however, for optimization full-wave simulation technique is used which are discussed in advanced texts [2].

In the design of CRLH-based microwave passive components and antennas, extraction of the equivalent circuit parameters L_R, C_R, L_L, and C_L of CRLH is essential; the details of this is given in advanced text [2] and has not been repeated here for brevity of presentation.

3.2.2 Microwave filter design with CRLH transmission line

Three CRLH resonators are used in the design of microwave band-pass filter in which first and third resonators are of the asymmetric type (see figure 3.3(a)) while the second one is symmetric type CRLH resonator (see figure 3.3(b)).

The target specification for the filter to be designed [4] is given in table 3.1.

Table 3.1. Filter specifications.

Parameter	Specification
Centre frequency	2.45 GHz
Bandwidth	100 MHz
Order	3
Insertion loss	<1.0 dB
Pass-band ripple	<0.1 dB
Return loss	<10 dB
Substrate	Roger RT/Duroid 5880
Dielectric constant	2.2
Height of the substrate	0.508 mm
Loss tangent	0.0009

Table 3.2. Final dimensions of CRLH-based BPF at 2.45 GHz.

	Value (mm)	
Items	Resonators 1 and 3	Resonator 2
W_{IDC}	0.466	0.2
l_{IDC}	5.6	3
Number of fingers	2	2
Gap between the fingers (g_{IDC})	0.18	0.36
W_{stub}	1	0.95
l_{stub}	8.8	12.67
Width of the port coupling	0.3	
Length of the port coupling	4.74	

The resonators have been optimized separately at 2.45 GHz in an electronic design automation (EDA) tool. At first, the designs of resonators need to be simulated in circuit simulator of the EDA tool (using approximate formulas) for calculating the initial values of the inductance and capacitance of the interdigital capacitor and stub inductor (as detailed under the heading 'design of unit cell of CRLH' in section 3.2.1) and then need to be optimized using the suitable optimization technique available in the EDA simulation tool. The optimized schematic is to be converted next to layout and final optimization is done using the electromagnetic (EM) simulation tool of the simulator. Doing all these steps one can get the final dimensions given in table 3.2 while the final layout looks like the one given in figure 3.5.

A typical CRLH TL-based band-pass filter (BPF) and a conventional (parallel-coupled/edge-coupled) microstrip BPF are given in figure 3.6(a) with their performance characteristics in figure 3.6(b)(i) and (ii). The performance and size comparison given in table 3.3 shows the superiority of CRLH-based filter both in terms of miniaturization and improved performance.

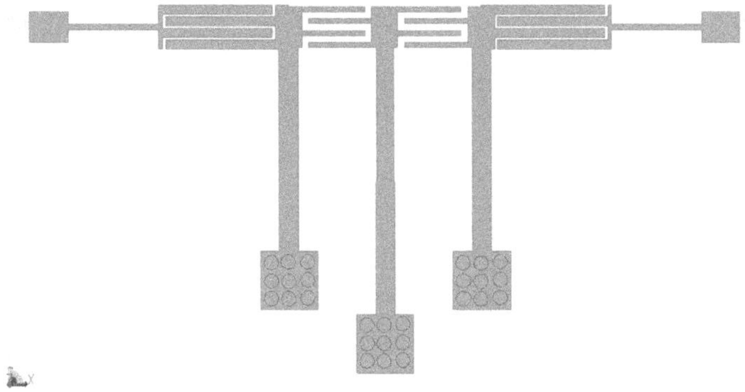

Figure 3.5. Final layout of CRLH TL band-pas filter (BPF) at 2.45 GHz.

(a)

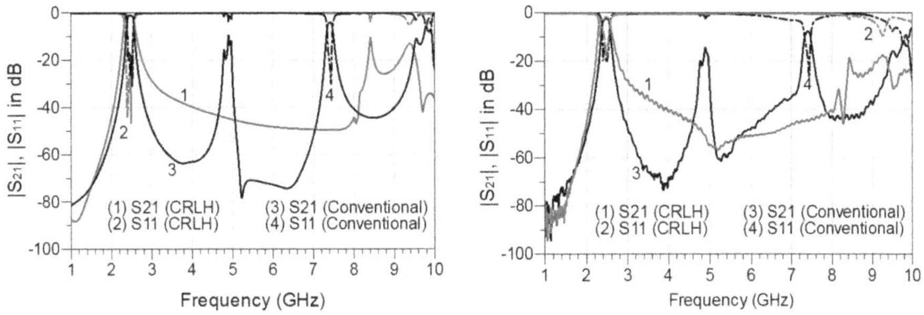

(i) (b) (ii)

Figure 3.6. (a) Fabricated CRLH-based BPF (top) and conventional (edge-coupled/parallel coupled) microstrip BPF (bottom). (b) Comparison of insertion loss (S_{21}) and return loss (S_{11}) of CRLH BPF with conventional BPF; (i) simulation results and (ii) measurement results.

3.3 CSRR-loaded microstrip patch antenna design

CSRR represents the complementary version of the well-known SRR (see figure 2.12(a)). In microstrip design, CSRR is the form of SRR which can be realized with planar fabrication technology. The gap portions of SRR are the

Table 3.3. Performance characteristics and size comparison of CRLH-based and conventional (parallel coupled/edge-coupled) microstrip BPF.

Performance characteristics/size	CRLH BPF	Conventional BPF
Insertion loss	1.6 dB	2.3 dB
Return loss	15 dB	18 dB
Harmonics suppression	At least up to 10 GHz	No harmonic suppression
Size*	4.3 cm × 3 cm	6.5 cm × 6 cm

* CRLH type BPF shows 67% size reduction (area wise) compared to the conventional one.

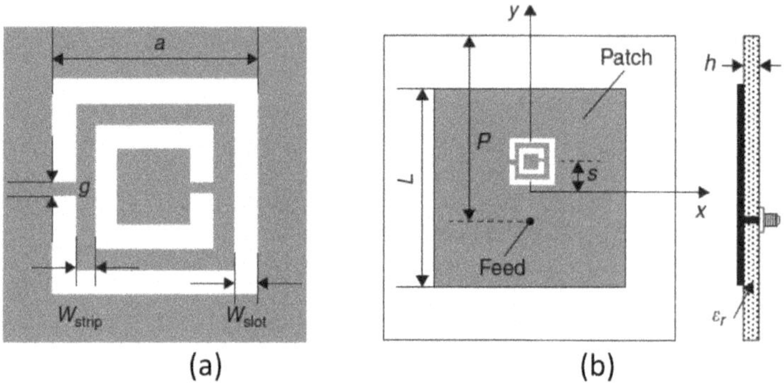

Figure 3.7. (a) Schematic of CSRR and (b) CSRR-loaded SMPA.

metallic portions for CSRR and vice versa. Electrically, CSRR is the *dual* of SRR (as per Babinet's principle), the latter being a magnetic inclusion while the former is an electric inclusion structure. The SRR can be considered as a resonant magnetic dipole that can be excited by an axial magnetic field, whereas the CSRR essentially behaves as an electric dipole with the same frequency of resonance (as that of the equivalent SRR) and can be excited by an axial electric field. As CSRR is electrically the dual of SRR, the equivalent-circuit inductance and capacitance of SRR will, respectively, correspond to the capacitance and inductance of CSRR. Since dual parameters are only related by constants, it is evident that, the resonant frequency which depends on the product of L and C will remain constant for a particular inclusion whether we treat it as an SRR or as a CSRR.

Schematic sketches of CSRR and CSRR-loaded square microstrip patch antenna (SMPA) are shown in figures 3.7(a) and (b), respectively; in figure 3.7(b) $\varepsilon_r = 2.2$ and $h = 0.508$ mm.

A conventional SMPA has been designed with metallic patch dimensions 18.45 mm × 18.45 mm printed on a substrate (Arlon Diclad 880 having $\varepsilon_r = 2.2$ and tan $\delta = 0.0009$) with size 50 mm × 50 mm and height $h = 0.508$ mm exhibiting resonance at 5.2 GHz [5]. It is fed with a standard 50 Ω coaxial cable having diameter of the inner and outer conductors 1.28 mm and 4.29 mm, respectively, with Teflon dielectric (having $\varepsilon_r = 2.1$). The impedance matching between the metal patch and the coaxial probe was done by optimizing the probe position P (see figure 3.7(b))

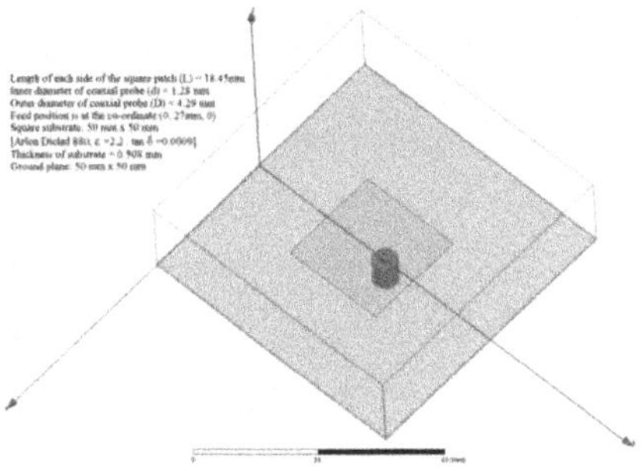

(a)

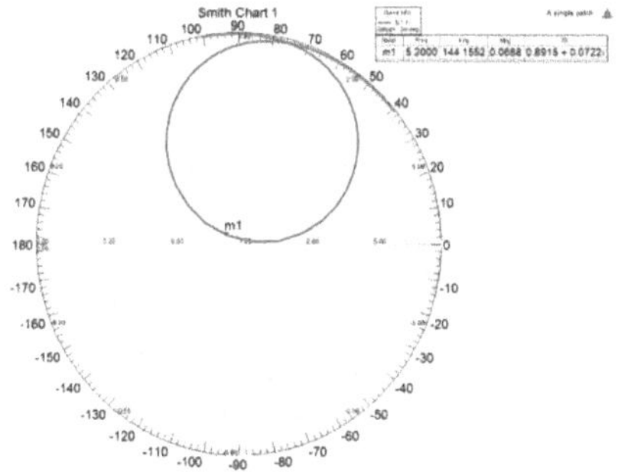

(b)

Figure 3.8. (a) Schematic of SMPA. (b) Smith chart showing the impedance matching.

using 3D electromagnetic field simulator. The SMPA schematic along with the Smith chart plot showing impedance matching is depicted in figure 3.8.

When CSRR resonator is loaded on the SMPA the performance optimization needs to be done by controlling the CSRR size and the parameter s (see figure 3.7(b)). Schematics of square patch loaded with square CSRR, having the line of cut in CSRR aligned orthogonal to the coaxial feed together with the surface current lines are shown in figures 3.9(a) and (b), respectively.

Other orientations of CSRR, one aligned with its line of cut along the line of the coaxial feed and facing the feed exhibits lossy nature, while another case when CSRR is aligned with its cut along the line of coaxial feed but rotated by 180° in clockwise direction exhibits no possibility of size reduction as it has resonance at 5.8 GHz. The CSRR-loaded patch given in figure 3.9(a) when optimized in terms of parametric study indicates that CSRR loading on patch brings down the resonance frequency below 5.2 GHz. To bring it back to 5.2 GHz with perfect impedance matching (see figure 3.9(c)) the patch size was found to be reduced to 14 mm × 14 mm with other optimized dimensions as indicated in table 3.4. It may be observed that a size reduction of more than 24% of the patch is realizable with CSRR loading on SMPA. A fabricated CSRR-loaded SMPA antenna and a conventional SMPA antenna structure is shown in figure 3.10.

A comparative performance of conventional SMPA antenna and CSRR-loaded SMPA antenna is given in table 3.5.

The characteristics of the designed antenna are shown in figure 3.11.

An analytical estimation of the resonant frequency of CSRR can be done as follows. Babinet's principle suggests that CSRR is the dual of SRR, so the resonant frequency of SRR and CSRR are expected to be the same [6]. The SRR is basically an LC resonator. The total inductance of SRR structure shown in figure 3.12 can be calculated as:

$$L_{\mathrm{eq}} = \frac{L_{\mathrm{out}} + L_{\mathrm{in}}}{2} = 11.4\ \mathrm{nH} \tag{3.2}$$

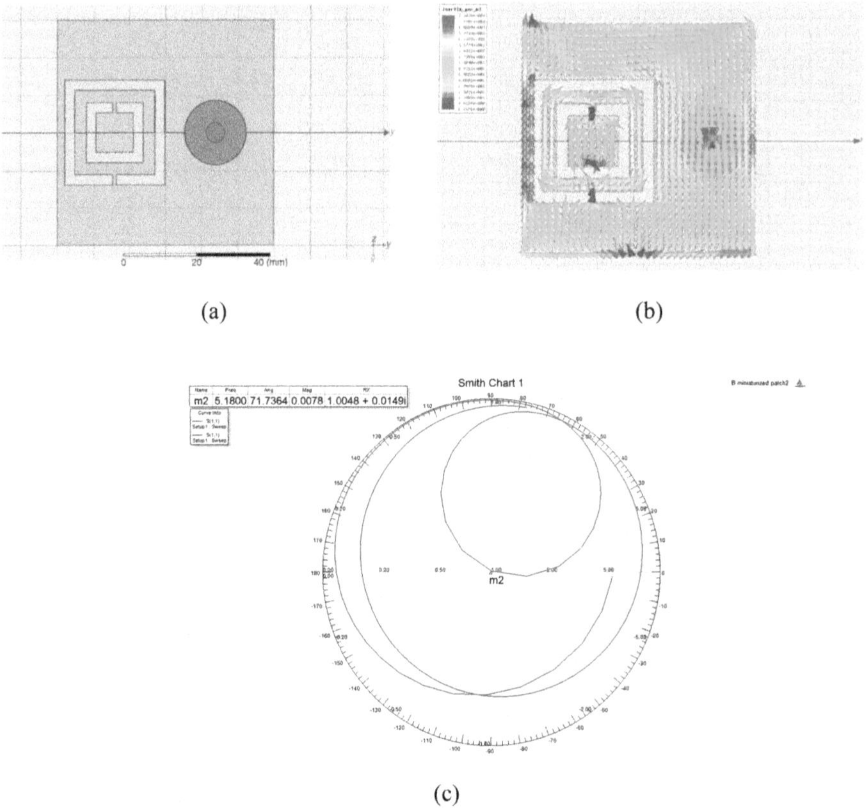

Figure 3.9. (a) Square patch loaded with CSRR, aligned orthogonal to the coaxial feed. (b) Surface current lines. (c) Smith chart plot showing matching between coaxial probe and the CSRR-loaded patch.

Table 3.4. Optimized dimensions of different parameters of CSRR-loaded patch antenna at 5.2 GHz.

Dimensions of different parameters for CSRR-loaded SMPA	
Parameters	Dimensions (in mm)
L	14
a	7.5
W_{strip}	0.8
W_{slot}	0.7
s	5.8
g	0.25
P	26.7

where L_{out} and L_{in} are calculated by considering each segment of the inner and outer rings as rectangular segments [7]. The evaluation of the equivalent inductance L_{eq} as the average of the L_{out} and L_{in} is justified. This is because the presence of small slits in both the ring results in circulation of current between the two rings. Thus the

Table 3.5. A comparative performance of conventional SMPA and CSRR-loaded SMPA antenna.

Antenna type →		Conventional SMPA		CSRR-loaded SMPA	
E-Plane	Obtained from→	Simulation	Measurement	Simulation	Measurement
	Maximum realizable gain	6.86 dB	6.11 dB	4.2 dB	4.17 dB
	Maximum cross-polarization	−56.35 dB	−31.1 dB	−7.43 dB	−10.36 dB
H-Plane	Maximum realizable gain	6.86 dB	6.114 dB	4.2 dB	4.187 dB
	Maximum cross-polarization	−51.06 dB	−20.4 dB	−7.43 dB	−6.39 dB
−10 dB Bandwidth		170.1 MHz	60 MHz	32.5 MHz	34 MHz
Efficiency		81.89%		50.73%	

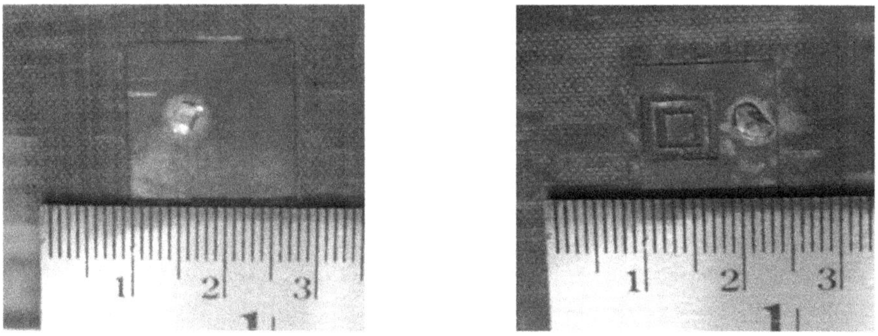

(a) (b)

Figure 3.10. Fabricated antennas. (a) Conventional SMPA. (b) CSRR-loaded SMPA (approximately 24% size reduction).

closed current path formed by traversing through both the rings is exactly like a current in a single loop.

Method of moments (MoM) has been used to find the capacitance per unit length between the two adjacent strips of SRR. Each strip is divided into N number of small rectangles such that surface charge distribution is nearly constant in a rectangle. The thickness of strips is very small (17–35 μm) compared to other dimensions and hence contribution due to this thickness in overall capacitance is neglected. Assuming unit potential on the strips, charge distribution on each rectangle is calculated from Poisson's equation which results in a $2N \times 2N$ matrix of $2N$ unknown variables. This matrix is solved using MATLAB to find the surface charge density on each rectangle. From surface charge density, total charge on a strip is calculated to

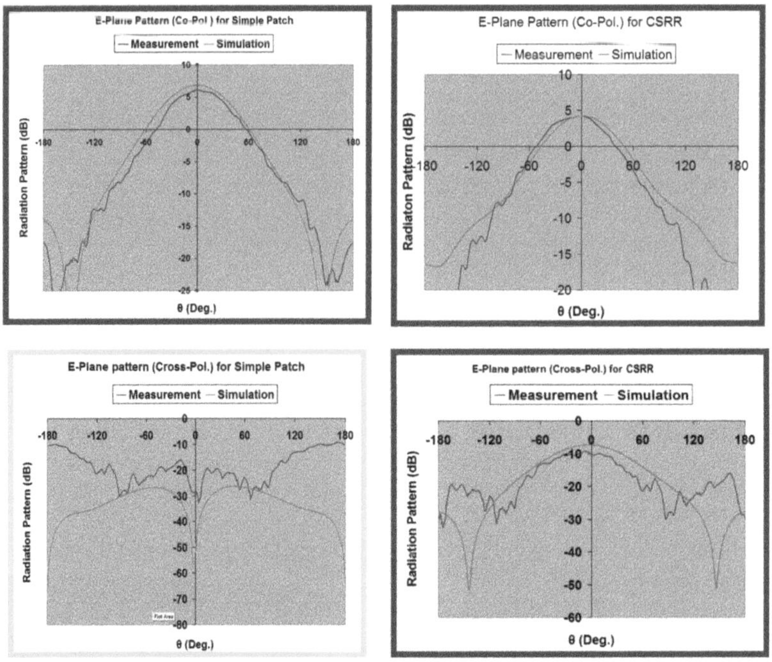

Figure 3.11. Characteristics of the designed CSRR-loaded patch antenna at 5.2 GHz.

find the capacitance for a given strip of width w, gap g and length l. It is observed that capacitance between two adjacent plates is a linear function of the length of strips. So, capacitance per unit length can be determined and used for further calculations. For our case of SRR, the capacitance calculated using MoM between the split rings is 0.016 pF where the width of the strip is 0.7 mm, gap between split rings is 0.8 mm. The half of SRR covers effective length of 10.5 mm and hence C_p (figure 3.12) will be 0.16 pF. The total capacitance for the full SRR structure is a series combination of two capacitors of value C_p. So the total equivalent capacitance for the SRR will be $C_{eq} = 0.084$ pF.

Using $L_{eq} = 11.4$ nH and $C_{eq} = 8.4 \times 10^{-14}$ F, the resonant frequency of the SRR, f_o is found to be 5.14 GHz. The CSRR-loaded patch antenna uses Arlon Diclad 880 substrate of dielectric constant 2.2 that has $\varepsilon_{eff} = (\varepsilon_r+1)/2 = 1.6$. Thus, the resonant frequency of the CSRR will be approximately $f_o/(\varepsilon_{eff})^{1/2} = 4.06$ GHz. One thing to be noted is that f_o is the resonant frequency of the SRR, whereas the metallic patch upon which the complementary form of SRR has been loaded is not an exact dual of SRR since the latter has been cut on a finite metal sheet (i.e. the patch). Due to this reason, some undue loading might have slightly changed the resonant frequency value of the CSRR. Possibly, this is the reason why the resonant frequency calculated analytically (4.06 GHz) and that obtained with simulation (3.03 GHz) of CSRR also differ.

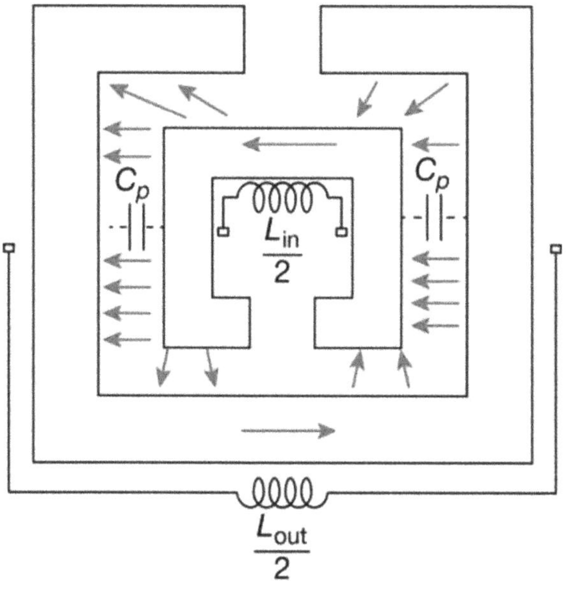

Figure 3.12. Equivalent circuit diagram for SRR to calculate resonant frequency.

3.4 Review of the developments of various other metamaterial-inspired antennas and passive components

3.4.1 Metamaterial-inspired antennas

3.4.1.1 Introduction

In recent years, the demand for miniaturization and integration of many functions of telecommunication equipment has been of great interest, especially for devices that are widely used in recent times such as smartphones using mobile communication systems, hand held tablets, GPS receivers, wireless internet devices and so on. To satisfy this requirement, the mobile device components must be compact and capable of multifunction with multifrequency band of operation. An antenna is one of them; it means that it must be conformal to the body of the device, reduced in size and capable of operating at multiple frequencies of mobile communication systems to be used by the so-called smart devices. Nowadays, there are many technical solutions applied in the antenna design to address these requirements. There are microstrip antenna technologies miniaturized by means of high-permittivity dielectric substrate, using shorting wall, shorting pins, some deformation, as the fractal geometry is, and so on. However, these methods have disadvantage such as narrow bandwidth and low gain. A new solution that is of great interest to designers is the use of electromagnetic metamaterials for antenna design. The use of metamaterials in antenna design not only dramatically reduces the size of the antenna but can also improve other antenna parameters such as enhancing bandwidth, increasing gain, or generating multiband frequencies of antenna's operation.

3.4.1.2 Zero-order antenna

In section 2.3.2.2 we have seen that metamaterial resonators, i.e. CRLH resonators are size-independent resonators with positive, negative, and zero order resonances. Such resonators can be used for the design of the so-called zero-order antenna. The zeroth-order resonator (ZOR) was seen to be essentially a CRLH TL structure with short/open termination operated at an arbitrarily-designed transition frequency, ω_0, where constant-phase and constant-magnitude field distributions are achieved along the structure (see figure 2.16), which corresponds to infinite propagation constant and therefore zero-order mode ($m = 0$) according to the resonance condition of equation (2.19b), $l = | m | \lambda/2$. Due to the unique nature of this zeroth-order mode, the size of the resonator does not depend on its physical length (unlike conventional resonator for which $l = n\lambda/2$, $n = 1, 2, \ldots$) but only on the amount of reactance provided by its unit cells. In terms of antenna, this represents a potential interest for miniaturization.

Any CRLH architecture can be used to implement this antenna, but via-less configuration [8] shown in figure 3.13 is useful as sometimes implementation of vias become problematic. In the via-less microstrip implementation of the CRLH zeroth order resonator the inductor L_L is connected to ground through a large-value capacitance C_g (see figure 3.13), providing a virtual ground; this essentially makes the equivalent circuit of the unit cell unchanged from the standard CRLH prototype. With such zero-order antenna a 75% footprint reduction is achievable compared to a conventional patch antenna of identical resonance frequency.

3.4.1.3 Dual-band ring antenna

The resonating antenna discussed above is based on $m = 0$. But in principle any mode $\pm m$ shown in figure 2.16(b) may be exploited for resonant radiation, and different modes exhibit different properties. An interesting choice of modes is that of an LH/RH pair ($-m$, $+m$), the same magnitude of m but negative sign for LH and positive sign for RH. As pointed out in section 2.3.2.2, the modes of a pair $\pm m$ are associated with propagation constants of identical magnitude, $| \beta_{+m} | = | \beta_{-m} |$ (same effective wavelength) and consequently exhibit similar field distribution and impedance characteristics. Therefore, they can provide *dual-band operation* with a single common feeding structure. In addition, the resulting resonator can be designed for

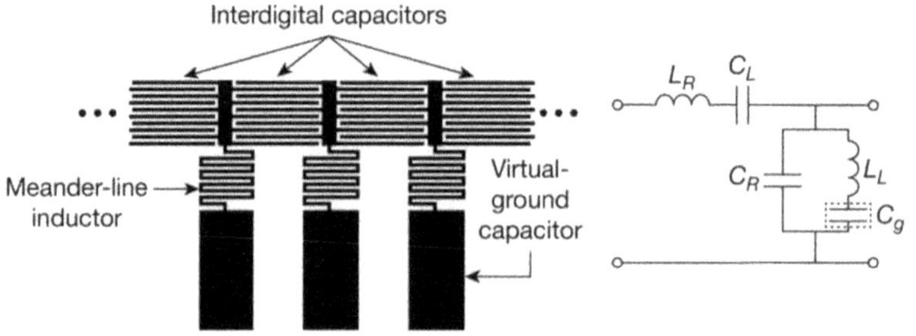

Figure 3.13. Via-less implementation of the CRLH zeroth order resonator for zero-order MTM antenna design.

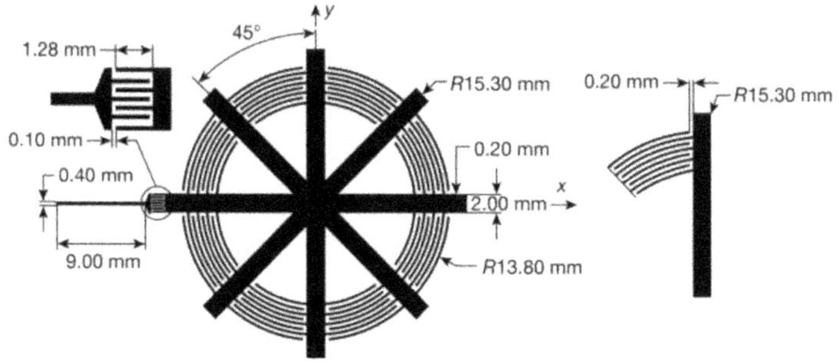

Figure 3.14. Microstrip implementation of eight-cell dual-band CRLH-based ring resonator antenna.

any arbitrary pair of frequencies by using CRLH design principle. A ring antenna that uses $\pm m$ dual-band resonance exhibits some interesting features as an MTM antenna.

Similar to the zero-order antenna, the ring antenna is also CRLH-based but it is a closed loop configuration. An eight-cell dual-band CRLH TL ring resonator antenna is shown in figure 3.14 with inductive stubs connected at the centre [9]. This antenna provides dual-band resonance based on $m = \pm 2$ with $m \neq 0$; because the resonance $m = 0$ does not exist in the absence of via and thus there is no danger of parasitic resonance.

3.4.1.4 Leaky-wave metamaterial antenna

A *leaky wave* (LW) is a travelling wave progressively leaking out power as it propagates along a wave-guiding structure. They are typically used as antennas, where the operating wave behaves as a travelling wave that progressively leaks out energy along the structure to produce a beam in the direction of phase coherence [10]. LW antennas are fundamentally different from resonating antennas, in the sense that they are based on a *travelling-wave* as opposed to the resonating-wave or *standing wave* mechanism of resonant-type antennas. Therefore, the size of LW antennas is not related to the operation frequency, but to directivity.

The conventional LW antenna is capable of beam scanning over only 90° extant but the speciality of MTM LW antenna is that it can scan over a total of 180° thus providing backward, forward and broadside radiation, see figure 3.15 [2, 11]. The sketch of a CRLH TL LW antenna illustrating the three radiation region may be observed from figure 3.15 where backward or *backfire radiation* ($\theta = -90°$) is achieved at the frequency ω_{BF} where $\beta < 0$, point A in figure 2.15 (LH range), *broadside radiation* ($\theta = 0°$) is achieved at ω_0 (i.e. at the transition frequency, f_0) where $\beta = 0$ (point B) and the forward or *endfire radiation* ($\theta = +90°$) is achieved at the frequency ω_{EF} where $\beta > 0$, point C in figure 2.15 (RH range).

The simulated radiation pattern for the PLTL based leaky wave antenna has been shown in figure 3.16 where the radiation beam scans a total of 180° by changing the frequency from 1.93 GHz to 4.0 GHz [12].

The full advantage of backfire-to-endfire (BE) leaky wave (LW) antenna can be taken by making the backward-wave LW antenna electronically tunable by varying

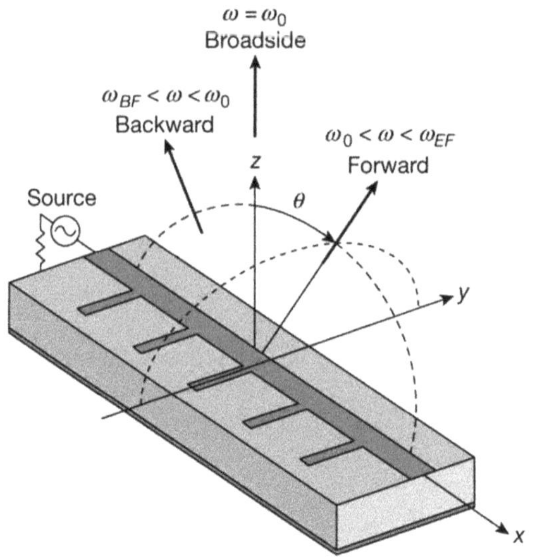

Figure 3.15. Sketch of CRLH TL-type LW antenna showing its special radiation regions.

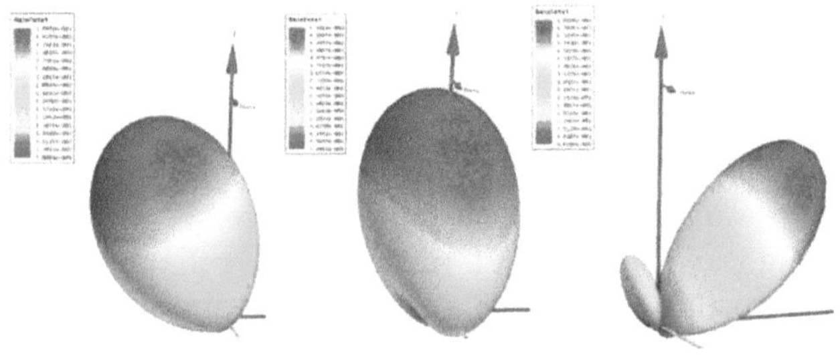

Figure 3.16. Simulated 3D radiation pattern at frequencies 1.93 GHz, 2.35 GHz and 4 GHz, respectively, for PLTL based leaky wave antenna.

the reverse bias of varactor diodes (at a fixed frequency) attached to it [11]. Such an antenna then achieves the capability of a continuous beam steering. In addition such an antenna provides the functionality of beam width control by using a non-uniform biasing distribution of the diodes, whereas beam width is conventionally controlled by changing the geometrical parameters of the structure or with a phased array of antennas. This antenna thus represents an attractive alternative to conventional phased arrays with its advantages of requiring only one radiating element, utilizing a very simple and compact feeding mechanism and not requiring any phase shifters.

3.4.1.5 Other microstrip antennas with metamaterial-inspiration

A lot of work has been done on metamaterial-inspired microstrip antenna; the detailed design of one type of this antenna [5] was given in section 3.3. In this section let us make a brief review of other important works done with this type of improvised microstrip patch antennas by researchers around the world.

The need for metamaterial-inspired microstrip patch antenna arises from the application point of view based on the demand generated from modern day communication system requirements. In fact, miniaturized integrated antennas are becoming a relevant need in several applications like mobile handsets, sensor network, biomedical imaging, wearable and RFID (radio frequency identification) systems, and so on. Several techniques have been proposed to realize electrically small microstrip patch antennas. For example, the use of slots on the patch surface [13], shorting pins [14], or high permittivity dielectrics may effectively lower resonant frequency for a fixed dimension of the patch [15]. However, the cross-polarization levels and the radiation pattern of these antennas might get worsened. Recently, metamaterial loading has emerged as an innovative technique to improve the performance of microstrip patch antenna leading to size miniaturization together with the challenging task of realizing the desired tradeoff between gain, bandwidth, impedance matching and radiation efficiency of the designed antenna. It may be mentioned herewith that conventional patch antenna size miniaturization is limited by the fact that for this 2D planar antenna the patch length l is approximately a half-wavelength ($l \approx 0.49\lambda$); and hence the requirement for miniaturisation with improvised metamaterial-inspiration.

A novel microstrip patch antenna loaded with meta-resonators (CSRRs) and metasurface (RIS—reactive impedance surface) resulted in a miniaturized antenna ($0.099\lambda_0 \times 0.153\lambda_0 \times 0.024\lambda_0$) at 2.5 GHz [16], see figure 3.17(a). The CSRR is incorporated on the patch to excite the antenna at a low CSRR resonance frequency while the RIS is inserted below the patch to miniaturize the antenna size and improve the antenna radiation performance. A dual-band antenna with orthogonal polarization was realized with it by simply adjusting the feeding position in order to excite the initial patch resonance. Another wideband antenna has been reported [17] in which microstrip patch antenna has been loaded with metamaterial unit cell composed of an interdigital capacitor and a CSRR, see figure 3.17(b). The designed antenna achieves a 55% reduction in patch size with metamaterial loading when compared with the

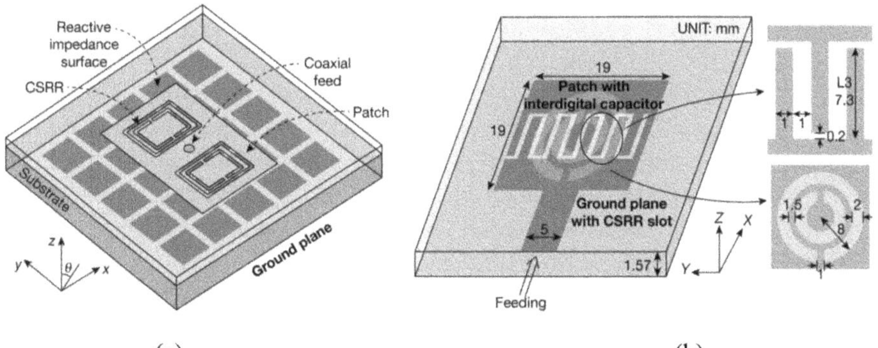

(a) (b)

Figure 3.17. Configuration of metamaterial-loaded patch antenna. (a) CSRR-loaded patch over a RIS. (b) Interdigital capacitor on top of the patch and a CSRR slot etched on the bottom side of the ground plane.

conventional patch antenna. Additionally, the interdigital fingers and the CSRR slot generate the TM_{01} mode which when combines with the normal TM_{10} mode of the patch antenna the bandwidth of the metamaterial-loaded antenna increases by 6.8%.

Another research reports a patch antenna with CSRR loading that is fabricated on a flexible substrate and folded in a cylindrical shape to form a nonplanar CSRR-loaded patch, featuring compact size, EMI (electromagnetic interference) shielding, and quasi-omnidirectional radiation pattern for a wireless endoscope application [18]. All the endoscope circuits could be wrapped around by the antenna, and thus the antenna serves as a packaging layer for the whole system as well. Yet another work reports a metamaterial-inspired patch antenna for WiMAX (worldwide interoperability for microwave access)/WLAN (wireless local area network) applications [19]. This consists of L-shape slotted ground microstrip patch antenna with CSRR embedded on patch structure so that the antenna can operate simultaneously at WiMAX (3.5 GHz) and WLAN (5.8 GHz). The metamaterial-inspired loading is exploited to create resonance for upper WLAN band while an L-shaped slot on the ground plane resonates at the WiMAX band, maintaining the antenna's small form-factor.

3.4.2 Metamaterial couplers

3.4.2.1 Introduction

Couplers are very well known passive components to microwave engineers. A directional coupler couples out a fraction of the input signal to the coupled-port which may be used for feeding the signal to very sensitive instruments like spectrum analyzer and so forth. The heart of a network analyzer is a very accurately designed directional coupler. The waveguide couplers are normally designed to have 10 dB, 20 dB or 30 dB coupling factor which, respectively, means that 0.1, 0.001, 0.0001 part of the input signal will be coupled to the coupled-port of the directional coupler. It may be noted that there exists some fundamental difference between the operation of waveguide-based directional couplers and microstrip-based coupled-line couplers making the so-called coupled port and isolated ports interchanged in the two cases, the details of which is available in standard textbooks on microwave engineering [3].

Conventionally used coupled-line couplers exhibit the advantage of broad bandwidth (typically 25%) but can achieve only loose coupling levels (typically less than 10 dB) in the case of edge-coupled configuration. In contrast, the branch-line couplers, such as the quadrature hybrid or the rat-race can achieve tight (3 dB) coupling but suffer poor bandwidth (typically less than 10%). A widely used planar coupler providing both broad bandwidth and tight coupling is the Lange coupler. But this coupler has the disadvantage of requiring cumbersome bonding wires with subsequent parasitic effects at high frequencies. But metamaterial-inspired couplers based on CRLH TL represent novel and unique alternatives to the Lange coupler because they are capable of providing arbitrarily tight coupling (up to virtually 0 dB!) over a broad bandwidth (more than 30%), while still being planar and without requiring bonding wires. There are two types of CRLH-based metamaterial couplers: symmetric impedance coupler and asymmetric phase coupler [2].

3.4.2.2 Symmetric impedance coupler

The CRLH-based symmetric impedance coupler consists of two identical CRLH TLs, the photograph of such a coupler [20] is shown in figure 3.18. This is an interdigital/stub-type CRLH backward edge-coupled directional coupler with 3 dB (realized with three cells) coupling factor. Because the CRLH TLs are quasi-TEM, the overall coupled-line structure is also quasi-TEM and the even/odd propagation constants can be considered to be approximately equal, $\beta_e \approx \beta_o$, consequently phase coupling will be negligible and the only possible coupling mechanism will be impedance coupling (IC) mechanism. However, it must be noted that the MTM IC mechanism is significantly different from the IC mechanism in conventional coupled-line couplers (CLCs).

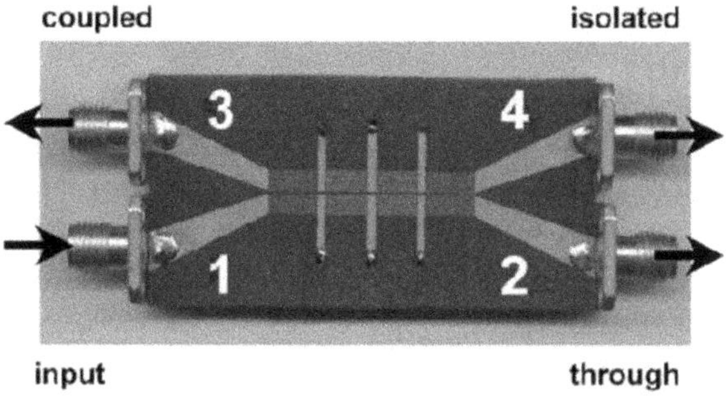

Figure 3.18. CRLH-based symmetrical impedance coupler with 3 dB coupling factor.

In this type of MTM coupler, any level of coupling can be achieved by choosing the number of CRLH cells and the interspacing of the two CRLH lines. Further, the arbitrary coupling level is not obtained at the expense of the bandwidth reduction, unlike the conventional CLC. With such couplers even 0 dB (!) coupling is possible to realize—though in that case the coupler will behave just as a DC block (because then just the capacitances of even/odd modes will come into play)—thus a 3 dB coupler is useful for practical purposes. The performance of this coupler has been quoted as [20]: an amplitude balance of 2 dB over a fractional bandwidth of 50%, from 3.5 to 5.8 GHz, the quadrature phase balance is 90°±5°, and average directivity is 20 dB.

3.4.2.3 Asymmetric phase coupler

When the two TLs constituting a coupled-line coupler are different, the structure is asymmetric. One possible interest of asymmetric CLC is that they provide broader bandwidth than their symmetric counterparts. The asymmetric phase coupler is constituted by the combination of a conventional purely right-handed (PRH) TL and of a CRLH TL, operated in its LH range [21]. But the symmetric impedance coupler we discussed above used two identical CRLH TLs operated in its LH range.

The photograph of the prototype of an asymmetric phase coupler is shown figure 3.19 in which the top line is a conventional microstrip line (PRH) and the bottom line is an interdigital/stub CRLH TL. The principle of operation of this MTM coupler may be understood as follows. As a signal is injected at port 1, the

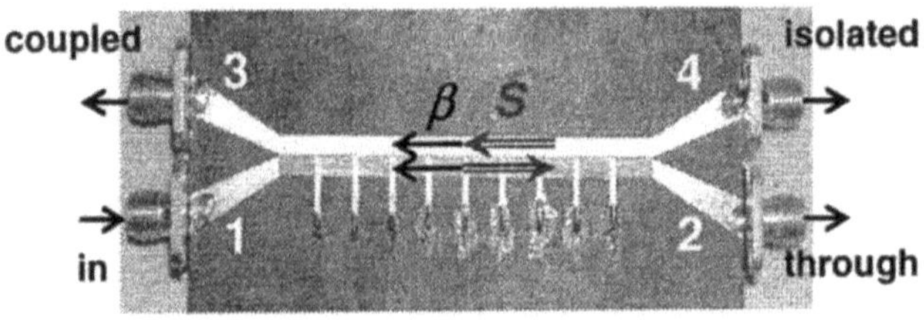

Figure 3.19. Metamaterial-based asymmetric phase coupler.

power propagates toward port 2, but phase propagation constant, β, propagates backward toward port 1 (because the bottom CRLH TL is operated in LH range). Therefore, because coupling to the other line (top line) with ports (3–4) occurs through transverse evanescent waves following the phase direction, phase and power propagate in the same direction toward port 3 in the PRH microstrip line. Consequently, *backward coupling* is functionally achieved, although the coupling mechanism is of phase coupling (PC) nature that is normally associated with forward coupling in conventional PC couplers. The performance of this coupler has been mentioned as [21]: −0.7 dB backward coupling over the range from 2.2 to 3.8 GHz with broad fractional bandwidth of 53% and an excellent directivity of 30 dB has been realized in the prototype measurement.

3.4.2.4 Other types of metamaterial-inspired couplers

Various other metamaterial-inspired couplers have also been reported in literature. Branch-line couplers are an important part of microwave integrated circuits and extensively used at microwave frequencies in the design of balanced mixers, image rejection mixers, balanced amplifiers, power combiners and power dividers, etc [3, 22]. The conventional branch-line coupler is composed of four quarter-wave transmission-line sections at the designed frequency, which results in a large occupied area, especially at low frequencies, making them difficult for incorporation in monolithic microwave integrated circuits (MMICs) because of its large dimension and demand for high chip cost. The present day's portable wireless communication system requires compact and cost effective components. Use of metamaterial-inspiration in the design of such couplers is found to meet these needs for miniaturization and with added functionalities. Two novel types of branch-line coupler design have been reported, *type* 1 and *type* 2, [23] that use a combination of regular microstrip (MS) and negative refractive-index (NRI) lines, see figure 3.20. The *type* 1 coupler uses regular microstrip lines (MS) for the low impedance branches and the NRI lines for the high impedance ones; see figure 3.20(a). The latter type (*type* 2) is the dual of the former (*type* 1) and utilizes NRI lines for the low-impedance branches and microstrip lines (MS) for the high impedance ones, see figure 3.20(b). For both couplers, power splits equally between the two output ports with a 0° phase shift (with respect to the input) at the coupled port. Furthermore, the *type* 1 MS/NRI branch-line coupler offers a negative phase quadrature (−90°) while *type* 2 provides a positive phase quadrature (+90°) at their through ports with respect to the input port.

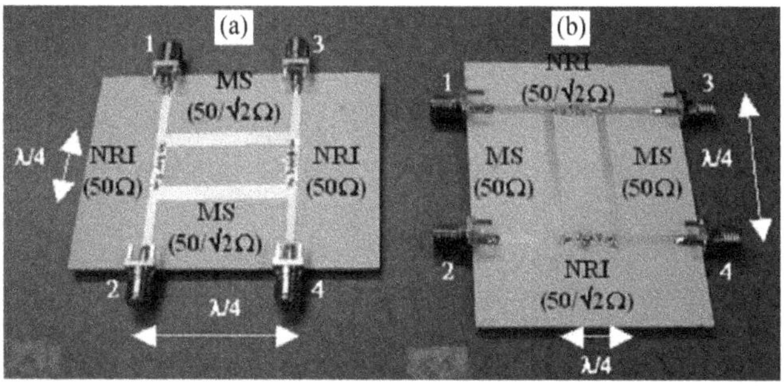

Figure 3.20. MS/NRI branch-line couplers using (a) *type* 1 (b) *type* 2.

These couplers exhibit similar power splitting characteristics as their regular counterparts, but allow the phase agility, i.e., phase compensation at the coupled port and choice of either positive or negative phase quadrature at the through port. Also they lead to a significant size reduction of one of the two orthogonal dimensions of the coupler.

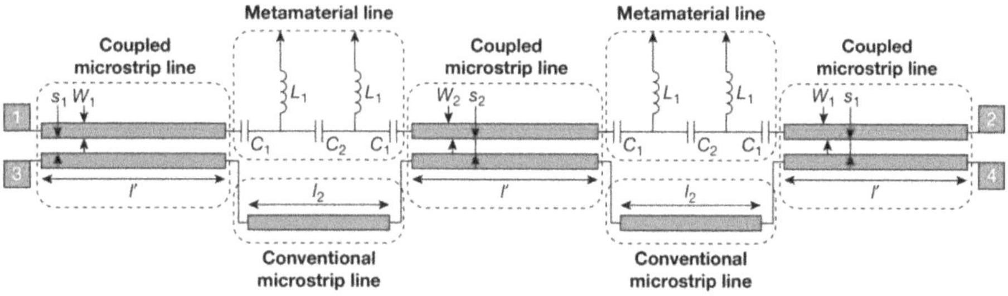

Figure 3.21. Schematic of microwave directional coupler using metamaterial lines. The design dimensions are: l': (−30° microstrip coupled line) = 7.7 mm, $W_1 = W_2 = 1.5$ mm, $s_1 = 1.1$ mm, $s_2 = 1.5$ mm, l_2: (−30° microstrip line) = 7.7 mm, $W = 1.5$ mm, metamaterial: +90° left-handed line ($C_1 = 4.3$ pF, $C_2 = 1.9$ pF and $L_1 = 4.7$ nH).

The directional coupler consisting of microstrip parallel-coupled line is widely used for its easy fabrication, whereas it shows poor directivity as the permittivity increases. In order to overcome the poor directivity, many researchers have reported several methods using wiggly lines, re-entrant coupling, additive lumped capacitances and so forth. Anyway, a tightly coupled microstrip line can hardly be implemented because parallel-coupled microstrip line has a loose coupling property inherently. In this backdrop, a novel directional coupler has been designed at microwave frequency making use of metamaterial [24] that can provide higher coupling factor, improved directivity, and broadband operation with size miniaturization. In this metamaterial-inspired directional coupler three sections of parallel-coupled microstrip line are used (each of $\lambda/12$ length instead of $\lambda/4$ line) along with two sections of asymmetrical delay lines. Each delay line consists of a right-handed TL ($\lambda/12$ length) and a lumped left-handed TL, as shown in figure 3.21. The designed three-section coupler has 18 dB coupling factor and provides a bandwidth of 38% with the centre frequency of 2.0 GHz having more than 20 dB directivity over the band [24].

3.4.3 Phase compensator and phase shifters

3.4.3.1 Introduction

In conventional positive refractive index (PRI) TLs, the phase lags in the direction of positive group velocity, thus incurring a negative phase shift with propagation away from the source. Thus the PRH TL is said to exhibit a phase delay. But as in an NRI line the phase and group velocities are antiparallel and the phase leads in the direction of group velocity it results in a positive phase shift with propagation away from the source. Thus a purely left-handed (PLH) TL is said to exhibit a phase advance. It therefore follows that the phase compensation can be achieved at a given frequency by cascading a section of NRI line with the section of a PRI line to synthesize positive, negative or even zero transmission phase over a short physical length. This idea of phase compensation was inherent in Veselago's flat-lens idea.

3.4.3.2 Phase compensator and its applications

A physical implementation of the concept of phase compensation with PRI–NRI combination TL is shown in figure 3.22 [25]. Here a slab of loss-less PRI material with positive index of refraction n_1 and thickness d_1 is cascaded with a loss-less NRI material with negative refractive index $-|n_2|$ and thickness d_2. Although not necessary, but for the sake of simplicity in the argument, let us assume that each of these slabs is impedance matched to the outside region (i.e. free space). Let a monochromatic uniform plane wave be incident normally on this pair of slabs. As this wave propagates through the slab, the phase difference between the entrance and exit faces of the first slab is obviously $n_1 k_0 d_1$, where $k_0 = \omega\sqrt{\varepsilon_0 \mu_0}$, while the total phase difference between the front and back faces of this two-layer structure is $|n_1| k_0 d_1 - |n_2| k_0 d_2$, implying that whenever phase difference is developed by traversing the first slab, it can be decreased or even compensated for by traversing the second slab. If the ratio of d_1 and d_2 is chosen to be $d_1/d_2 = |n_2|/|n_1|$ at a given frequency, then the total phase difference between the front and back faces of this two-layer structure will become zero. This means that the NRI material slab acts as a phase compensator in this cascaded PRI–NRI structure.

It may, however, be noted that such phase compensation/conjugation does not depend on the sum of thickness, $d_1 + d_2$, rather it depends on the ratio of d_1 and d_2.

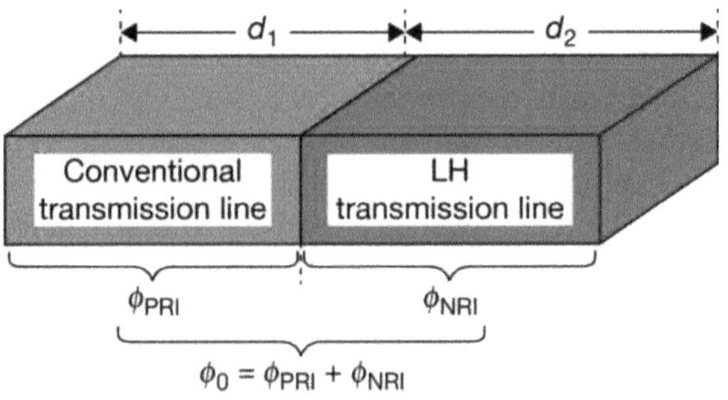

Figure 3.22. Phase compensator with cascaded PRI and NRI slabs.

So in principle, $d_1 + d_2$ can be of any value as long as d_1/d_2 satisfies the above condition. Therefore, even though this two-layer structure is present, the wave traversing this combined structure would not experience any phase difference between the input and output faces. This feature can lead to several interesting ideas in device and component design, e.g., in implementing thin sub-wavelength resonators [26], series-fed linear arrays [27] and so forth. It should be pointed out that these phase-shifting/phase-compensating lines offer an advantage when compared with conventional delay lines. Due to their compact, planar design they lend themselves easily to integrating with other microwave components and devices. The MTM phase-shifting lines are therefore well suited for broadband applications requiring small versatile, linear devices.

Let us now discuss here the application of zero-degree phase-shifting line realizable with cascaded PRI–NRI TL for the design of a series-fed linear array.

A conventional series-fed linear array is designed by using TL-based feed network that uses a one-guide-wavelength (λ_g) long meander line in between successive antennas of the array, see figure 3.23(a), that incurs a phase of -2π. The interelement spacing d_{IE} is maintained to be less than half a free-space wavelength ($d_{IE} < \lambda_0/2$) to avoid capturing grating lobes in the visible region of the array pattern. However, such a feed network is frequency dependent and thus narrow band and hence a change in the operating frequency will cause squinting of the emerging beam from its purely broadside direction. In addition, the fact that the lines are meandered causes the radiation pattern to experience high cross-polarization levels, particularly in CPW (co-planar waveguide) implementation, as a result of parasitic radiation due to scattering from the corners of the meandered lines. But if we use non-radiating zero-degree phase shifting lines realized with cascaded PRI–NRI TL instead of conventional meandered line, see figure 3.23(b) then most of the above-mentioned problems will be resolved and the successive antennas of the array will be fed in-phase that will help to radiate at broadside and the antenna will also operate over broad bandwidth [27].

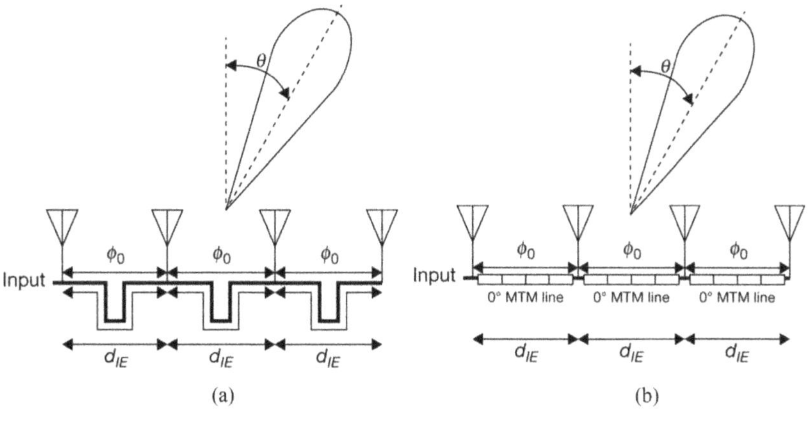

Figure 3.23. Series-fed linear array. (a) Conventional -2π TL using meandered feed lines. (b) Improved one using 0° MTM feed lines.

3.4.3.3 Metamaterial-inspired phase shifter

Phase shifters are essential in microwave systems, including phased array antennas and radars. Most of the phase-shifters used today are based on PIN diodes, GaAs varactors or field-effect transistor (FET) switches [28]. However, these technologies suffer from relatively high power consumption and high losses, especially at millimetre-wave frequencies. On the other hand, the advanced microwave systems used today put stringent requirements on phase-frequency performances, power consumption, size, and cost. This makes it necessary to look for new technologies and circuit architectures. In this respect, ferroelectric materials, considered since 1962 [29], offer cost effective solutions along with small sizes and low power consumption. Successful integration of high Q ferroelectric varactors on semi-conductor substrates also demonstrates the high density integration potential of this technology. Additionally, recent advances in metamaterials, i.e. left-handed TLs, offer phase shifter circuit solutions [30–32] with improved phase-frequency performances. Let us discuss in the following a compact linear lead/lag metamaterial phase shifter for broadband applications [30].

The unit cell of the phase-shifter is shown in figure 3.24 whose length is d_0. The structure consists of a host TL medium (RH-medium) periodically loaded with discrete lumped element components, L_0 and C_0 (constituting the LH-medium, dual of RH-medium). The structure recalls the CRLH shown in figure 2.14(b), and a CRLH is understood to possess both negative and positive phase shift capabilities which are indicated in the dispersion diagram shown in figure 2.15. By adjusting the NRI-medium lumped element values, L_0 and C_0, the phase shift can be tailored to a given specification. For a TL medium with periodicity d_0 and phase shift $\varphi_{TL} = \beta_{TL} d_0$, the total phase shift per unit cell, $|\phi_0|$, is given by:

$$|\phi_0| = \phi_{TL} + \frac{-1}{\omega\sqrt{L_0 C_0}} \tag{3.3a}$$

The second term on right-hand side of equation (3.3a) follows from table 2.2 for LH-medium. In the said equation the matching conditions for two media:

$$Z_0 = \sqrt{L_0/C_0} = \sqrt{L/C} \tag{3.3b}$$

need to be maintained. Thus equations (3.3a) and (3.3b) can be used to determine unique values of L_0 and C_0 for *any* phase shift, φ_0, given a TL section with intrinsic phase shift φ_{TL} and characteristic impedance Z_0.

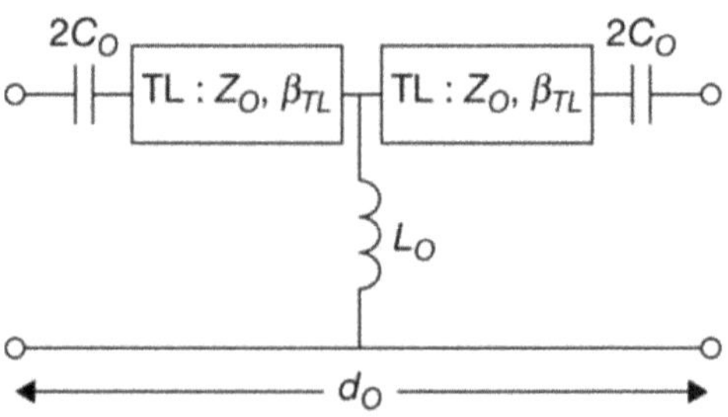

Figure 3.24. Unit cell of 1D metamaterial phase shifter.

The phase shifter discussed above offers some significant advantages over conventional delay lines. The phase shifter is compact in size and exhibits a linear phase response around the design frequency. It can incur either a negative or positive phase shift, as well as 0° phase depending on the values of the loading elements, while maintaining a short physical length. In addition, the phase shift incurred is independent of the length of the structure. Due to its compact, planar design, the structure lends itself easily for integration with other microwave components and devices. It is therefore ideal for broadband applications requiring small, versatile, linear phase shifters.

Other types of MTM-based phase shifters have also been reported in published literature. A CRLH phase shifter using ferroelectric varactors as tunable element exhibits a differential phase shift with flat frequency response around the centre frequency [31]. The ferroelectric varactors are realized in parallel plate version. Under 15 V DC bias applied over each varactor, the differential phase shift is flat around 17 GHz and has an absolute value of 50°. A fully integrated metamaterial-based phase shifter has also been reported [32] that replaces the TL sections by their lumped *L–C* equivalent circuit in a suitable form. This enables integrating the entire phase shifter on a single MMIC chip, resulting in a compact implementation. The IC phase shifter achieves a tunable phase over −35° to +59° at 2.6 GHz with less than −19 dB return loss from a single stage occupying 550 μm × 1300 μm.

3.5 Innovative trends for passive components and antenna developments

3.5.1 Introduction

In addition to metamaterial-inspiration, other innovative trends of artificial material development have emerged that have the potential of designing passive components and antennas with promising and otherwise unavailable characteristics. One such electromagnetic artificial material is the so-called *high impedance surfaces* (HIS) [33, 34] which were briefly mentioned in section 1.2.2 of chapter 1 and discussed in detail in section 3.5.2.2 of this chapter. HIS may be considered as the forerunner of metasurfaces and has great significance in the future developments of metasurface technology. These are 'textured' electromagnetic surfaces that can be used to alter the properties of metal surfaces to realize a variety of functionalities. For example, specific textures can be designed to change the surface impedance for one or both polarizations, to manipulate the propagation of surface waves, or to control the reflection phase. These surfaces are used to design new boundary conditions for building electromagnetic structures, such as for varying the radiation pattern of small antennas. They can also be tuned, enabling electronic control of their electromagnetic properties. Tunable impedance surfaces can be used as simple steerable reflectors or steerable leaky-wave antennas [35].

3.5.2 Textured electromagnetic surfaces

3.5.2.1 The simplest structure

The simplest example of a textured electromagnetic surface is a metal slab with quarter-wavelength deep corrugations [36], as shown in figure 3.25(a). In this structure

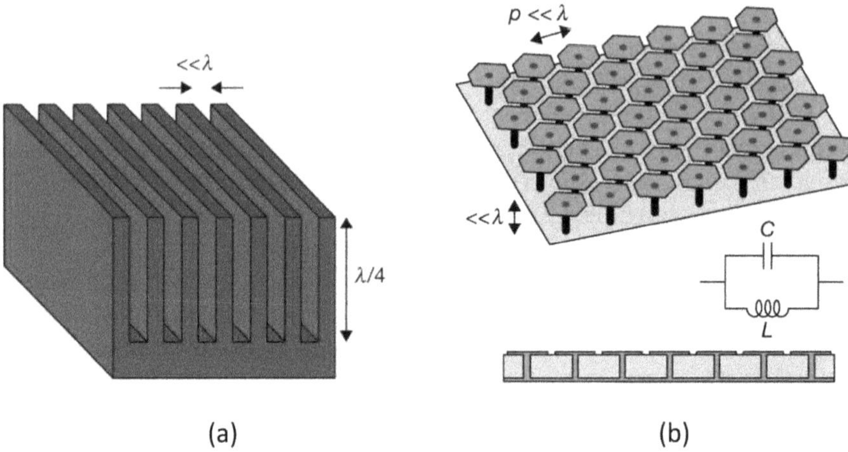

Figure 3.25. Textured electromagnetic surfaces. (a) Traditional corrugated surface. (b) A 2D lattice of plates attached to a ground plane by metal-plated vias. (Reprinted/adapted from [35] with permission from De Gruyter, copyright Li *et al* (2018).)

the corrugations behave as quarter-wave long TLs, in which the short circuit at the bottom of each grove is transformed into an open circuit at the top surface. This provides a high-impedance boundary condition for electric fields polarized perpendicular to the grooves and low impedance for parallel electric fields. Such textured impedance surface can be used for manipulating the radiation patterns of horn antennas or controlling the edge diffraction of reflectors. 2D structures have also been built, such as shorted rectangular waveguide arrays which are typically one quarter-wavelength thick in order to achieve a high impedance boundary condition.

3.5.2.2 Compact high-impedance surface and its potential applications

Compact high-impedance structures have been developed which can also alter electromagnetic boundary condition of a metal surface but that are much less than one quarter-wavelength thick [33, 35]. They typically consist of an array of sub-wavelength metal protrusions on a flat metal sheet. The protrusions are arranged in 2D lattice plates attached to a ground plane by metal-plated vias (to form a continuous conductive metal texture), resembling mushrooms or thumbtacks protruding from the metal surface below, as shown in figure 3.25(b) [35]. The plates provide capacitance and inductance, and it has high electromagnetic impedance near its shunt LC resonance. For greater capacitance, multilayer circuit boards with overlapping plates can be used. Such textured electromagnetic structure provides a high impedance boundary condition for both polarizations and for all propagation directions.

As the period (p) of the protrusions is small compared to the wavelength (λ) of interest, see figure 3.25(b), the material may be analyzed with effective medium theory, with its surface impedance defined by effective lumped-circuit parameters that are determined by the geometry of the surface texture. A wave impinging on the material causes electric fields to span the narrow gaps between the neighbouring metal patches, and this can be described as an effective sheet capacitance C. As current oscillates between the neighbouring patches, the conducting plates through the vias and the ground plane provide a sheet inductance L. These form a parallel resonant circuit that dictates the electromagnetic behaviour of the material. Its surface impedance is given by the expression: $Z_s = j\omega L/(1 - \omega^2 LC)$, with resonance frequency: $\omega_0 = 1/\sqrt{LC}$. Below the resonance, the surface is inductive and supports TM surface waves while above resonance, the surface is capacitive and supports TE surface waves. Near ω_0 the surface impedance is much higher than the impedance of free space, and the material does not support bound surface waves. By the way, a surface wave is a wave that is bound to a surface and decays exponentially ($e^{\alpha z}$) away from the surface, where α is the decay constant.

In addition to its unusual surface wave properties, the high-impedance surface also has unusual reflection-phase properties. They reflect with a phase shift of zero, rather than π, as with an electric conductor. It is known that in an electrical conductor, which is known as a low-impedance surface, the ratio of electric field to magnetic field is small and the electric field has a node at the surface while the magnetic field has an antinode. Conversely, in a high impedance surface (i.e. textured metal surface) the electric field has an antinode at the surface but the magnetic field has a node. Thus the high impedance surface is sometimes termed as *artificial magnetic conductors*. Because of this unusual boundary condition, the high impedance surface can function as a new type of ground plane for low profile antennas. The image currents in the ground plane are in-phase with the antenna current, rather than out of phase, allowing radiation elements to lie directly adjacent to the surface while still radiating efficiently. For example, a dipole lying flat against a high-impedance ground plane is not shorted as it would be on an ordinary metal sheet. Further, as the reflection phase for normally incident wave: $\varphi = \mathrm{Img}[\ln\{(Z_s - \eta_0)/(Z_s + \eta_0)\}]$, where Z_s is the surface impedance of the textured surface and η_0 is the impedance of free space; is determined by the frequency of the incoming wave with respect to the resonance frequency—such a surface can perform as a distributed phase shifter. When the resonance frequency is swept from low to high values around the resonance frequency the phase varies from $-\pi$ to $+\pi$.

In addition to such unusual reflection-phase properties, these materials have a surface wave bandgap, within which they do not support bound surface waves of either transverse electric (TE) or transverse magnetic (TM) polarization. They may be considered as a kind of electromagnetic band gap (EBG) structure or photonic crystal for surface waves. Although bound surface waves are not supported, leaky TE waves can propagate within the bandgap, which can be useful for certain applications.

By incorporating tunable materials or devices into textured surfaces, their capabilities are expanded to include active control of electromagnetic waves [34, 35]. This can be accomplished using mechanical structures such as movable plates or electrical components such as varactor diodes. An electrically tunable impedance surface can be built by connecting neighbouring cells with varactor diodes, which have voltage-tunable capacitance. Changing the bias voltage on the diodes adjusts the capacitance and tunes the resonance frequency. To supply the required voltage to all the varactors, half of the cells are biased alternately and the other half are grounded in a checkerboard pattern, see figure 3.26. At the centre of each biased cell, a metal via passes through a hole in the ground plane and connects to a control line located on a separate circuit layer on the back of the surface. The varactors are oriented in opposite directions in each alternate row, so that when a positive voltage is applied to the control lines, all the diodes are reverse biased. By individually addressing each cell, the reflection phase can be programmed as a function of position across the surface.

Thus with a tunable textured surface of this sort, one can build devices such as programmable reflectors that can steer or focus a reflected microwave beam [37]. These can provide a low-cost alternative to traditional electrically scanned antennas (ESAs) where phase shifters and complicated feed structures are replaced by a planar array of varactor diodes and a free-space, quasi-optic feed. Despite being low-cost, these steerable reflector antennas are ruled out for some applications because they require free-space feed and are not entirely planar.

An alternative to the above is to use a leaky-wave design [38], where a surface wave is excited directly in the surface and then radiates energy into the surrounding space as it propagates. This method involves programming the surface with a periodic impedance function that scatters the surface wave into free space. The period of the surface impedance can be varied to change the phase-matching condition between the surface wave and the space wave and thus steer the radiated

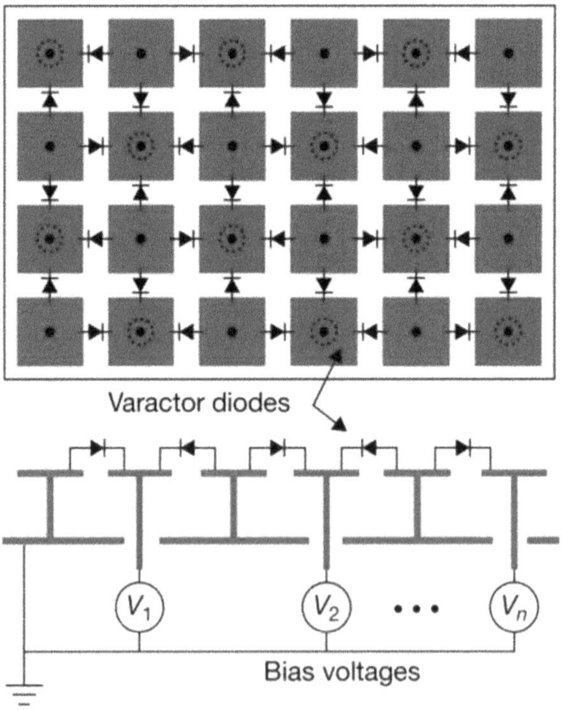

Figure 3.26. A tunable high-impedance surface. (Reprinted/adapted from [35] with permission from De Gruyter, copyright Li *et al* (2018).)

wave. The beam can be electronically steered over a wide range in both the forward and backward directions. Backward leaky waves can also be understood as resulting from bands of negative dispersion, similar to those in other negative-index materials.

3.6 Metamaterial-inspired microwave active devices

3.6.1 Introduction

Some efforts have also been made by researchers to use metamaterial-inspiration for improved design of both solid-state active devices like oscillators and power combiners at microwave frequency and also improving the design of vacuum tube devices at microwave and terahertz (THz) frequencies. We will give a short account of all these in the following.

3.6.2 Solid-state active devices using metamaterial-inspiration

A Japanese group of Mitsubishi Electric Corporation has reported an oscillator with low phase-noise of -121 dBc Hz^{-1} for 1 MHz offset from the operating frequency at Ku band in which an HBT (heterojunction bipolar transistor) device has been used in conjunction with a reflection-type zeroth order CRLH resonator [39]. Since the zeroth order resonator operates with the propagation constant (β) equal to zero, see section 2.3.2.2 of chapter 2, it operates as an infinite wavelength resonator so that its size is small and also the conductance loss of the resonator is very small making it possible to a design size-miniaturized oscillator with loaded Q very high and hence the realization of low phase-noise.

Power combiners are required to combine the power output from a number of active devices, especially the solid-state devices, as the output power from a single solid-state device is normally inadequate for many airborne applications [3]. In conventional design, the devices used for combining their power are so connected via TL that they add their power together in-phase. Metamaterial-inspired design of power combiners is found to be more compact in size and with better performance characteristics.

A novel power combining method for tunnel diode oscillators using the infinite wavelength phenomenon observed in CRLH metamaterial TL has been reported by Itoh *et al* [40, 41]. They have reported two schemes for power combiners based on metamaterial-inspiration and compared their performance. The first combining scheme uses a series combiner composed of the so-called zero-degree CRLH lines while the other one uses an open-ended CRLH TL as a zeroth order resonator. The power combining method using zero-degree line series combiner may be used for tunable oscillator, whereas the zeroth order resonator may be used for high Q oscillations.

The CRLH TL is understood to possess either the propagation properties of a PRH TL—with phase delay—or a PLH TL—with phase advance—depending on the frequency of operation. The transition frequency between the RH and LH regions is a point at which the propagation constant is zero, i.e., $\beta = (2\pi/\lambda) = 0$. Thus at this transition frequency an infinite wavelength can exist, see figures 2.15 and 2.16(b). Since the wavelength is infinite, both the phase and amplitude of a wave

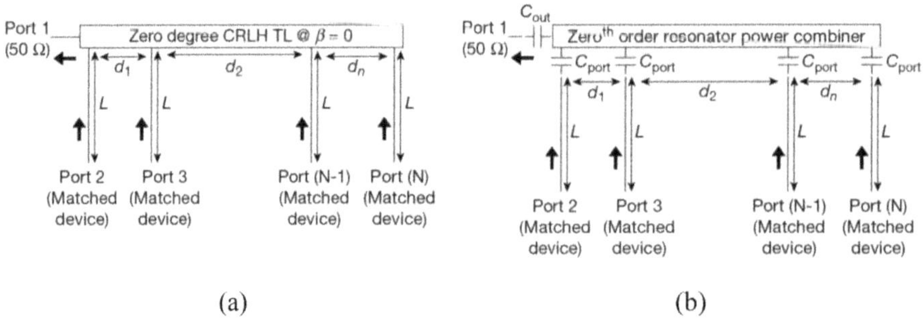

Figure 3.27. Metamaterial-inspired power combiner schemes. (a) CRLH zero-degree line combiner. (b) CRLH zeroth order resonator combiner.

propagating along the line are independent of position and if the line is used as a resonator (with open or short termination) it supports a stationary wave.

The schematic of CRLH-based zero-degree line power combiner is shown in figure 3.27(a). The output port (Port 1) of zero-degree line is impedance matched to 50 Ω while the other ports (Port 2 to Port *N*) are connected via TL of length $L = \lambda_0/4$ (where λ_0 is the wavelength of the fundamental frequency of operation) to tunnel diode oscillators. In other words, the optimum impedance of the device (tunnel diode) is matched to the impedance of the zero-order line with quarter-wave impedance transformer. All the ports, except Port 1, are connected in series with respect to the zero-degree line; hence it is a series combiner. It may be noted that the distance between each port: d_1, ... d_n can be arbitrary, and still provide in-phase power combining due to the fact that $\beta = 0$ at the operating frequency. This eases constraints on the combiner layout and oscillator spacing.

The zeroth order resonator based power combiner, see figure 3.27(b), has the output port (Port 1) and all the remaining ports (Port 2 to Port *N*) loosely coupled to the resonator structure via capacitors while the other side of 2 to *N* ports are connected via quarter-wave TL to match with the optimum impedance of the tunnel diode device. The resonant characteristics of the zeroth order resonator help to provide high *Q* and hence a better phase-noise characteristic is realizable. A maximum power combining efficiency of 131% is obtained with this latter power combiner and the operating frequency of both types of the tunnel diode oscillator is 2 GHz.

3.6.3 Vacuum tube devices with metamaterial-inspiration

In parallel with the global efforts in improving the performance of microwave lenses, phase shifters, directional couplers, antennas, sensors, absorbers, system components, etc, by metamaterial (MTM) assistance, the efforts have also been made to improve the performance of vacuum electron devices (VEDs) by MTM assistance. For the past decade, metamaterial-assisted high-power microwave VEDs have started to gain attention worldwide due to their device miniaturization as well as their gain, power, and efficiency enhancement capabilities. Vacuum tube devices like klystron, travelling wave tube (TWT) backward-wave oscillator, gyrotron and so forth are known to enjoy more power producing and handling capabilities than their conventional solid-state device counterparts. However, metamaterial-inspiration when included in the tube design is found to improve the structure interaction impedance (K) leading to further improved device output power, gain, efficiency, etc. There has been a recent surge in the exploration of MTM assistance in VEDs and their interaction structures which have alleviated the shortcomings of conventional VEDs. In the following we will discuss in brief the present scenario of metamaterial-inspired vacuum tube developments.

In a VED, the kinetic or potential energy of an electron beam is transferred to electromagnetic waves supported by a material medium, which may be termed as the beam–wave interaction structure. A material, treated as a medium for electromagnetic waves to interact with, can be classified in terms of the permittivity ε and permeability μ of the medium as: (i) double-positive (DPS) (ii) epsilon negative (ENG) (iii) double-negative (DNG) and (iv) mu negative (MNG) as may be seen from figure 1.4 of chapter 1. If appropriately engineered, an MTM can be designed to exhibit not only the characteristics of DNG material, but also those of DPS, MNG and ENG materials. We have already seen how to realize ENG, MNG, and DNG MTMs, respectively, with TW, SRR and its variants, and with combination of TW-SRR and so forth in chapter 2.

Let us now make a brief review of the status of the studies on metamaterial-inspired VEDs that are based on the electromagnetic analysis and simulation of MTM-assisted interaction structures of VEDs and underlying beam–wave interaction mechanisms [42].

Purushothaman and Ghosh [43] investigated an SWS (slow wave structure) used in TWT consisting of a helix supported by discrete ENG, MNG, and DNG–MTM rods in a metal envelope (figure 3.28(a)) with the help of electromagnetic field analysis in the tape-helix model that takes into account the axial harmonic effects of

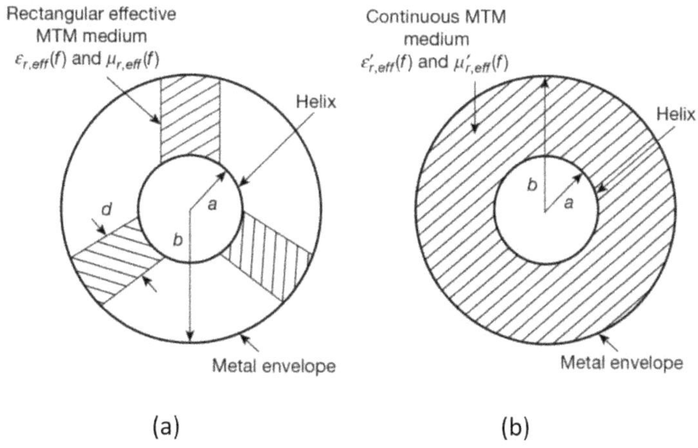

Figure 3.28. (a) Helix of radius a with discrete MTM support rods, typically of rectangular cross-section, enclosed in a metal envelope of radius b. (b) An equivalent structure model consisting of a helix surrounded by an MTM tube in a metal envelope. (Reprinted/adapted from [42] Taylor and Francis Group LLC (Books) US http://www.tandfonline.com.)

the helix turns. In their study, they have assumed that the properties of the MTM rods are isotropic, homogeneous and frequency independent. They have also considered the real values of ENG, MNG, and DNG. As far as the dispersion control at enhanced *K* of the SWS is concerned, they found that the ENG- and MNG–MTM helix-support rods are superior to the DNG–MTM rods.

It has been a conventional method to widen the bandwidth of a helix-TWT by controlling the helix dispersion with low *K* values by metal vane loading of the metal envelope though the method entails the risk of arcing in the tube. Guha and Ghosh [44] have proposed and simulated, using CST Microwave Studio, a novel DPS–MTM loaded helix SWS in which I-shaped metallic strip printed on both the azimuthal faces of the conventional dielectric (BeO, $\varepsilon_r = 6.5$) helix-support rods, see figure 3.29, has been used.

Contrary to the narrow band MTMs like SRR, CSRR etc, the double-positive (DPS), non-resonant MTMs, such as the I-shaped MTMs, behave as non-resonant MTM with double-positive constitutive parameters. However, unlike the conventional DPS materials, such DPS MTMs provide more freedom in tailoring the constitutive properties of the MTMs exhibiting extremely low-loss, broadband response [44]. The proposed structure exhibits the dispersion characteristics similar to those of the conventional vane-loaded helix SWS with a higher *K* leading to the enhanced device gain and efficiency with relaxed possibility of arcing.

Conventional folded-waveguide (FW) SWSs for TWTs are extensively used for high frequency applications as transverse dimensions of the waveguide remain bigger than helix SWSs, for which they have better power handling capability. However, typically FW-SWSs exhibit extremely low *K* values at their wide operating bandwidth. In this regard, Rashidi and Behdad [45] studied ENG–MTM (with frequency independent effective medium property) inclusion in FW–SWS for TWTs (figure 3.30(a)) where, due to the ENG–MTM insertion, operating frequency gets

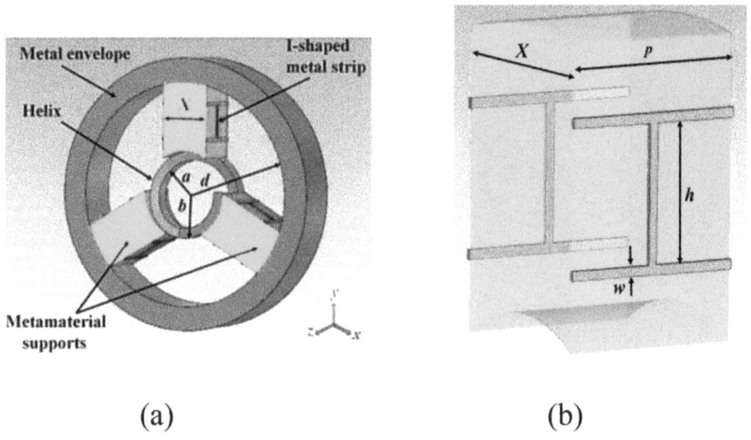

Figure 3.29. (a) Single-turn helical SWS with I-shaped MTM support rods ($a = 0.7$ mm, $b/a = 1.29$, $d/a = 3.33$, and pitch/a = 1.51). (b) I-shaped metal strips imprinted on a host dielectric (BeO, $\varepsilon_r = 6.5$) substrate ($p = 1.057$ mm), $h = 0$.

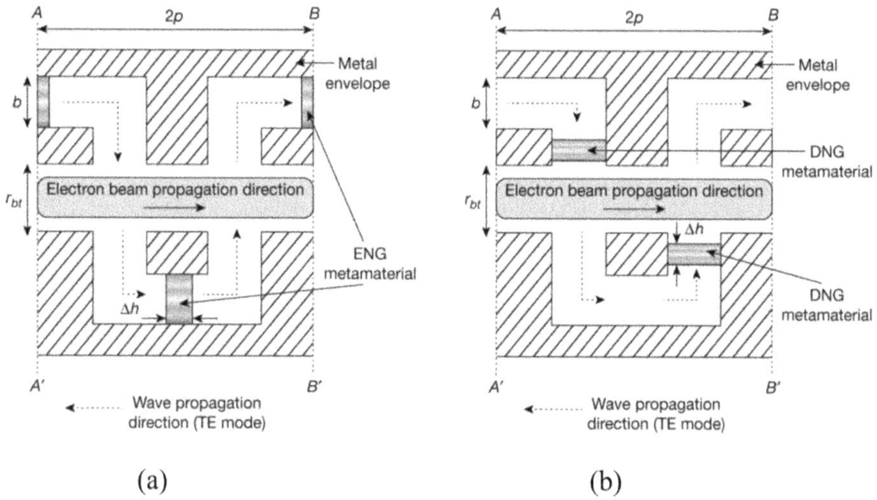

Figure 3.30. Unit cell of an FW loaded by (a) ENG–MTM, and (b) DNG–MTM inserts in a rectangular waveguide of height b with a beam tunnel diameter r_{bt}, showing also the axial periodicity p of the insert, and the thickness Δh of both the types of MTM loading. (Reprinted/adapted with permission from [42] Taylor and Francis Group LLC (Books) US http://www.tandfonline.com.)

up-shifted caused by the increase of phase velocity, obtained using CST Microwave Studio, owing to the introduction of ENG–MTM insert acting as a shunt inductance. However, if the transverse dimensions of FW with an ENG–MTM insert are increased by a scale factor, this up-shifting of operating frequency can be prevented and can be restored back to that corresponding to the case of FW without ENG–MTM loading, however, at the cost of the bandwidth. Interestingly, the larger transverse dimensions of the ENG–MTM loaded FW would allow for a larger beam tunnel diameter, and consequently, a larger beam current, larger beam power and a higher device RF output power for a given frequency of operation. This advantage of ENG–MTM assistance is of special significance in designing high-frequency TWTs, such as the millimetre-wave TWTs, for which, in the conventional design, the beam diameter and current become restricted due to the smaller transverse dimensions of the SWS. Further, added to this advantage of the ENG–MTM assistance is enhanced K of FW making it potentially employable as SWS of a high-gain, high-efficiency ENG–MTM-assisted FW-TWT. The TWT designers face the difficulty of fabrication as well as thermal management in designing and developing the conventional TWTs in high-frequency, such as in the millimetre-wave regime, due to their reduced transverse dimensions. Clearly, the ENG–MTM loading of FW-TWT can alleviate this problem by providing scaled-up larger transverse dimensions as required for preventing the operating frequency band from shifting to higher frequencies. Further, a higher K of an ENG–MTM-loaded FW would enhance the gain and efficiency of an ENG–MTM-assisted FW-TWT [45].

The DNG–MTM loaded FW has also been studied [45], see figure 3.30(b). It has been observed that in a DNG–MTM loaded FW-TWT, the MTM parameters provide an additional control over the gain-frequency response, unlike in a conventional FW-TWT without MTM loading in which only the structural dimensions control such a response. Further, the dependency of frequency corresponding to the maximum power transfer from the electron beam to electromagnetic waves on the beam accelerating potential makes the DNG–MTM-loaded FWT rather frequency tuneable. However, the limited frequency regime in which the MTM loaded FW exhibits the DNG property makes the DNG–MTM loaded FW-TWT essentially bandwidth-limited and dependent on the properties of the DNG–MTM used. The need of the hour is to reduce the inherent ohmic losses associated with the MTM and widen the DNG frequency regime of MTM to make the DNG–MTM loaded FW-TWT a more useful VED.

The resistive-wall amplifier (RWA) is a growing technique in VED technology in which the gain in the device is obtained through the interaction between the charge of electron beam and that on the resistive metal wall that is induced by the beam such that the charge induced on the wall acts on the electron beam so as to cause larger and larger electron bunches to be formed, thereby resulting in the spatial growth of signal. Rowe *et al* [46, 47] studied the effect of MTM assistance on the performance of RWA tube devices at first by considering MTM as isotropic [46] and also as anisotroipic [47]. Rowe *et al* [46] proposed two different configurations of RWAs assisted by an MTM that were supposedly isotropic: (i) one using a loss-less ENG–MTM as the support material for the resistive layer (figure 3.31), where region I is the electron beam region, region II is a vacuum region, region III is a resistive layer region and region IV is a loss-less MTM region, and (ii) yet another using a lossy ENG–MTM as both the resistive and support mediums integrated into a single layer, the regions III and IV being merged to a single region. The MTM exhibits here a plasma-like response with a positive permeability and a negative

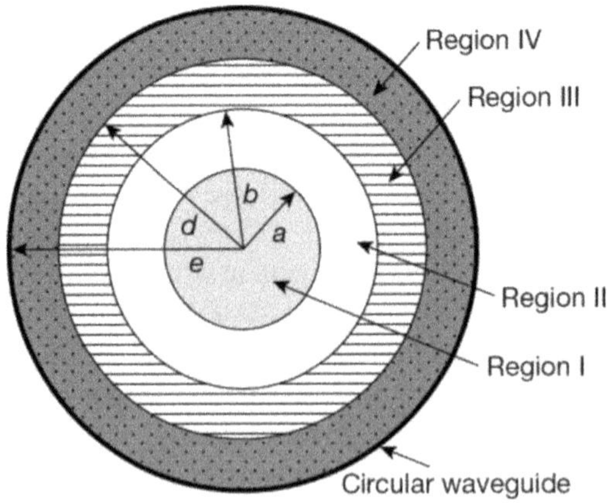

Figure 3.31. Cross-sectional views of an MTM-assisted RWA showing (i) beam region (Region I), (ii) vacuum region (Region II), (iii) resistive layer region (Region III), and (iv) loss-less MTM region (Region IV). Here $r = a$ represents the beam radius; $r = b$ and $r = d$ represent the inner radii of the resistive layer and MTM regions, respectively; and $r = e$ represents the inner wall radius of the circular waveguide. (Reprinted/adapted with permission from [42] © Taylor and Francis Group LLC (Books) US http://www.tandfonline.com.)

permittivity, responsible to maintain the inductive wall admittance, which, in turn, yields high gains of the device over relatively broad bandwidths with moderate beam filling factors [46]. Rowe *et al* [47] through their analysis also found that the anisotropic MTM liner provided higher inductive wall admittance as well as an increased peak gain rate within a narrow bandwidth as compared to an isotropic MTM liner.

Apart from these few representative works extensive research is going on with improved performance klystron [48], backward-wave oscillator [49], and so forth both at microwave and terahertz frequency domain [50]—a concise review of that, including experimental prototype design, is available in reference [42]. As a concluding remark it may be mentioned that newer MTM-assisted interaction structures and devices need to be explored for higher frequencies and higher powers as well as for the size reduction or compactness which might make tube devices of the future for more varied applications which have not yet been thought of.

References

[1] Kar S and Ghosh M 2012 Advances in metamaterial research *Asian J. Phys.* **21** 1–16

[2] Caloz C and Itoh T 2006 *Electronic Metamaterials: Transmission line Theory and Microwave Applications* (New York: Wiley Interscience)

[3] Kar S 2022 *Microwave Engineering—Fundamentals, Design, and Applications* 2nd edn (Hyderabad: Universities Press)

[4] Kar S, Kumar A, Majumder A, Ghosh S K, Saha S, Sikdar S S and Saha T K 2016 CRLH and SRR based microwave filter design useful for microwave communication *Int. J. Comput. Electr. Autom. Control Inf. Eng.* **10** 629–33

[5] Kar S, Ghosh M, Kumar A and Majumder A 2016 Complementary split-ring resonator-loaded microstrip patch antenna useful for microwave communication *Int. J. Electron. Commun. Eng.* **10** 1321–5

[6] Falcone F, Lopetegi T, Laso M A G, Baena J D, Bonache J, Beruete M, Marques R, Martin F and Sorolla M 2004 Babinet principle applied to the design of metasurfaces and metamaterials *Phys. Rev. Lett.* **93** 197401

[7] Wu R, Kuo C and Chang K K 1992 Inductance and resistance computations for three-dimensional multiconductor interconnection structures *IEEE Trans. Microw. Theory Tech* **40** 263–71

[8] Sanada A, Murakami K, Awai I, Kubo H, Caloz C and Itoh T 2004 A planar zeroth-order resonator antenna using a left-handed transmission line *Proc. 34th European Microwave Conf. (Amsterdam, 12–14 October 2004)* 1341–4

[9] Otto S, Rennings A, Caloz C, Waldow P, Wolff I and Itoh T 2005 Composite right/left-handed λ-resonator ring antenna for dual frequency operation *Proc. IEEE AP-S USNC/URSI National Radio Science Meeting (Washington, DC, 3–8 July 2005)*

[10] Balanis C A 2008 *Modern Antenna Handbook* (New York: Wiley)

[11] Lim S, Caloz C and Itoh T 2005 Metamaterial-based electronically controlled transmission-line structure as a novel leaky-wave antenna with tunable radiation angle and beamwidth *IEEE Trans. Microw. Theory Tech* **53** 161–73

[12] Majumder A, Das S, Kar S, (Guide-cum-Consultant) and Kumar A 2013 Technical Report on Left Handed Maxwell's Systems—Experimental Studies *Report number: Report on*

Project No. 2009/34/37/BRNS/3176/ Dated-12-2-2010) [https://researchgate.net/publication/324889314_Technical_Report_on_Left_Handed_Maxwell's_Systems_-Experimental_Studies/figures].
[13] Maci S, Gentili G B, Piazzesi P and Salvador C 1995 Dual-band slot-loaded patch antenna *Proc. Inst. Elect. Eng. Microw. Antennas Propag* **142** 225–32
[14] Porath R 2000 Theory of miniaturized shorting-post microstrip antennas *IEEE Trans. Antennas Propag.* **48** 41–7
[15] Pozar D M and Schaubert D H 1995 *Microstrip Antennas: Analysis and Design of Micristrip Antennas and Arrays* (New York: IEEE Press)
[16] Dong Y and Itoh T 2011 Miniaturized patch antennas loaded with complementary split-ring resonators and reactive impedance surface *Proc. 5th European Conf. on Antennas and Propagation (EUCAP) (Rome, 11–15 April 2011)* 2415–8
[17] Ha J, Kwon K, Lee Y and Choi J 2012 Hybrid mode wideband patch antenna loaded with a planer metamaterial unit cell *IEEE Trans. Antennas Propag.* **60** 1143–7
[18] Cheng X, Senior D E, Kim C and Yoon Y-K 2011 A compact omnidirectional self-packaged patch antenna with complementary split-ring resonator loading for wireless endoscope applications *IEEE Antennas Wirel. Propag. Lett.* **10** 1532–5
[19] Malik J and Kartikeyan M V 2012 Metamaterial inspired patch antenna with L-shape slot loaded ground plane for dual band (WiMAX/WLAN) applications *Prog. Elecromagn. Res. Lett.* **31** 35–43
[20] Caloz C, Sanada A and Itoh T 2004 A novel composite right/left-handed coupled-line directional coupler with arbitrary coupling level broad bandwidth *IEEE Trans. Microw. Theory Tech* **52** 980–92
[21] Caloz C and Itoh T 2004 A novel mixed conventional microstrip and composite right/left-handed backward-wave directional coupler with broadband and tight coupling characteristics *IEEE Microw. Wirel. Compon. Lett.* **14** 31–3
[22] Pozer D M 2005 *Microwave Engineering* 3rd edn (New York: Wiley) ch 7 pp 333–7
[23] Islam Rand Eleftheriades G V 2004 Phase-agile branch-line couplers using metamaterial lines *IEEE Trans. Wirel. Compon. Lett.* **14** 340–2
[24] Chun Y H, Hong J S, Moon J Y and Yun S W 2006 High directivity directional coupler using metamaterial *Proc. 36th European Microwave Conf. (Manchester, UK)* 329–31
[25] Engheta N and Zilokowski R W (ed) 2006 *Metamaterials—Physics and Engineering Explorations* (Piscataway, NJ: IEEE Press, Wiley Interscience)
[26] Engheta N 2002 An idea for thin, subwavelength cavity resonators using metamaterials with negative permittivity and permeability *IEEE Antennas Wirel. Propag. Lett.* **1** 10–3
[27] Qie M, Simcoe M and Eleftheriades G V 2002 High gain meander-less slot arrays on electrically thick substrates at mm-wave frequencies *IEEE Trans. Microw. Theory Tech* **50** 517–28
[28] Koul Kand Bhat B 1991 *Microwave and Millimeter Wave Phase Shifters* (Norwell, MA: Artech House)
[29] Didomenico M Jr and Pantell R 1962 X-band ferroelectric phase shifter *IRE Trans. Microw. Theory Tech* **10** 179–85
[30] Antoniades M A and Eleftheriades G V 2003 Compact linear lead/lag metamaterial phase shifters for broad band applications *IEEE Antennas Wirel. Propag. Lett.* **2** 103–6
[31] Kuylenstierna D, Vorobiev A, Linner P and Gevorgian S 2006 Composite right/left handed transmission line phase shifter using ferroelectric varactors *IEEE Microw. Wirel. Compon. Lett.* **16** 167–9

[32] Abdalla M, Phang K and Eleftheriades G V 2007 Printed and integrated CMOS positive/negative refractive-index phase shifters using tunable active inductors *IEEE Trans. Microw. Theory Tech* **55** 1611–23

[33] Sievenpiper D, Zhang L, Romulo F, Broas J, Alexopoulos N G and Yablonovitch E 1999 High-impedance electromagnetic surfaces with a forbidden frequency band *IEEE Trans. Microw. Theory Tech* **47** 2059–74

[34] Sievenpiper D 1999 High-impedance electromagnetic surfaces *PhD Dissertation* (Los Angeles, CA: Department of Electrical Engineering: University of California)

[35] Li A, Singh S and Sievenpiper D 2018 Metasurfaces and their applications *Nanophotonics* **7** 989–1011

[36] Elliot R 1954 On the theory of corrugated plane surfaces *IRE Trans. Antennas Propag* **2** 71–81

[37] Sievenpiper D, Schaffner J, Song H J, Loo R and Tangonan G 2003 Two-dimensional beam steering reflector using an electronically tunable impedance surface *IEEE Trans. Antennas Propag* **51** 2713–22

[38] Sievenpiper D 2005 Forward and backward leaky wave radiation with large effective aperture from an electronically tunable textured surface *IEEE Trans. Antennas Propag* **53** 236–47

[39] Mizutani H, Tsuru M, Kawakami K, Nishino T and Miyazaki M 2008 Ku band oscillator with a reflection type zeroth order resonator *The Institute of Electronics, Information and Communication Engineers (IECIE) Technical Report* ED2006-2008, MW2006.161 (2007-1) (in Japanese)

[40] Dupuy A, Leong K M G K and Itoh T 2006 Power combining tunnel diode oscillators using metamaterial transmission line at infinite wavelength frequency *IEEE MTT-S Int. Microwave Symp. Digest* 751–5

[41] Itoh T, Leong K and Dupuy A 2009 Power combiners using meta-material composite right/left hand transmission line at infinite wavelength frequency *US Patent* 20080001684A1 Issued on: 27-01-2009

[42] Guha R, Wang X, Varshney A K, Duan Z, Shapiro M A and Basu B N 2021 Review of metamaterial-assisted vacuum electron devices *Metamaterials: Technology and Applications* ed P K Choudhury (Boca Raton, FL: CRC Press, Taylor and Francis Group) ch 13 pp 351–81

[43] Purushothaman N and Ghosh S K 2013 Performance improvement of helix TWT using metamaterial *J. Electromagn. Waves Appl* **27** 890–900

[44] Guha R and Ghosh S K 2022 Dispersion control of a helix slow-wave structure by I-shaped metamaterial loading for wideband traveling-wave tubes, *URSI-RCRS*, IIT-BHU, Varanasi, India, 2020. *Also*: Dispersion control and size enlargement of helical slow-wave structure by double-positive metamaterial assistance *IEEE Trans. Electron Devices* **69** 771–6

[45] Rashidi A and Behdad N 2014 Metamaterial-enhanced travelling wave tubes *Proc. IEEE Int. Vac. Electron. Conf. (IVEC) (Monterey, CA, April 20–22 2014)* 199–200

[46] Rowe T, Booske J H and Behdad N 2015 Metamaterial-enhanced resistive wall amplifiers: theory and particle-in-cell simulations *IEEE Trans. Plasma Sci.* **43** 2123–31

[47] Rowe T, Behdad N and Booske J H 2016 Metamaterial-enhanced resistive wall amplifier design using periodically spaced inductive meandered lines *IEEE Trans. Plasma Sci.* **44** 2476–84

[48] Wang X, Li S, Zhang X, Jiang S, Wang Z, Gong H, Gong Y, Basu B N and Duan Z 2020 Novel S-band metamaterial extended interaction klystron *IEEE Electron Device Lett.* **41** 1580–3
[49] Hummelt J S, Lewis S M, Shapiro M A and Temkin R J 2014 Design of a metamaterial-based backward-wave oscillator *IEEE Trans. Plasma Sci.* **42** 930–6
[50] Duan Z, Guo C, Guo X and Chen M 2013 Double negative-metamaterial based terahertz radiation excited by a sheet beam bunch *Phys. Plasmas* **20** 093301

Chapter 4

Terahertz and optical metamaterial

The beacon of life should be the effulgence that streams out of one's soul

4.1 Introduction

Metamaterial development initially started in the year 2000 at microwave frequency to realize negative refractive index using thin-wire (TW) and split-ring resonator (SRR) combination with the then available conventional printed circuit technology. The race for the optical negative-index materials (NIM) was on, and virtually every year saw a decrease in the operating wavelength where a negative refractive index was observed—from centimetres in 2000 to optical wavelengths in 2005 [1, 2]. This was possible because of the progress of fabrication technology, especially nanofabrication technology, resulting in newer application possibilities in the terahertz (THz) and optical frequency domain. The ubiquitous split-ring metamolecule, commonly used as magnetic inclusion metamaterial structure at microwave frequency, can be scaled down in size from MHz to THz [3–8], with various shaping of SRR (see figure 4.1) [2], but this scaling breaks down at higher frequencies as the metal no longer behaves as a conductor for wavelengths shorter than 1.5 μm, i.e. beyond 200 THz range. Thus, going from microwaves to optics poses a real challenge to the metamaterial researchers as far as the physics and technology is concerned. At optical frequencies some noble metals like silver and gold behave differently in that they do not exhibit conductivity in the usual sense, but instead exhibit plasmonic resonance (i.e. coupling of optical signals with collective oscillation of conduction electrons at these metal surfaces) due to the negative real part of their permittivity. Thus, the physical mechanisms and the distribution of the electromagnetic field's currents in the structures at microwave and optical frequencies are radically different. Resonant effects used to tailor μ and ϵ in optical metamaterials are essentially plasmon resonances. For instance, there is no need for employing electrical engineering terms to define the so-called LC resonances which are but the fundamental plasmonic modes of the nanostructure. Thus, optical metamaterials are closely related with the

doi:10.1088/978-0-7503-5532-2ch4

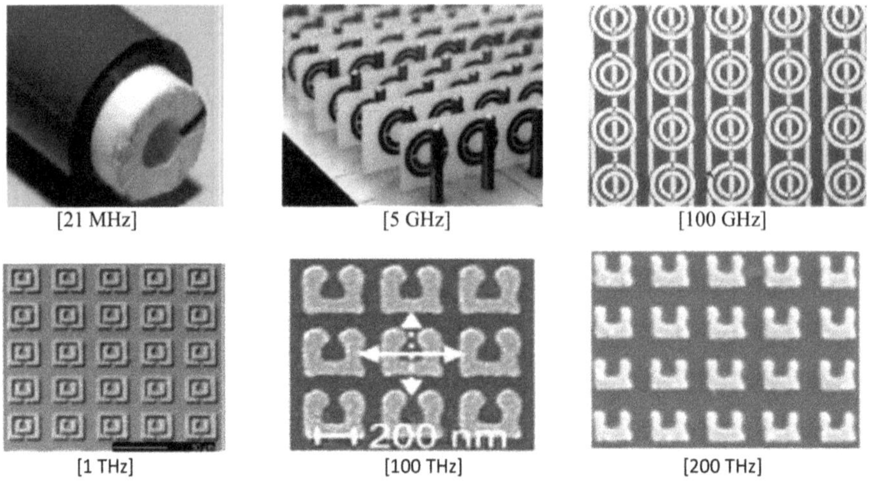

Figure 4.1. MHz to THz metamaterial structures. (Reproduced/adapted from [2], copyright (2006) with permission from Elsevier.)

progress of plasmonics. Along with this physical conceptual issue, the fabrication limitations and difficulties of making nanometer size SRRs along with metal wires (SRR-TW combination) led to the development of alternative designs using sub-wavelength nanoparticles for the design of metamaterial inclusion structures at optical frequencies. Thus, at optical frequency regime coupled-nanorod/nanostrip structures and the so-called fishnet structures and their various improvisations, designed with nano-fabrication technology, are used to realize metamaterial property.

Light is the ultimate means of sending information to and from the interior structure of materials—it packages data in a signal consisting of photons (the fundamental quantum of light) of zero mass that travels with an unmatched speed. However, light is, in a sense 'one-handed' when interacting with atoms of conventional materials. This is because from the two field components of light—electric and magnetic—only the electric hand efficiently probes the atom of a material, whereas the magnetic component of light is normally weak. This is because magnetic coupling to an atom is proportional to the Bohr magneton $\mu_B = e\hbar/2m_ec = \alpha e a_0/2$, (where e is the electronic charge, $\hbar$ is the reduced Planck's constant, m_e is the electronic mass, a_0 is the Bohr radius) and the electric coupling is ea_0. The induced magnetic dipole also contains the fine-structure constant $\alpha \approx 1/137$ so that the effect of light on the magnetic permeability is α^2 weaker than on the electric permittivity. This also explains why all naturally occurring magnetic resonances are limited to relatively low frequencies (in fact the magnetic susceptibility of all magnetic materials tails off towards the THz and microwave frequencies) while the electric resonances occur at much higher frequencies (optical to ultraviolet ranges). As the magnetic response is a precursor for realizing negative refraction, it is of critical importance to address the fundamental problem of engineering artificial optical magnetism in optical metamaterial design.

The problem of low coupling to magnetic field component of electromagnetic signal was overcome by mimicking the principle of magnetism used by Mother Nature. In 1999 J B Pendry was able to realize artificial magnetism at microwave (gigahertz) range with the help of his ubiquitous SRR. However, at optical frequency such artificial magnetism is realizable with some other innovative techniques which will be discussed in section 4.3.2. Metamaterials, that is, artificial materials with radically different properties, can allow both field components of light to be coupled to meta-atoms, enabling entirely new optical and THz properties with exciting applications. In fact, metamaterials are opening up a new gateway in the optical and THz frequency domain with unprecedented electromagnetic properties and functionality unattainable from naturally occurring materials. The structural units of metamaterials can be tailored in shape and size, their composition and morphology can be artificially tuned and inclusions can be designed and placed at the desired location to achieve new and varied functionalities that were not possible with natural materials so far.

The optical community has been excited for more than a decade or so with the stunning early example of the demonstration of negative refractive index at optical frequency, something long thought impossible. Optical invisibility cloaks have intrigued the popular press. Other promising practical capabilities of metamaterials, not available from conventional optical materials, include broadband polarising filters, near perfect absorption and so forth. An important field of optics that has emerged with great success with the advent of metamaterilas/metasurface is the 'transformation optics'—which can transform the wavefront at the will of the designer. In fact, by varying the shape of the structures or the configuration of the sub-wavelength building blocks of the metamaterial/metasurface one can create 'gradients' in optical properties that can manipulate light in practically all possible ways. Another attraction of metamaterials is their ability to produce strong circular polarization, which can be difficult with conventional bulk optics.

At THz frequency, the high absorption of metamaterials is of special attraction where natural materials generally have low absorption capability. Thus metamaterial-based THz absorbers are now possible to develop that can be made insensitive to polarization and to angles of incidence too. THz metamaterials have also demonstrated other interesting properties like high-speed switching and modulation, and tuning of resonance behaviour by manipulating the inclusion structures. In many cases the metamaterial-based THz devices outperform their conventional counterparts, and in fact for many applications such conventional counterparts do not even exist at THz frequencies. For example, strong magnetic responses in THz frequency regime have been engineered with a composite material containing split-ring structures made of copper. Such strong magnetic responses at THz frequency do not occur in naturally available materials.

4.2 THz—a promise for various emerging applications including those in metamaterials

4.2.1 Introduction

Electromagnetic response of natural materials is not evenly distributed across the electromagnetic spectrum. At frequencies of a few hundred gigahertz and lower, electrons are the principal particles which serve as the workhorse in designing the devices (sources, detectors and so forth)—this has given birth to the discipline of electronics. On the other hand, at infrared through optical/UV wavelengths, the photon is the fundamental particle of choice and thus we have the discipline of photonics. In-between these two response regimes there exists a region comparatively devoid of material response, commonly referred to as the 'terahertz gap' (0.1–10 THz) [9, 10], i.e. having a frequency of a trillion cycles a second, see figure 4.2. This corresponds to the millimetre and sub-millimetre wavelengths between 3 mm (EHF band) and 0.03 mm (long-wavelength edge of far-infrared light).

The THz frequency domain has remained practically unused by technologists for a long time, till the end of the 20th century, except for very few uses in radio astronomy, due to the non-availability of suitable devices, circuits in this frequency spectrum. But if this T-ray technology is properly developed this could lead to innovative imaging and sensing technologies that hold enormous potential in biomedicine and security. In fact, from the beginning of the 21st century THz frequency domain has been observing exponential growth of its use with the advancement of both RF and optical technologies. As mentioned earlier, the THz frequency spectrum falls between the higher RF (that uses electronics) and the visible spectrum (that uses photonics), see figure 4.2. Although enormous efforts have been focussed on the search for 'terahertz' materials or alternative techniques to enable the construction of device components, much work still remains to be done. There is a wide range of natural phenomena that could be probed with terahertz (THz) devices. Specifically, a THz detector would be useful for imaging in areas such as biology [11, 12] and security [13, 14].

The most important applications of THz frequency domain at this time are for concealed weapon detection (CWD) and stand-off detection (meaning: 'to see with its radio eye' a few meters/tens of meters away) of explosives and illegal drugs for

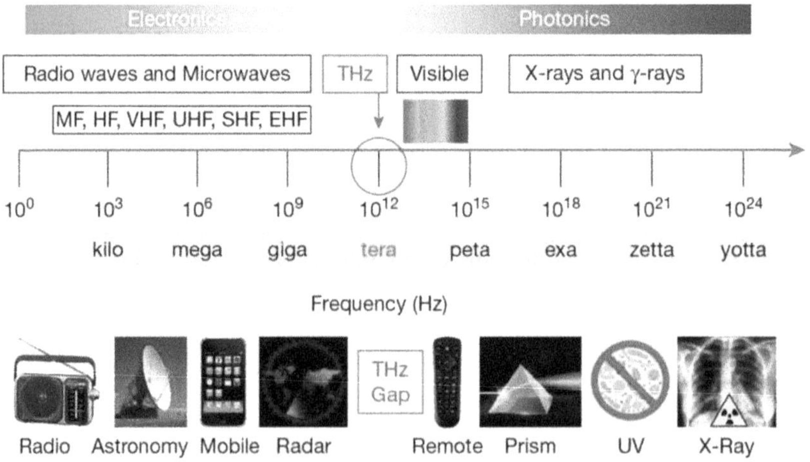

Figure 4.2. Electromagnetic spectrum showing the location of terahertz (THz) frequency range and some applications in various other frequency ranges.

airport and other security purposes. A T-ray imager can clearly distinguish between talcum powered and explosives like RDX in stand-off detection. T-ray also offers hope for early diagnosis of cancer, including skin cancers [15] like melanoma, because they track molecular signatures of the malignancy that are not easily seen with other types of scans. All these applications of THz have been possible due to the special property of THz having the penetration capability of RF (even in zero visibility) and spectroscopic capability (i.e., chemical-discriminating capability) of optical frequencies. Further, THz being much higher in frequency compared to millimetre-waves the THz imaging provides much better fidelity (i.e. image resolution) compared to millimetre-wave imaging. And last but not the least, THz is a non-ionizing radiation unlike x-rays, the latter being highly ionizing; and thus THz is a harmless electromagnetic radiation useful for medical imaging purposes.

To address the requirement of various THz devices, the newly developed engineered electromagnetic materials, i.e. 'metamaterial', are expected to play a meaningful role within the so-called void of natural materials in the THz gap—this is what we are going to discuss now.

4.2.2 THz metamaterial

The terahertz metamaterial (MTM) research also started picking-up pace from the first decade of this century. THz MTMs have received much attention, since conventional and natural materials possesses weak response to THz electromagnetic waves. THz MTMs are beyond this limitation and can interact with THz waves strongly by tailoring unit cells to typical periodicities of tens or hundreds of micrometers [16]. At lower THz frequency and below 200 THz the scaling of SRR structure used at microwave/millimetre-wave frequencies works fine. Yen *et al* [6] in 2004 generated artificial magnetic moments at THz frequency (0.6–1.6 THz) by exploiting the resonance of SRR particles proposed originally by J B Pendry in 1999 [17]. It is known that when such structure is excited by a time-varying magnetic field with a component perpendicular to the plane of the rings of SRR, a strong artificial magnetic moment is generated at resonance of the rings, thus exhibiting a resonant-type macroscopic magnetic polarization. As the structure is resonant in nature it can provide a modest fractional bandwidth, but such a bandwidth may be sufficient in absolute terms for THz applications. Also, the operation frequency of SRR MTM is known to be tunable by proper adjustment of the structural design parameters. In 2005, Moser *et al* [18] demonstrated LH resonance between 1 and 2.7 THz with SRR-TW MTM structure consisting of a 2D array of Ni and Au SRR and rods embedded in an AZ P4620 photoresist matrix.

The realization of tailorable artificial electric and magnetic materials in the THz range may represent an important step toward bridging the gap between microwave/ millimetre-wave and optical frequencies. It has the possibility of developing novel devices for biological and security imaging, biomolecular fingerprinting, remote sensing, and guidance in near-zero visibility weather conditions [6]. Such THz MTM and derivatives may also play an important role in the development of engineering devices, such as THz sources, harmonic generators, photo-mixers, and others [19].

4.2.3 THz metamaterial applications

4.2.3.1 Metamaterial-based THz absorbers

MTMs are found to be very useful for the development of perfect absorbers (PAs) that resemble the so-called 'black body radiators' in electromagnetic spectrum from GHz through THz to optical frequencies. MTM based absorbers are useful to enhance the efficiency in capturing solar energy [20] and applied to plasmonic sensor [21], bolometer [22], wireless power transfer [23] and perfect light absorber [24, 25].

Despite the existence of the famous THz gap, below which the response of a device is dominantly electronic, while above which it is dominantly photonic, the basic principle of operation of MTM PA in the whole frequency range is nearly identical to each other, except for small details, for example, localized surface-plasmon polaritons actually come into play in the IR/visible range. Therefore, a simple scale down of the structure operating at GHz is possible for operating at THz frequencies, however, at IR-to-optical frequency the MTM PA design is more involved, which has been discussed in section 4.3.4.2.

Ruck *et al* [26] classified perfect EM-wave absorbers into two groups: (i) broadband absorbers; and (ii) resonant absorbers. So far, the resonant absorbers can have perfect absorption in a narrow bandwidth, while only non-resonant techniques have been known to be employed for broadband absorption. It is, of course, possible to make resonant absorbers operating in a broad range of frequencies by stacking several layers; however, this kind of broadband absorber is very thick and bulky.

As already discussed earlier, the artificially engineered metmaterials are metal–dielectric composites whose electromagnetic properties originate from oscillating electrons in unit cells comprised of highly conducting shaped metals (copper/silver/gold). The sub-wavelength unit cells are shaped in order to exhibit electrical and magnetic response (in the case of plasmonic metamaterial with TW/CW and SRR, respectively) so that their combined structure can replicate material properties which is tailorable by the designer. Resonant structures that couple strongly to either the electric [27] or magnetic [28] fields have been demonstrated at THz.

Since its inception in 2001 with the successful experimental realization of negative refractive index at microwave frequency, significant growth in metamaterial research has been directed towards the investigation of negative refractive index phenomena and their applications like cloaking, sub-wavelength imaging and so forth. To realize such structure for these applications, it is important to minimize losses over the operating frequency range, which is associated with the imaginary part of the complex part n_2 of the refractive index $[\tilde{n}(\omega) = \sqrt{\tilde{\varepsilon}(\omega)\tilde{\mu}(\omega)} = n_1 + jn_2$, where complex electric permittivity $\tilde{\varepsilon}(\omega) = \varepsilon_1(\omega) + j\varepsilon_2(\omega)$ and complex magnetic permeability $\tilde{\mu}(\omega) = \mu_1(\omega) + j\mu_2(\omega)]$; hence one strives for $n_2 \to 0$. So, for realizing negative refractive index the real part of n needs to be such that $n_1 < 0$, while for cloaking application one needs $0 < n_1 < 1$. But for applications like—*absorbers*—it would be desirable to maximize the loss in the metamaterial. Design of metamaterial-based absorbers at THz frequency would be of particular importance because it is difficult to find naturally occurring materials with strong absorption coefficients at THz frequency domain.

A resonant metamaterial absorber at microwave frequency was first reported by Landy *et al*, in 2008 [29]. This had the advantage of small size and was quite thin compared with the conventional absorbers. They realized theoretically and experimentally a perfect MTM absorber (97% absorption around 11 GHz) that was composed of conducting electric ring resonator on the front side and cut wire on the rear one, separated by a dielectric substrate—a similar structure which was later used by Tao *et al* [30] for the design of THz PA.

Since the work of Landy *et al*, and Tao *et al*, a large volume of research work has taken place both at microwave [31–35] and THz [31, 36–38] frequency for the design of MTM PA. In 2010, Li *et al* [32] realized a dual-band metamaterial absorber in the microwave region. The triple-band MTM PA at GHz was theoretically and experimentally investigated by Shen *et al* [33]. Park *et al* [34] theoretically and experimentally reported polarization-independent quadruple-band MTM PAs in the GHz range. Tuong *et al* [35] also reported polarization-insensitive and polarization-controlled dual-band absorption in the GHz range. The EM (electromagnetic) wave in the GHz range is used primarily for cell phone and military radar bands. In order to be applied to these application areas, therefore, MTM PAs operating in the GHz range should overcome the problems relevant to the size of the unit cell, the flexibility, the thermal stability, etc.

Shen *et al* [36] observed absorption spectrum with MTM structure having three near-unity (more than 95%) absorption peaks at frequencies of $f_1 = 0.50$, $f_2 = 1.03$ and $f_3 = 1.71$ THz. Since the MTM PAs fabricated on rigid substrates are limited in applications not only on curved surfaces but also on varying surfaces, ensuring flexibility in MTM PAs is one of the challenging tasks for practical applications. Shan *et al* [37] demonstrated a flexible dual-band THz MTM PA fabricated on a thin polyimide (PI) substrate. The fabricated MTM PA is claimed to be 'ultrathin' because the thickness of PI spacer layer is only 25 μm. Experimental results show that the absorber has two resonant absorption frequencies (0.41 THz and 0.75 THz) with absorption of 92.2% and 97.4%, respectively. They also demonstrated that their MTM PA was insensitive to polarization, and had high absorption (over 90%) for wide incident-angle range from 0° to 45°. The absorption was also better than 90% even in the case of wrapping it to a curved surface. Apart from the flexibility, tunability is also one of the challenging tasks in MTM PAs. Zhao *et al* [38] demonstrated highly-flexible and tunable MTM PA at THz frequency (0.2–2.0 THz).

The research progress with THz MTM PAs clearly reveals that the future of THz MTM PAs is very optimistic, because researchers in this field have exerted enormous efforts in exploiting new application fields and in realizing sophisticated properties of THz MTM PAs. It is desirable that in practical application of THz MTM PAs the EM-wave response should be broadband, omni-directional, polarization-independent, and flexible. It will be especially challenging if the combination of two or more features, for instance, the combination of omni-directionality and polarization independence, is achieved. In spite of these hardships, multi-dimensional physical properties, such as broadband, flexibility, polarization independence and tunability are expected to be realized in the near future.

Anyway, the first generation THz metamaterial absorber was the report by Tao *et al* [30] which could achieve a resonant absorptivity of 70% at 1.3 THz, but similar to Landy *et al*'s GHz MTM PA [29] this was also a highly selective absorber over a narrow band at THz frequencies. However, it would be useful to have a brief discussion of the principle of operation of this basic MTM PA here.

The THz absorber reported by Tao *et al* [30] consists of a combination of electric ring resonator (ERR) and split-wire matrix (as shown at the middle of figure 4.3(a)). In the same figure the split-wire array is shown in the top left and an individual unit cell of the absorber is also shown in the bottom. A unit cell of ERR (deposited on polyimide), split-wire (deposited on GaAs wafer), and ERR-split-wire combination is shown in figure 4.3(b) with *E*, *H*, and *k* vector of the incident THz signal.

A single unit of the absorber consists of two distinct metallic elements: an ERR and a split (cut)-wire structure, see figure 4.3(a) bottom picture. The ERR consists of two single-split rings sitting back to back. The two inductive loops are of opposite handedness and thus couple strongly to a uniform electric field, and negligibly to magnetic fields [27, 28]. The magnetic component of electromagnetic signal couples to both the centre section of the ERR and the split-wire structure, thus generating antiparallel currents resulting in resonant $\mu(\omega)$ response. The magnetic response can therefore be tuned independently of the electric resonator by changing the geometry of the split-wire structure and the distance between elements. In this structure, the metallic pattern on the front layer (the ERR) is for the minimized reflection of electromagnetic (EM) wave by impedance matching with the incident medium, and the metallic layer on the back side (split-wire) is for blocking the transmission. However, the PA MTMs discussed above are for specific frequency and thus have limited applications.

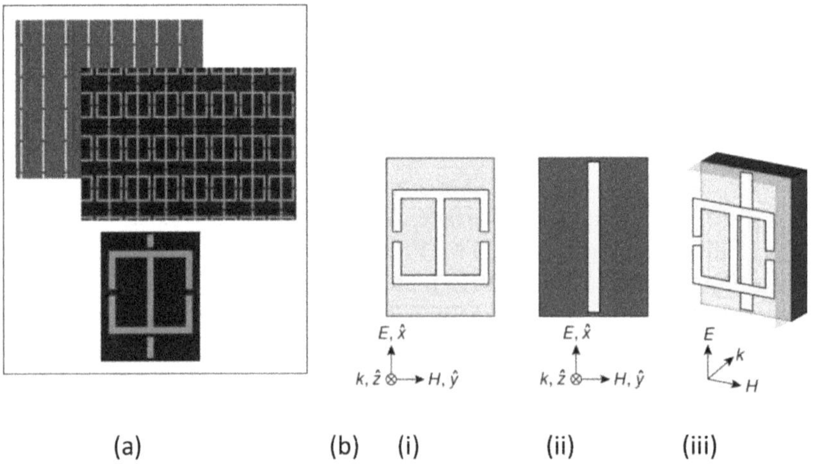

Figure 4.3. THz absorber structure. (a) ERR (electric ring resonator) and split-wire combination matrix (middle), split-wire array (top left), unit cell of the absorber (bottom). (b) (i) A unit cell of ERR (deposited on polyimide), (ii) split-wire (deposited on GaAs wafer), and (iii) ERR-split-wire combination. (Reprinted/adapted with permission from [30], copyright The Optical Society of America.)

This necessitated the research on multi/broadband, polarization-independent, incident-angle-independent and frequency tunable MTM perfect absorbers.

It may be noted herewith that the performance of a THz radiation detector depends on the efficiency of converting radiation energy to an output signal. Therefore, maximizing the THz radiation absorption efficiency is integral to the development of a functional THz detector/imager. It is difficult to find strongly absorbing materials at THz frequencies that are compatible with standard photo-lithography. Thus, a potential application of these THz metamaterial absorbers might aid in the design of thermal detectors. A strong absorption coefficient is also necessary to have a small thermal mass. This is important for optimizing the temporal response of thermal detectors.

4.2.3.2 THz sensors

For sensor applications, metamaterial-based THz sensors are very useful for medical and other applications. Using THz signal, the sensing action is determined in terms of the resonances of various molecular vibrations in the sample to be sensed. As metamaterial magnetic inclusion structure (SRR/LR) is a highly resonant structure, it aids in enhancing the resonances and hence leads to better sensing with metamaterial assistance. The greater the Q (quality factor) of the metamaterial resonators, the better is the sensing capability. Effective sensing is evaluated by measuring the resonant response differences of the metamaterial unit cells thereby identifying and detecting minute amounts of differences in chemical and biochemical substances. In the structure shown in figure 4.4 [39], the Q-value of 30 has been realized with maximum refractive index sensitivity of 788 GHz/RIU or 1.04 × 105 nm/RIU. It may be observed that the SRR structure used here is of square shape and has two cuts forming the so-called LR (or double-split resonator) that does not suffer from bi-anisotropy problem (caused by asymmetric placement of slits in the two rings resulting in imbalance of the current in the two rings) prevalent with single-split conventional SRR structures. Absence of bi-anisotropy in double SRR or more aesthetically called LR helps to achieve LC resonance of high quality factor.

The use of a metallic structure in metamaterial design has inherent drawbacks; hence, the Q-factor gets compromised. If we use perfect absorbers designed with

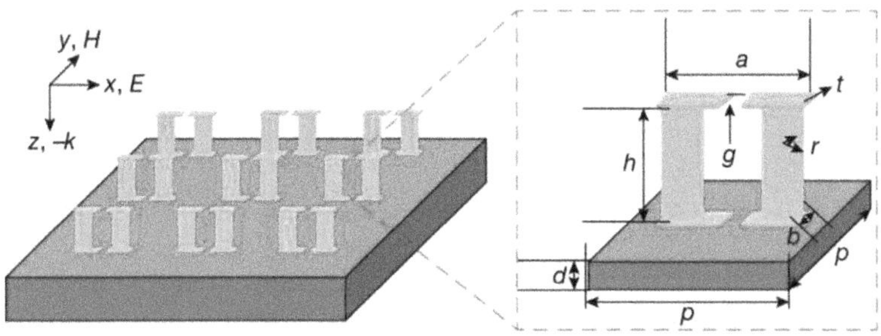

Figure 4.4. 3D schematic drawing of the metamaterial sensor and a single enlarged unit cell with its geometrical dimensions. (Reprinted/adapted with permission from [39], copyright The Optical Society of America.)

metamaterial, then we can get a higher Q-factor. Realization of high Q-factor resonance in these structures leads to enhanced sensor sensitivity to detect minute frequency shifts. Bi-material sensors with metamaterial absorbers have been used for sensor design, see figure 4.5 [40]. This consists of a sensing element (absorber) that converts the incoming THz radiation to heat, which is eventually transmitted by conduction to two symmetrically located bi-material legs connected to the host substrate (heat sink) with supporting structures of lower thermal conductance (anchors). Temperature rise caused by the absorption of incident THz radiation results in deformation of bi-material legs. The deformation can be probed by different approaches, among which the optical readout is a simple technique that requires a reflective surface, which is embedded into the absorber. Such sensors have responsivity values as high as 1.2° μW^{-1}, have time constants as low as 200 ms, have minimum detectable power on the order of 10 nW, and can operate with low power THz sources.

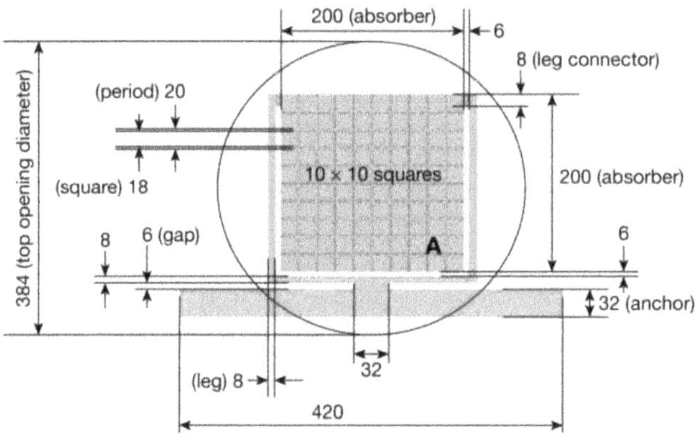

Figure 4.5. Structure of bi-material THz sensor using metamaterial absorber. (Reprinted/adapted with permission from [40], copyright The Optical Society of America.)

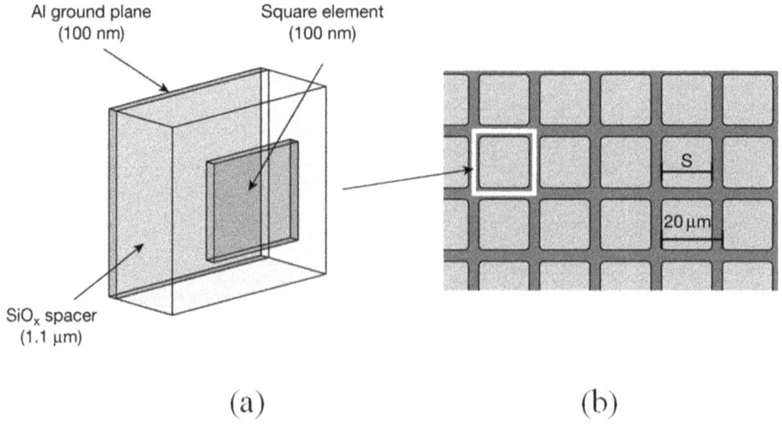

Figure 4.6. Metamaterial absorber; (a) unit cell, and (b) 1D periodic array. (Reprinted/adapted with permission from [40], copyright The Optical Society of America.)

A 'perfect' absorber can be constructed with the proper design of structural parameters [40]. The challenge remains in the design of a metamaterial film that is thin enough to provide low thermal capacitance, while providing structural strength, low stress and a flat reflective surface for an optical readout. Figure 4.6 shows that a typical metamaterial absorber consists of a periodic array of Al square elements separated from an Al ground plane by a SiO_x layer, a single unit cell being shown in figure 4.6(a). Such a combination allows matching to the free-space impedance at specific frequencies, eliminating the reflection. Figure 4.6(b) shows a periodic array of this metamaterial absorber [40].

4.3 Optical metamaterial

4.3.1 Introduction

The optical metamaterial has opened-up an incredibly new paradigm that promises groundbreaking new functionalities which were not possible in the past with conventional optics using natural material. In addition to invisibility and imaging with unlimited resolution now we have gain-assisted, controllable and nonlinear metamaterials at optical frequency with varied applications in designing flat laser or laser spaser, high-speed carbon nanotube (CNT) switches, high-quality optical sensors, and even in light harvesting. However, all these have been possible with some unique features provided by the characteristics inherent with the artificially structured material, the matamaterial. In fact, it may be observed that in natural media the optical response is determined by quantum energy-level structures of atoms or molecules, which are set by Nature while the electromagnetic properties of metamaterials even at optical frequency are determined in terms of the resonant characteristics of the sub-wavelength plasmonic resonators from which they are constructed and that too are in designer's control.

A general question naturally arises, irrespective of the frequency range of operation: why are the NIMs not available in Nature but can be realized with artificially structured metamaterial? The answer for this is simple and trivial and based on the way the natural material responds to an incident electromagnetic stimulus. In natural material, the resonant characteristics exhibited in the electromagnetic spectrum at electrical and magnetic domain are not overlapping. In fact, we have noted in the introduction part of this chapter that the electric resonance for natural materials takes place at much higher frequency than the magnetic resonance frequency. But in artificially structured materials like metamaterials this limitation can be overcome (at microwave and THz frequency domain) by designing at the designer's will the unit cells or 'meta-atoms'—with TW for electric response and SRR for magnetic response—that can exhibit both electric and magnetic resonant responses in an overlapping frequency range, thus making the realization of negative refractive index possible.

Another question that haunts us is: can we reach the optical domain just by scaling down the metamaterial inclusions (electric and magnetic ones) with structural (i.e. unit cell) scaling-down methodology? The answer is NO, and thus we have to resort to innovative metamaterial structures for operation in the optical domain. In the microwave frequency range the TW negative-permittivity structure and SRR negative-permeability structures are quite fine to realize practical NIM structures. The approaches for moving into the higher frequencies or shorter wavelengths were initially based on the usual concept of scaling down the unit cell size. The idea that is used for this purpose is that the magnetic resonance frequency of SRR is inversely proportional to its size. Using single SRR, this approach works up to 200 THz [41, 42]. However, this scaling breaks down for higher frequencies for the single SRR case because, for wavelengths shorter than the 200 THz range, the metal starts to strongly deviate from an ideal conductor [42]. For an ideal metal with infinite carrier density, hence infinite plasma frequency, the carrier velocity and the kinetic energy are zero, even for finite current in a metal coil. For a real metal, though the velocity and kinetic energy becomes finite, for a small SRR, at optical frequencies, the frequency approaches a constant and becomes independent of the SRR size—and the scaling law is not applicable like those at microwave or THz frequencies. This scaling limit combined with the fabrication difficulties of making nanometre-scale SRRs along with metallic TW led to the development of alternative designs that are more suitable for higher THz (>200 THz) and optical regimes.

At this juncture, for a quick recapitulation, the complete electromagnetic spectrum highlighting the optical, i.e. visible, light range has been given in figure 4.7.

It may be mentioned herewith that as per the definition by the International Commission of Illumination, the optical/visible range extends from 380 nm to 780 nm in wavelengths (in general it is 400 nm, violet and purple light, to 700 nm, deep red light, is considered to be the optical range; just for comparison, a human hair is about 100 000 nm in diameter); below this range is infrared and above this range is ultraviolet.

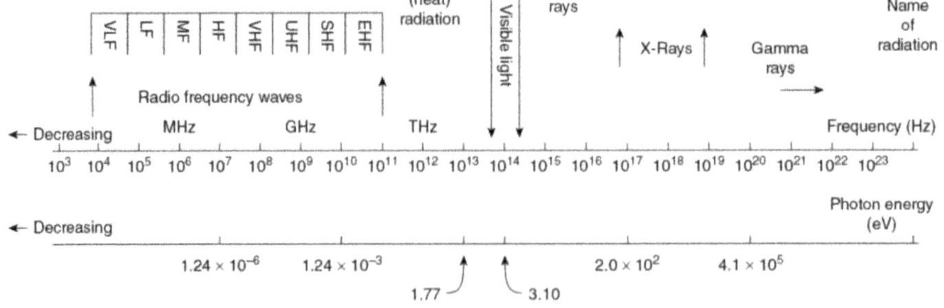

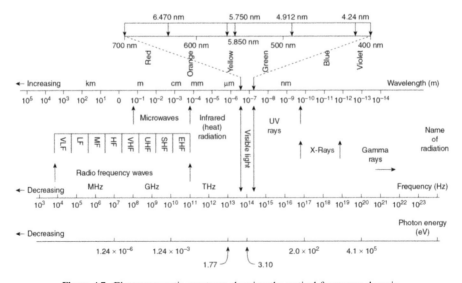

Figure 4.7. Electromagnetic spectrum showing the optical frequency domain.

4.3.2 Optical metamaterial structures

Design techniques of metamaterial at optical frequency are quite different from those at microwave-to-THz frequencies. The first experimental demonstrations of negative refractive index in the optical range were accomplished nearly at the same time for two different metal–dielectric geometries. In one case, pairs of metal nanorods separated by a dielectric layer (see figure 4.8(a)) [1], known as coupled-nanorod structure, were used while in the other the metal–dielectric–metal layers were used having voids (see figure 4.8(b)) [43] popularly known as the fishnet structure.

In the so-called coupled-rod design, consisting of a pair of metal nanorods, the time-varying electric field of the incident electromagnetic signal applied parallel to both the rods induces parallel current in both the rods when an electric resonance occurs with $\varepsilon < 0$. Again, the magnetic resonance with $\mu < 0$ results due to antiparallel currents in the two rods/strips induced by the perpendicular magnetic field component of the incident electromagnetic field. The magnetic response will be dia- or paramagnetic depending on whether the wavelength of the incoming magnetic field is shorter or longer than the magnetic resonance of the coupled rods. The metal rods can also be thought of as inductors, where the gaps at the ends form two capacitors. The overall result is a resonant LC circuit with a current loop that operates at optical frequencies. A detailed analysis of coupled-rod structure as optical metamaterial is given in the advanced text [44]. Anyway, in the coupled-rod/strip structure it is difficult to get overlapping regions of $\varepsilon < 0$ and $\mu < 0$. The particular design shown in figure 4.8(a) used an array of pairs of parallel 50 nm thick gold rods separated by 50 nm of SiO_2 spacer and was fabricated with electron-beam lithography (EBL) via a standard lithography—deposition—lift-off process. The coupled strip-based NIM exhibited a negative refractive index of $n_{\text{eff}} \approx -0.3$ at 1.5 μm [1].

The fishnet structure, figure 4.8(b), combines magnetic coupled strips (to provide $\mu < 0$) with continuous electric strips (to provide $\varepsilon < 0$) over a broad spectrum. Hence, the overlapping frequency zone for simultaneously negative ε and μ is easily obtained at the optical frequency. The so-called fishnet structure can be thought of

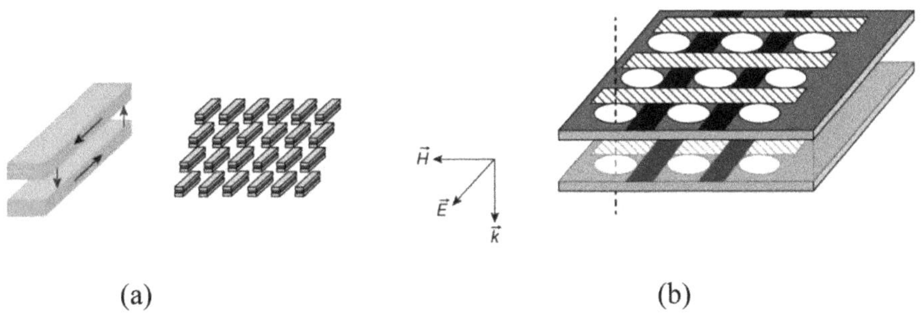

Figure 4.8. Optical metamaterial structures. (a) Coupled-nanostrip. (Reprinted/adapted with permission from [1], copyright The Optical Society of America.) (b) Fishnet. (Reprinted/adapted with permission from [43], copyright (2005) by the American Physical Society.)

as the inverse of the pair of nanorod/strip resonant structure discussed above. Let us look at an array of pairs of metal nano circles separated by a dielectric, which are similar to the pair of rods of figure 4.8(a). The inverse of this design would be paired circularly shaped voids in metal films. To make such a structure, we can begin with two films of metal separated by a dielectric. Then circularly shaped voids etched in the two metal films will form the inverse of the original structure of paired metal circles. Both such samples, in accordance with Babinet's principle, should have similar resonance behaviour if the orientation of the electric and magnetic fields are interchanged. In the actual design, interference lithography (IL) with 355 nm UV source was used to define a 2D array of holes in a multilayer structure (60 nm thick Al_2O_3 dielectric layer between two 30 nm thick gold films on a glass substrate). The active regions for the electric (dark regions) and magnetic (hatched regions) responses are indicated. A minimum negative refractive index of $n_{\text{eff}} \approx -2$ was obtained around a 2 μm wavelength [43]. The fishnet structure with circular voids had a relatively small figure of merit: $F = 0.5$, which was later improved with elliptical voids having a rather large figure of merit ($F = 2$).

In the summer of 2008, Xiang Zhang's group at UC Berkeley made the first successful 3D optical metamaterial [45]. It was made of 21 alternating sheets of 30 nm thick silver metal and 50 nm thick glasslike substance (non-conducting magnesium fluoride), the structure, dubbed as fishnet, with rectangular holes that resemble waffles or sieves, see figure 4.9. The structure was fabricated with focussed-ion beam (FIB) milling. When the light travels through the fishnet, the alternating layers act as nano-circuits bending light in an unusual or 'wrong' way. The refractive index of the structure was found to vary from $n \approx -0.63$ at 1200 nm to $n \approx -1.23$ at 1775 nm.

A separate group in Zhang's lab accomplished a similar feat using silver nanowires in a solid base [46]. The metamaterial was formed from silver nanowires which were electrochemically deposited in porous aluminium oxide. The metamaterial exhibited negative refraction for all incident angles in the visible region and since in this structure the negative refraction occurs far from any resonance, it has low loss and can provide broadband propagation at visible frequencies.

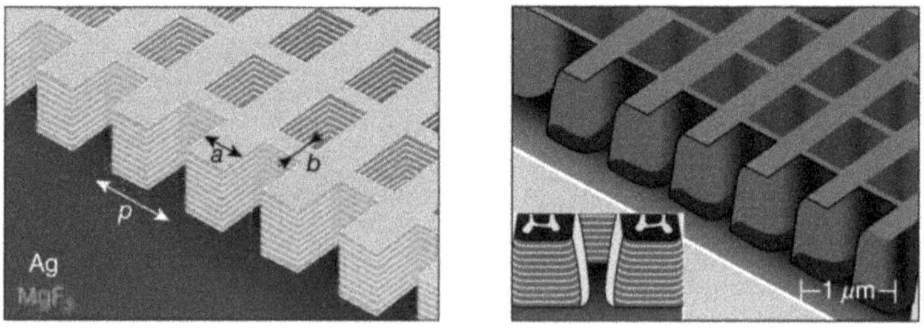

(a) (b)

Figure 4.9. 3D Fishnet metamaterial showing (a) fabricated 21-layer structure with unit cell (p = 860 nm, a = 565 nm, b = 265 nm). (b) SEM image with the side etched, showing the cross-section. The structure consists of alternating layers of 30 nm silver (Ag) and 50 nm magnesium fluoride (MgF), as in diagram (a).

In early 2007, a metamaterial with a negative index of refraction for a light wavelength just outside the frequency of the colour red was announced. The material had an index of −0.6 at 780 nm [47]. The fabricated structure had good large-scale homogeneity; however, as the feature size required for visible range NIMs is to be sub-100-nm, the side walls happened to be quite rough.

It would be worthwhile at this point to discuss the construction and the general principle of operation of fishnet metamaterial which is the most popular NIM structure useful at optical frequencies [48].

The fishnet can be thought of as a combination of two separate constituents, which allow the permeability and permittivity of the structure to be engineered independently. The first component of the fishnet metamaterial is thick metallic strips oriented along the direction of the magnetic field of the incoming light which are separated by a dielectric layer (figure 4.10(a)). This structure functions as an inductor and capacitor resonator wherein antisymmetric currents are created in the metal strips, which give rise to an induced magnetic polarization in the structure (figure 4.10(e)). This provides a magnetic response, or artificial permeability, that can achieve negative values near resonance. The second constituent consists of thin metallic wires oriented in the direction of the electric field of the incoming light (figure 4.10(b)). These wires functions as a diluted metal with decreased plasma frequency, providing a negative effective permittivity that can be engineered separately from the permeability. These two constituents are combined to form the fishnet metamaterial where both the negative permeability and permittivity can be engineered at a particular wavelength (figure 4.10(c) and (d)); figure 4.10(d) being a unit cell of the fishnet metamaterial structure. An alternative explanation of the negative refraction arising from multilayer fishnet metamaterials is based on the

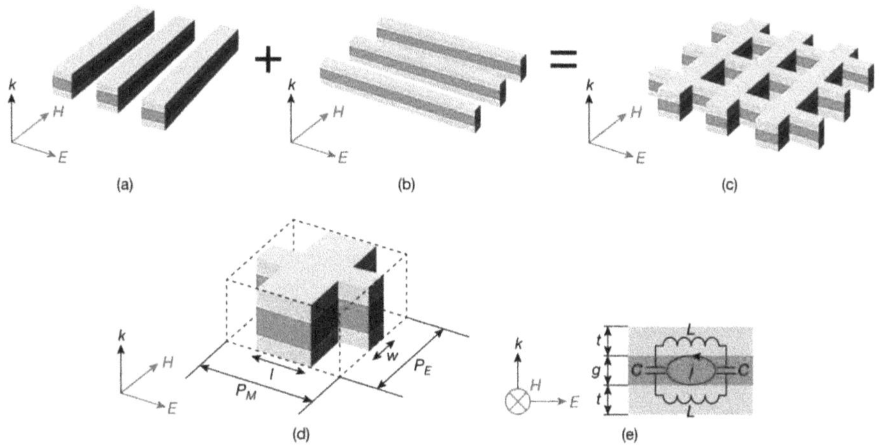

Figure 4.10. Fishnet metamaterial construction schematic showing a single functional layer (metal/dielectric/metal): (a) and (b), respectively, represent the magnetic and electric response structures (grey layer is made of metal and green layer is a dielectric), (c) complete fishnet metamaterial which is a combination of the structures shown in (a) and (b), while (d) is the unit cell geometry of the metamaterial, (e) depicts the LC circuit and current loop that form the magnetic response in the fishnet metamaterial.

spoof plasmon supported by each perforated metallic film as explained by Pendry *et al* [49].

4.3.3 Metamaterial design technologies at optical frequencies

Practical realization of optical NIM demands very small periodicities (about 300 nm and less) and tiny feature sizes (about 30 nm) to ensure that they behave as effective medium. Thus the fabrication of optical NIM is challenging since we aim at high-precision, high-throughput, and low-cost manufacturing processes. Feature sizes for metamaterials operating in the infrared or visible range can be smaller than the resolution of the state-of-the-art photolithography (due to the diffraction limit), thus requiring nanofabrication processes 100- or sub-100-nm resolution. Due to the limitations of current nanolithography tools, fabricated metamaterials do not really enter the real meta-regime where the unit cell size is order of magnitude smaller than the wavelength. However, the feature size is typically small enough compared to the wavelength so that one can describe such material, at least approximately, as a medium with effective permeability, μ_{eff} and effective permittivity, $\varepsilon_{\mathrm{eff}}$ (which is in contrast to photonic crystals where the lattice period matches the wavelength).

Further, the choice of the metal for optical NIM is a crucial one because the overall losses are dominated by losses in the metal components. Because silver and gold has lowest losses at optical frequencies these are the first choices of metals for NIMs in the optical regime. Since the refractive index is a complex number: $n = n' + n''$, where the imaginary part n'' characterizes light extinction (losses), a convenient measure for optical performance of NIM is the figure of merit (FOM), defined as the ratio of the real and imaginary parts of n, $F = | n'|/n''$ [50]. Aside from the proper metal choice, loss compensation by introducing optically amplifying materials can be considered for achieving low-loss NIMs [51].

There are various fabrication technologies for the design of 2D and 3D optical NIMs. For 2D design, the techniques used are: EBL, FIB milling, IL, nanoimprint lithography (NIL). For 3D design multiple-layer formation can be done by lift-off procedure, layer-by-layer technique or deep anisotropic etching. But for the design of complex 3D nanostructures various techniques are used such as direct electron-beam writing, FIB chemical vapour deposition (FIB-CVD) and so forth. But a fabrication method that is most promising to date for true 3D optical NIMs is based on two-photon photopolymerization (TPP) technique.

Electron-beam lithography, as opposed to photolithography, is a maskless lithography method that utilizes an electron gun from a scanning electron microscope to pattern nanoscale features on a substrate surface with an electron beam (e-beam). EBL was first reported as a technique for patterning substrate materials as early as the 1960s. EBL does not rely on a pre-existing patterned mask, but can write the pattern directly from stored data. However, similar to photolithography, substrates for EBL are coated with a film (called the resist), exposing the resist and selectively removing either exposed or non-exposed regions of the resist (called developing) generates the desired pattern on the substrate. Due to the high resolution provided by EBL it is still the first choice for fabricating small area metamaterials demanding nano-level patterning as in near-optical/optical NIMs. Writing larger areas requires long e-beam writing times and hence escalating the operating cost. This approach is suitable for proof-of-principle fabrication of optical NIMs as was discussed earlier [1, 43].

Focussed-ion beam milling technique can be used for rapid prototyping of metamaterials. In FIB, a focussed beam of gallium ions is used to sputter atoms from the surface or to implant gallium atoms into the top few nanaometers of the surface, making the surface amorphous. Because of the sputtering capability, the FIB is used as a micro-machining tool, to modify or machine materials at the micro- or nanoscale. This technique was used for fabricating magnetic metamaterials based on SRRs [52]. Scaling of the SRR structure requires sub-100-nm gap sizes (down to 35 nm) for a 1.5 μm resonance wavelength. For such small features, EBL-based fabrication requires time-consuming tests and careful optimization of writing parameters and processing steps, leading to relatively long overall fabrication times. In contrast, the rapid prototyping of complete structure can be fabricated via FIB writing in times a short as 20 min [52]. The process is based on FIB writing and corresponds to an inverse process where the FIB removes metal (20 nm of Au) deposited on a glass substrate. After FIB writing the structure is ready and no further post-processing steps are required. Both the EBL and FIB are serial process and are not considered to be feasible for the large-scale metamaterial fabrication required for real-life applications. Both these techniques have the limitations of low throughput.

Interference lithography is a type of optical lithography which is powerful technique for wide array of samples for nanotechnology. Integrated circuit (IC) technology uses optical lithography for large-scale manufacturing in industry. The IL fabrication technology is based on the superposition of two or more coherent optical beams forming a standing wave pattern. Being a parallel process, IL provides a low-cost large-area (up to ~cm^2) mass-production capability. Moreover, multiple exposure, multiple beams, and mix-and-match synthesis with other lithographic techniques can extend the range of IL capability [53]. IL offers high structural uniformity combined with considerable, but not total, pattern flexibility while its resolution is now approaching the 20 nm scale. NIMs have been fabricated with negative refractive index over wavelengths 1.56–2 μm, 1.64–2.2 μm and 1.64–1.98 μm with circles, ellipse and rectangles (fishnet), respectively, with IL fabrication technology [54]. Given the simplicity and robustness of making a high-quality, single layer of metamaterial using IL, one can envision further investigations aiming at piling 2D layers to create 3D structures. Such a transition to 3D fabrication will turn parallel IL process into a step-by-step procedure that would require alignment of subsequent layers. Even though for compact versions of IL such a transition might result in a time-consuming fabrication process due to multiple alignments, proper technique development would make it possible to optimize and automate the alignment procedure. Thus, IL can be considered to be one approach for making 3D optical NIMs [55].

Nanoimprint lithography is another promising direction for the fabrication of production-compatible, large area, high-quality optical NIMs at low processing cost and time [56]. NIL accomplishes pattern transfer by the mechanical deformation of the resist via a stamp rather than a photo- or electro-induced reaction in the resist as in most lithographic methods. Thus the resolution of the technique is not limited by the wavelength of the light source, and the smallest attainable features are given solely by stamp fabrication. Since NIL requires stamp fabrication by other nano-fabrication techniques (like EBL) it is ideal for parallel production of already optimized metamaterials, when the preliminary test structures were patterned via EBL. Thus, NIL can be seen as a large-scale, low-cost process of making EBL-written structures that offers solutions to the intrinsic EBL drawbacks. NIL has high throughput and is a good candidate for fabricating 2D optical NIMs. Two types of NIMs may be mentioned that were fabricated with NIL technology. One structure is an ordered 'fishnet' array of metal–dielectric–metal stacks that demonstrated negative permittivity and permeability in the same frequency region and hence exhibited a negative refractive index of $n_{\text{eff}} = -1.6$ at a wavelength near 1.7 μm [57]. The other metamaterial structure was an ordered array of fourfold symmetrical L-shaped resonator (with a minimum feature size of 45 nm) that were shown to exhibit negative permittivity and a magnetic resonance with negative permeability near wavelengths 3.7 μm and 5.25 μm, respectively [58].

The possibility of creating a truly 3D periodic NIM or metamaterial at optical frequency is a critical requirement for real-life applications like cloaking, sub-wavelength imaging, optical antennas, ultra-compact optical circuits and so forth. Our commonsense world is a 3D world, in geometrical sense, and we are able to see things with optical (i.e. light) signal in the visible spectrum. Thus for truly macro-applications of metamaterial like making an aeroplane or a ship invisible with metamaterial cloaking device (a device that can make things invisible) we practically need a periodic 3D optical metamaterial. However, it may be noted that geometrical 3D which we understand and find all around us (with length, breadth and height) is not the periodic 3D (which needs to be the case in metamaterial, photonic crystal etc). The 3D periodic structure reminds us of the 'checker cube' we are familiar with; an interesting toy that demands our patience to get it correctly colour-matched in all directions. For clarification, a representative diagram of 1D, 2D, and 3D periodic structure is shown in figure 4.11.

Complex 3D metal–dielectric nanostructures can also be fabricated by electron-beam writing (EBW) [59] and FIB-CVD [60]. These methods offer 3D fabrication

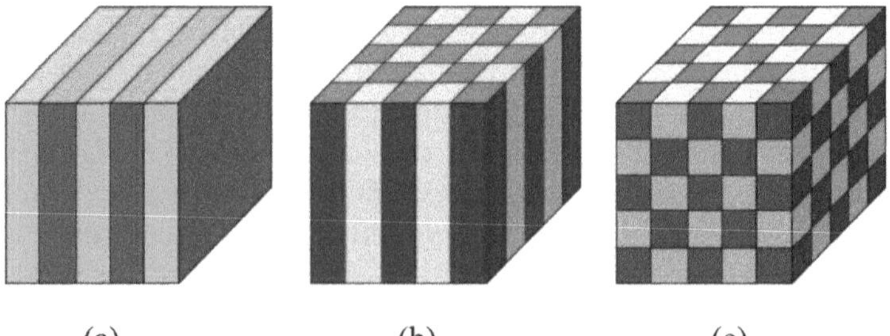

Figure 4.11. Representative diagrams of periodic structures: (a) 1D, (b) 2D, and (c) 3D.

that is not possible using traditional layered optical and EBL techniques. The use of EBW was demonstrated for building structures of multiple layers with linewidth resolution of 80–100 nm using electron beam to cause direct sintering of 2–10 nm nanoparticles [59], while various free-space Ga- and W-containing nanowirings were successfully fabricated by FIB-CVD, also called 'free-space-nanowiring' fabrication technology [60]. Though these techniques offer unique possibilities for making very complex 3D structures, they all suffer from severe material limitations in terms of what materials can be patterned/deposited. Moreover, such methods are complex and time consuming. Thus, they can be used only for making first prototypes or single structures for proof-of-principle studies.

The *two-photon-photopolymerization* technique has become the most successful technique for designing 3D NIMs at IR and optical frequencies. TPP involves the polymerization of a material via a nonlinear, multiphoton process that only occurs at the focal point of a tightly focused laser beam, thereby providing 3D control over the location of polimerization. TPP enables the fabrication of complex objects with sub-diffraction resolution (down to 100 nm) [61] since the absorption of light in the material occurs only at the focal region of the laser beam. In addition to direct single beam laser writing of complex structures into a polymer matrix, large-scale 3D polymer structures for future real-life applications can also be realized via 3D, multiple beam TPP technique [62].

3D, multilayered, polymer and metallic structures can also be realized by NIL [63]. This method offers high reproducibility over large areas (millimetre scale) and has been used for fabricating 3D cubic arrays of gold cubes and 3D woodpile-like polymer structures for photonic crystals-based devices. Combined with multimaterial deposition, these approaches might also be adapted for future NIM fabrication.

To reach real NIM applications, several tasks have to be fulfilled such as loss reduction, large-scale 3D fabrication and new isotropic designs. Careful material choice (for example, new crystalline metals with lower absorption instead of traditional silver and gold) and process optimization (reduced roughness and high uniformity of the materials) can help on the way to creating low-loss optical NIMs. Another possibility is to introduce a gain material into the NIM, thus compensating for the losses. Even though it is still a long way to truly 3D, isotropic, NIMs at optical frequencies, several fabrication approaches do seem to be feasible. With emerging techniques such as nanoimprint, contact lithography, direct laser writing and possibly new types of self-assembly, it seems likely that truly 3D metamaterials with meta-atom sizes much smaller than the wavelength can be created. In the next generation of optical NIMs, for any chosen manufacturing approach, the careful choice of materials and process optimization will be required including the cost of production in order to obtain high-quality structures commensurate with the desired blueprint.

4.3.4 Applications of optical metamaterial

4.3.4.1 Hyperlens

It is known that in biological laboratories to study living cells, as well as in the high-tech industry to create integrated circuits and so forth, optical microscopes are commonly used. But the optical microscopes are constrained by the so-called 'diffraction limit'—a fundamental limit in optics that allows a resolving power to the extent of the wavelength of the optical signal received from the object—making them unsuitable in resolving nanometer-level detail in an optical image of the object. However, to capture details down to a few nanometers, scientists currently use scanning electron or atomic force microscopes, which create images by scanning objects point by point. In contrast to the optical microscope, which can snap an entire frame of an image in a single shot, scanning electron microscopes can take up to several minutes to get an image. Since the object must remain immobile and in a vacuum during the image taking process, imaging by scanning electron microscope is restricted to non-living samples.

The propagating waves are used by optical lenses to form an image, but the so-called evanescent waves contain the finer details (the sub-wavelength details) of an object. Capturing the information carried by the evanescent waves is the Holy Grail of optical imaging. But the evanescent waves decay too quickly, typically within a quarter-wavelength of the signal, and thus the conventional lenses are unable to capture the sub-wavelength details of the object at the image location. However, J B Pendry in 2000 proposed the so-called superlens/perfect lens [64] that uses a metamaterial plane slab with a view to realize the 'Holy Grail' of optical imaging beating the diffraction limit. The sub-wavelength imaging, possible with the superlens concept of metamaterial, is also gaining enough enthusiasm that it might one day make it possible to image individual strands of DNA, thereby bringing about a revolution in medical research. But the problem with the superlens is that it demands very close proximity of the object and image from the metamaterial plane slab, i.e. it is in principle applicable for near-field imaging only. The first superlens with a negative refractive index provided resolution three times better than the diffraction limits and was demonstrated at microwave frequencies [65]. Subsequently, the first optical superlens (the optical lens that exceeds the diffraction limit) was created and demonstrated [66]—but the lens did not rely on negative refraction; instead a thin silver film was used to enhance the evanescent modes through surface-plasmon coupling.

The *hyperlens* shows a new way to beat the diffraction limit, which would allow biologists to not only see a cell's nucleus and other smaller components, but to study the dynamic movement and behaviour of individual molecules in living cells in real time. Hyperlens consists of a metamaterial formed by a curved periodic stack of silver (Ag) and aluminium oxide (Al_2O_3) deposited on a half-cylindrical cavity fabricated on a quartz substrate [67]. When an object is illuminated, its evanescent waves travel through the lens. As the wave vectors move outward, they are progressively compressed. This compression allows the image of the objects to be magnified by the time it reaches the outer layers of the hyperlens. At this point, it can be captured by a conventional optical lens and projected outward onto a far-field plane a meter or so away. The superlens may be termed as 'near-sighted'—the projected image only exists near the surface of the lens and that too with no magnification. This limits the practical applications for the superlens, since the camera needs to be within an object's 'near-field' range. To make a lens useful for far-field imaging beating the diffraction limit one must convert evanescent waves to propagation waves, which is what the hyperlens does. The hyperlens shown in figure 4.12 has been demonstrated with resolution down to 125 nm at 365 nm working wavelength [67]. Such a hyperlens might have possible applications in nanotechnology photolithography and a host of other applications yet undreamt of.

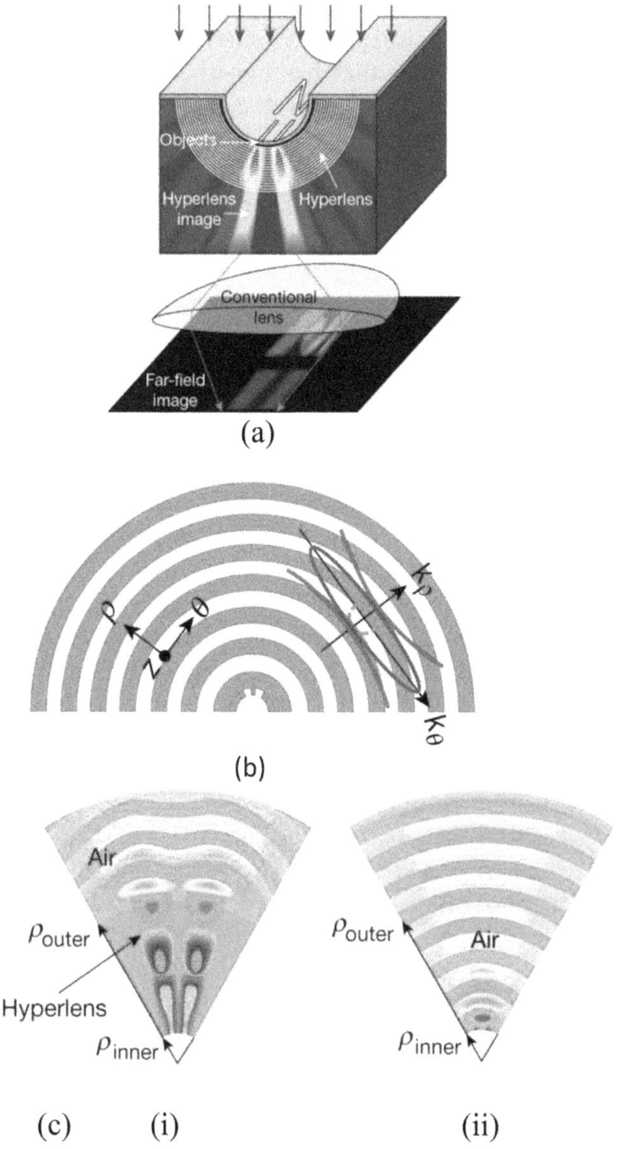

Figure 4.12. (a) Hyperlens based on hyperbolic metamaterial to magnify diffraction limited objects. (b) Dispersion characteristics of hyperbolic metamaterial. (c) Far-field resolved image of a sub-wavelength object (i) in hyperbolic metamaterial (ii) in air.

The physics behind the design and development of hyperlens may be understood on the basis of the report by Jacob *et al* [68]. In terms of spatial harmonics, it is known that the so-called low- and high-k (wavevector) components of the electromagnetic wave carry the coarse and finer details of an object via the propagating wave and the evanescent wave, respectively. The perfect lens/superlens proposed by J B Pendry can only provide an amplified version of the high-k waves and thus further manipulation of such waves in the far-field with the use of conventional optical system is not possible. The iso-frequency surfaces of a cylindrical anisotropic metamaterial (where ϵ and μ are expressed using tensors) with $\varepsilon_r < 0$ and $\varepsilon_\vartheta > 0$ exhibit a dispersion characteristics that has hyperbolic nature, see figure 4.12(b); thus for a fixed frequency, the wavevector k can take an arbitrarily large value (unlike an isotropic material where k has bounded vales). Such cylindrical hyperbolic metamaterial can support high-k waves and can transform them into low-k waves as the waves move outwards in the radial direction. Such an anisotropic multilayer (with alternate layer of metal and dielectric) metamaterials in a curved geometry is used in the design of the hyperlens shown in figure 4.12(a). In this design the layer thickness (h) of each of metal and dielectric is less than the operating wavelength (λ) and when $h \ll \lambda \leqslant r$ one can treat such metamaterial as effective medium with: $\varepsilon_\theta = (\varepsilon_{\rm m} + \varepsilon_{\rm d})/2$ and $\varepsilon_{\rm r} = 2\varepsilon_{\rm m}\varepsilon_{\rm d}/(\varepsilon_{\rm m} + \varepsilon_{\rm d})$, where $\varepsilon_{\rm m}$ and $\varepsilon_{\rm d}$ denote the dielectric permittivities of the metal and dielectric layers, respectively.

On the basis of the transformation optics, the conversion of the high-k waves to low-k waves with the curved contour of the hyperbolic metamaterial may be understood as follows. It may be mentioned that owing to the conservation of angular momentum, the tangential component of the high-k's are progressively compressed as the waves travel along the outward radial direction—thus for a subwavelength object placed at the inner boundary, a magnified image is obtained at the outer boundary; the magnification at the output surface is given simply by the ratio of the radii at the two boundaries. The magnified image, once larger than the diffraction limit, will be resolved in the far-field. Figure 4.12(c) shows two subdiffraction-limited line sources separated by a distance of ~80 nm clearly resolved by using hyperbolic metamaterials (left), but not by air alone (right). In the simulation, the radii of the hyperlens at the inner and outer boundaries are $\rho_{\rm inner} = 240$ nm and $\rho_{\rm outer} = 1200$ nm, respectively. The permittivity for metal and dielectric in the hyperbolic metamaterial-based hyperlens is $-2.3 - j0.3$ and 2.7.

It must be noted that even though a hyperlens can project a magnified image of a diffraction limited object up to a meter away, the object that is being imaged still needs to be placed in the near-field zone of the lens. But this is definitely a major first step in reaching the goal of making a far-field optical nanoscope. Efforts are also there to demonstrate another type of far-field superlens by using surface grating to convert evanescent waves into propagating waves [69]. The metamaterial far-field superlens, composed of a metal–dielectric multilayer and a 1D sub-wavelength grating, can work over a broad range of visible wavelengths. Anyway, only time will tell whether the 'hyperlens' or the 'far-field superlens' will be used first for practical applications for far-field imaging using metamaterials beating the diffraction limit.

4.3.4.2 Metamaterial for absorber design at IR and optical frequencies

The most challenging problem in realization of metamaterial perfect absorbers (MTM PAs) operating in the IR and the visible ranges is the size of the unit cell of patterned array. The size of unit cell should be much smaller than the wavelength of incident EM wave to fulfil the strict sub-wavelength condition of MTMs. Because the wavelength of IR and optical EM waves is in the range of a few hundreds of nanometre (see figure 4.7), the periodicity of the patterned array should be a hundred or a few tens of nm. Various types of magnetic inclusion structures of MTM that have been used commonly from MHz to THz frequency range are Swiss rolls, SRRs, U-shaped resonators and so forth, see figure 4.1. For IR/optical MTM coupled-nanorod/strip and fishnet structure, see figure 4.8, are the basic ones. Current level of techniques of nanofabrication might be able to handle the range of unit-cell size they demand; however, the uniformity of the fabrication of individual unit cells is still challenging. Because of this, difficult progress in MTM PA operating in the IR and the optical ranges demands innovative designs.

After the first demonstration of the design of MTM PA in the GHz range [70] there were many attempts to fabricate MTM PA operating in the IR and/or the optical ranges. Avitzour *et al* [71] were the first researchers who attempted to design MTM PA operating in the IR range. The design of the absorber was based on a perfectly impedance-matched NIM; however, they employed quite a complicated structure.

A highly-efficient conformal dual-band metamaterial was realized by Jiang *et al* [72]. Figure 4.13 shows the schematic diagram of the structure with blow-up view of the unit cell. This conformal MTM PA maintains its absorption properties when integrated even onto curved surfaces of arbitrary shapes.

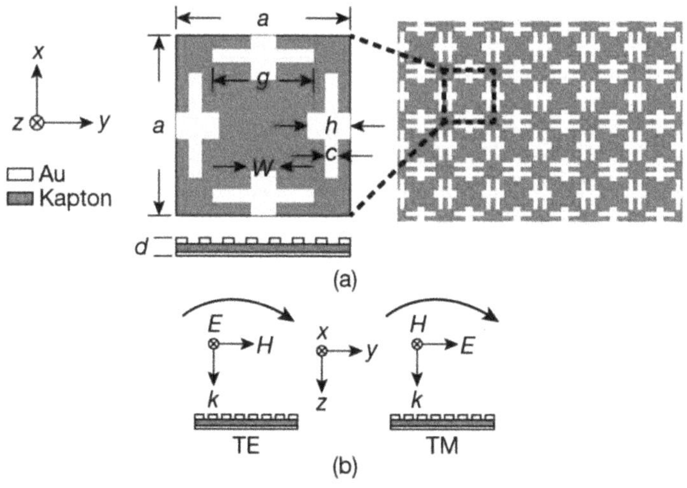

Figure 4.13. (a) Doubly periodic array of H-shaped nano-resonators showing blow-up view of one unit cell on the right. (b) Orientation of the incident fields with respect to the MTM PA. (Reproduced/adapted with permission from [72], copyright (2011) American Chemical Society.)

Bossard *et al* [73] employed a strategy to integrate several nanoscale resonators, resonating at different frequencies, into a unit cell. By doing so, they were successful at realizing the broadband absorption in the mid-IR range. In figure 4.14 a schematic diagram of the structure along with the top view of a unit cell is shown. The measured average absorption was found to be greater than 98%, which was maintained over a wide range of angle of incidence (±45°) for mid-IR wavelengths between 1.77 and 4.81 μm.

Aydin *et al* [74] designed, fabricated and characterized a broadband and polarization-independent resonant MTM PA by utilizing ultrathin plasmonic super absorbers. A broadband resonant light absorption over the entire visible spectrum (400–700 nm) was achieved with an average measured absorption of 0.71 and simulated absorption of 0.85. A high-performance, wide-angle, polarization-independent dual-band MTM PA was realized by manipulating the ratio of disk size to the periodicity and the disk size itself.

Cheng *et al* [75] designed, fabricated and characterized a dual-band MTM PA made of Ag/SiO_2/Ag tri-layer on top of Si wafer, see figure 4.15. The top layer is a square array of circular Ag disks.

Pitchappa *et al* [76] reported the tunability of a near-IR microelectromechanical system (MEMS) switchable complementary MTM PA. Its schematics of unit cell and switching principles are depicted in figure 4.16. The micro-actuator, which will eventually tune the response of MTM PA, can be switched 'on' and 'off' by the applied voltage.

Recent progresses in MTM PAs operating in the IR and the optical ranges clearly reveal that in the near future this kind of MTM PAs can be realized. Like the case of THz MTM PAs, it is desirable that in practical application of this kind of MTM

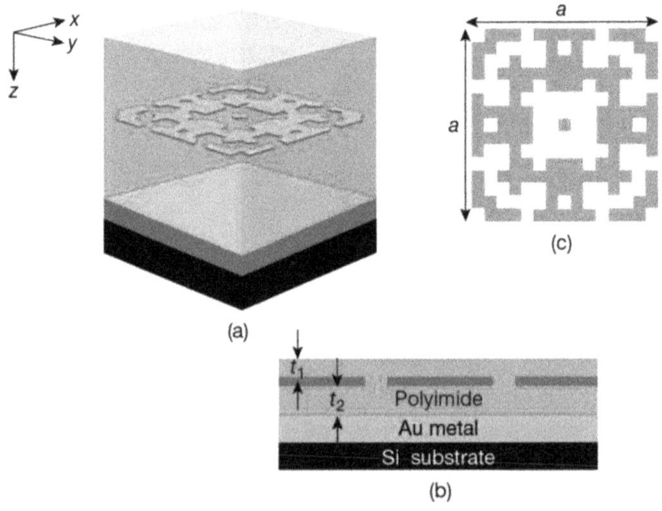

Figure 4.14. (a) Schematic diagram of the Au-based MTM PA structure, (b) side-view and (c) top view of one unit cell of the structure. (Reproduced/adapted with permission from [73], copyright (2014) American Chemical Society.)

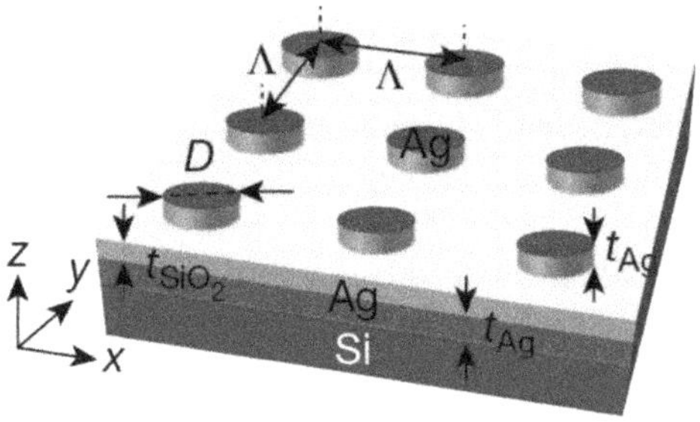

Figure 4.15. Schematic diagram of the metallic-disc structure for the dual-band MTM PA. (Reprinted/adapted with permission from [75], copyright The Optical Society of America.)

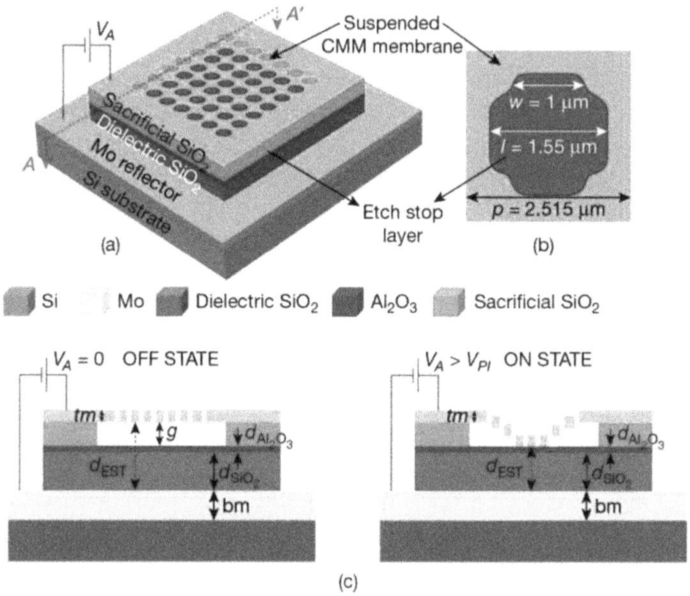

Figure 4.16. (a) Schematic diagram of near-IR MEMS switchable complementary metamaterial absorber (MSCMA). (b) Top view of unit cell within the MTM geometry with dimensions indicated. (c) Schematic cross-sectional view along AA of MSCMA in OFF state and ON state. (Reproduced/adapted from [76], with permission AIP Publishing.)

PAs, the EM-wave response should be broadband, omni-directional, polarization independent, and flexible. If this kind of MTM PAs becomes practically workable in the field, a revolutionary change in the field of, for instance, harvesting of energy delivered from the Sun may become possible in the coming future.

4.3.5 Other emerging applications

4.3.5.1 Introduction

Metamaterial was considered to be 'material-like' (i.e. one with tailorable constitutive properties), during the initial phase of metamaterial research.

Metamatetials became a big issue of concern when this new tree of knowledge (with its roots embedded deep in the soil of microwave technology, at the initial stage) brought forth the 'forbidden fruit' of negative-index media. Aggressively contested when it first appeared, the concept of negative index is now widely accepted with a number of possible applications. However, a paradigm shift in metamaterial research philosophy has taken place subsequently, since 2010, and metamaterial is now considered as a 'device': where the hybridization with functional agents brings new functionality and the response becomes gain-assisted, nonlinear, switchable and so forth. This next stage of technological revolution with metamaterials has led to the possibility of development of active, controllable, and nonlinear metamaterials surpassing natural media as platforms for optical data processing and quantum information applications [77, 78].

In developing active gain-assisted metamaterials, the main goal is the compensation of losses that dampen the coupled oscillations of electrons and light (known as plasmons) in the nano structures. These losses render photonic negative-index media useless from a practical point of view. One solution is to combine metamaterials with electrically and optically pumped gain media such as quantum dots [79], semiconductor quantum wells and so forth, embedded into the metal nanostructures. Electrically and optically pumped semiconductor gain media and emerging technology of carbon monolayers (graphene) could be expected to provide the loss compensation from optical to terahertz spectral ranges. Another grand goal is to develop a gain-assisted plasmon laser, or 'lasing spaser' device [80]: a flat laser, in which the emission of the spaser (Surface-Plasmon Amplification by Stimulated Emission of Radiation) is fueled with the plasmonic excitations in an array of coherently emitting metamolecules. In contrast to conventional lasers operating at wavelengths of suitable natural molecular transitions, the lasing spaser does not require external resonators and its emission wavelengths can be controlled by metamolecule design. We now also have 'metamaterial-based carbon nanotube': that exhibits an order-of-magnitude higher non-linearity compared to bare CNT leading to high-speed switching possible at optical frequency. Further, it is now possible to realize improved sensor design and light-harvesting using the new functionalities provided by expanding paradigm of metamaterials. Finally, 'transparent metamaterial': made entirely of dielectric materials, might make it possible for interconnecting circuits using photons instead of electrons to process and transmit data and also paving the way for light sources at the single-photon level.

4.3.5.2 Lasing SPASER

The idea of combining gain media with metamaterials has attracted the attention of the metamaterial research community since 2010. This new generation of metamaterials has the capability of playing a key role in developing novel laser sources. Indeed, it is now experimentally verified that hybridizing a gain medium (semiconductor quantum dots or quantum well structures) with a plasmonic metamaterial can lead to a multi-fold-intensity increase and a narrowing of their photoluminescence spectra. The luminescence enhancement is a clear manifestation of the quantum Prucell effect, and it can be controlled by a metamaterial's design.

This is an essential step towards the development of metmaterial-enhanced gain media and the 'laser spaser': a 'flat' laser as mentioned above.

The use of metamolecules in developing 'lasing spaser' is a dominant example of the 'gain-assisted' plasmonic metamaterial application in photonics [80]. In spaser, also known as plasmonic laser, the light-quanta-photons of the laser are being replaced with electronic excitations at the surface of metals called surface plasmons, which can have atomic-scale dimensions [81]. However, a spaser produces very little light, which is not collimated into a narrow beam. But, if the emission can be fuelled by plasmonic excitations in an array of coherently emitting metamolecules (designed with magnetic inclusion structure, such as SRRs) supported by a gain medium (quantum-dot-doped dielectric), which can overcome the radiation losses and Joule losses in the metallic structure of metamolecules, we have the 'lasing spaser' [80, 82], see figure 4.17. In contrast to conventional lasers that operate at wavelengths of suitable natural atomic or molecular transitions, the emission wavelength of lasing spaser can be controlled by metamolecule design. Being the thinnest (~100 nm) laser, the lasing spaser promises new applications ranging from displays to high-speed communications.

It is noted that adding gain media to the metamaterial has resulted in compensation of joule losses that otherwise dampens plasmons in metal nanostructures. Lowering of losses is crucial for performance of metamaterial-based negative-index

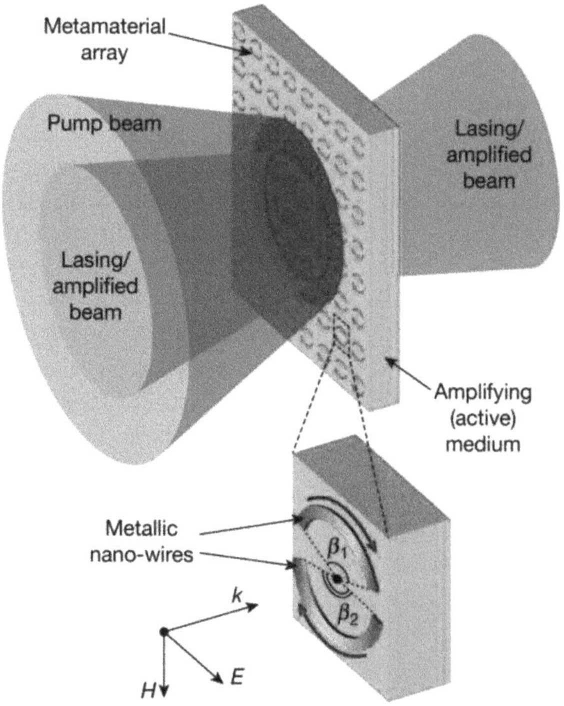

Figure 4.17. Schematic of metamaterial-fueled lasing spaser. (Reprinted/adapted with permission from [78], Taylor and Francis Group LLC (Books) US http://www.tandfonline.com.)

devices, waveguides, spectral filters, delay lines and, in fact, practically any application of metamaterials.

The combination of artificial classical electromagnetic resonators (SRRs), forming the metasurface, plays the role of the active medium in the lasing spaser, just as an assembly of essentially quantum inversely populated atoms plays the same role in a conventional laser. These identical plasmonic resonators impose the frequency at which the device will lase, drawing energy from a supporting gain substrate. In a conventional laser, the direction of emission is dictated by the external resonator, and its coherence is underpinned by the stimulated emission of atoms in the gain medium. In a lasing spaser, the direction of emission is normal to the plane of array, where strong trapped-mode currents in the plasmonic resonators oscillate in-phase.

4.3.5.3 Metamaterial-based CNT switches and MEMS tuners

CNTs are 1D systems, with diameter of a few nanometers and length on the micron scale with unique nonlinear optical properties. Single-walled CNTs rolled from a graphene sheet are direct gap semiconductors with absorption spectra dominated by exciton lines. Possible technological uses include nanoscale light sources, photodetectors, and photovoltaic devices.

Switchable and tunable metmaterials are another very rapidly expanding domain of research with metamaterials. Indeed, the development of nanophotonic all-optical data processing circuits depends on the availability of fast and highly responsive nonlinear media that react to light by changing their refractive index and absorption. In all media where the functionality depends on electronic or molecular anharmonicity or nonlinearities, stronger responses often come at the expense of longer reaction times, a constraint that is practically impossible to break. The plasmonic nonlinearity of metals constituting metamaterial nanostructures is extremely fast and could provide terahertz modulation, but requires high intensities to operate. Combining conventional nonlinear media with metamaterial nanostructures is a powerful way of engineering an enhanced response. Metamaterials can slow light, thereby increasing the interaction time with nonlinear medium embedded in it, or they can help by concentrating the local field and thus enhancing a nonlinear response. Prime contenders emerging for such hybridization with metmaterials are semiconductors and semiconductor multiple-quantum well structures used as substrates for metallic framework, liquid crystals, conjugated polymers, CNTs [77].

Recent experiments show that the nonlinear response of CNTs can be strongly enhanced by adding a metamaterial layer. Single-wall semiconductor CNTs deposited on metamaterials exhibit an order-of-magnitude higher nonlinearity than the existing strong nonlinear response of the nanotubes themselves, due to a resonant plasmon–exciton interaction making ultra-high-speed switching possible at optical frequency [83]. Also, the metamaterial environment allows one to spectrally tailor the nonlinear response.

When the high-speed switching is not the prime objective, metamaterials can be reliably and reversibly controlled by microelectromechanical (MEMS) actuators that can reposition parts of the metamolecules in an array, allowing the tunability of transmission and reflection spectra, which has been convincingly demonstrated for terahertz and far-infrared metamaterials [77]. Reconfigurable optical metamaterials require moving components on the scale of a few tens of nanometers (NEMS actuators) which can be fabricated on bimorph membrane and are tunable by temperature.

4.3.5.4 Sensor and light-harvesting applications

Sensor applications represent another rapidly growing area of metamaterials research. Since metamaterials can exhibit a strong localization and enhancement of fields, they can be used to improve the sensor selectivity of detecting nonlinear substances and to enable detection of extremely small amounts of analytes [84]. For instance, asymmetric SRRs supporting high-quality Fano resonances or metamaterial arrays of nanaoscale antennas are well suited to detect low-concentration analytes such as sugar, hydrogen etc, through variations of their transmission and reflection characteristics [77]. Interdisciplinary boundary between metamaterials science and sensing technology has become a fertile ground for new scientific and technological development. We have already discussed two metamaterial-based THz sensors in section 4.2.3.2 of this chapter. In recent times, a high refractive index THz sensor has been researched [85] which is expected to be a highly sensitive protein sensor useful for diagnosis of severe contagious diseases such as Ebola, severe acute respiratory syndrome, including COVID-19. In fact, biosensing technologies based on metamaterials have attracted significant attention from the microwave-to-optical frequency because of their cost-efficient and label-free biomolecule detection. According to operating frequency of sensing biomolecule and component, the metamaterial-based sensors are classified into three types: microwave biosensor, terahertz biosensor and plasmonic biosensor [86]. A number of review articles are available in literature on metmaterial-based sensors [86, 87] which the reader may refer to for further details.

Plasmonic metamaterial nanostructures can also be used to improve light-harvesting solutions, permitting a considerable reduction in physical thickness and improved efficiency in solar photovoltaic absorber layers [77]. Solar cells are known to be a low-cost, renewable, and carbon-emission-free energy source that converts solar energy into electrical energy. Solar energy is the most promising among other sources of renewable energy because of its abundance in Nature and cleanness. Photovoltaic solar energy conversion with solar cells normally suffers from relatively low conversion efficiency. This is due to the wavelength mismatch between the narrow wavelength band associated with the energy gap in the semiconductor and the wide band of the Sun's (black body) emission curve. Thus, in solar cells the photon management is expected to contribute improving spectrally and spatially the absorption of the incident sunlight. In this respect, numerous advanced micro- and nano-optical principles have been tried including: photonic crystals [88], metallic nanoparticles [89], and metamaterials [90, 91]. A metamaterial absorber that will be used in designing the photovoltaic cell should have broadband of operation so that it can absorb solar energy over a broad frequency spectrum, thus increasing the conversion efficiency from solar energy to electrical energy.

4.3.5.5 Transparent metamaterial

Unlike conventional metamaterials (that use noble metals like gold or silver on dielectric substrate), a class of new metamaterials, known as *transparent metamaterials*, is made entirely of dielectric materials or insulators and non-metals [92–94]. In fact, everything will be in a silicon platform, allowing integration of electronic and photonic devices on the same chip. The absence of metal in the metamaterial design will save light from getting unnecessarily lost as heat in the photonic device and interconnections [92]. Such metamaterials can make it possible for computer chips and interconnecting circuits to use photonics, instead of electronics, to process and transmit data, representing a potential leap in performance. In computers and consumer electronics, we still use copper wires between the different parts of the chip. But, if we can confine light to the same size as a nanoscale copper wire, we can have much faster clock speed, and hence, enormously fast data processing—the transparent metamaterial can make this possible [92, 93] (figure 4.18). The innovation in transparent metamaterial (TMTM) lies in modifying the phenomenon of total internal reflection (TIR)—the principle normally used to guide light in an optical fibre. Here the optical momentum of evanescent waves is controlled as opposed to conventional photonic devices, which manipulate propagating waves [93]. This dramatically reduces cross-talk (mutual coupling), the figure of merit of photonic integration, among close-by devices compared to any dielectric waveguide (slot, photonic crystal, etc), making dense photonic integration possible [94]. The introduction of engineered anisotropy in permittivity tensor ($\varepsilon_x = 4.8$ and $\varepsilon_z = 11.9$) in the momentum of cladding (realizable with TMTM) brings about the phenomenon of relaxed-TIR [92].

Quantum information technologies, such as quantum cryptography, quantum information storage and optical quantum computing, demand effective stable sources of single photons and nanostructures to control the quantum dynamics (of these photons) [78, 95]. Transparent metamaterial can aid in enhancing

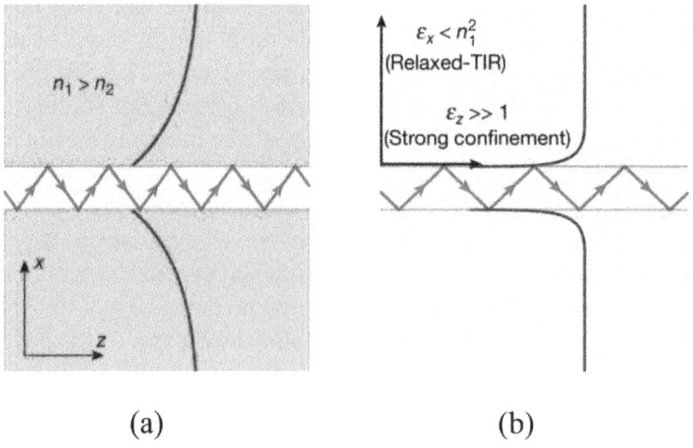

Figure 4.18. (a) Conventional TIR; most power is in cladding that decays slowly. (b) Relaxed-TIR in TMTM; light decays fast in cladding. (Reprinted/adapted with permission from [92], copyright The Optical Society of America.)

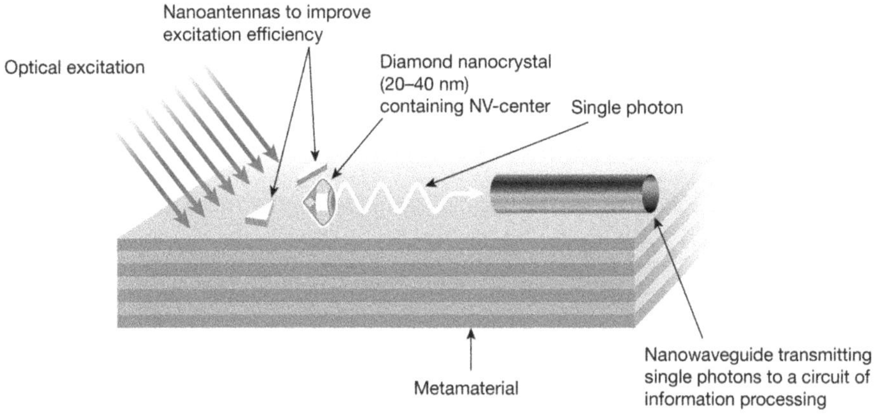

Figure 4.19. Principle of single-photon source with metamaterial cladding. (Reprinted/adapted with permission from [78], Taylor and Francis Group LLC (Books) US http://www.tandfonline.com.)

the single-photon radiation over a broad spectral range, the principle of which is discussed in figure 4.19.

Here single-photon generation is based on coupling diamond nitrogen-vacancy (NV) centres (or silicon-vacancy centres) with a metamaterial having hyperbolic dispersion characteristics in which the enhancement of single-photon emission is due to the presence of metamaterial (that may be two orders of magnitude or so) [96]. Such single-photon sources may also have applications in nanochemistry to control chemical reactions at the level of individual molecules, biochemical analysis to determine the dynamics of molecular configuration, decoding DNA and so on [97].

References

[1] Shalaev V M, Cai W, Chettiar U K, Yuan H K, Sarychev A K, Drachev V P and Kildishev A V 2005 Negative index of refraction in optical metamaterials *Opt. Lett.* **30** 3356–8

[2] Padilla W J, Basov D N and Smith D R 2006 Negative refractive index metamaterials *Mater. Today* **9** 28–35

[3] Wiltshire M C K, Pendry J B, Young I R, Larkman D J, Gilderdale D J and Hajnal J V 2001 Microstructured magnetic materials for RF flux guides in magnetic resonance imaging *Science* **291** 849–51

[4] Smith D R, Padilla W J, Vier D C, Nemat-Nasser S C and Schultz S 2000 Composite medium with simultaneously negative permeability and permittivity *Phys.Rev. Lett.* **84** 4184

[5] Gokkavas M, Guven K, Bulu I, Aydin K, Penciu R S, Kafesaki M, Soukoulis C M and Ozbay E 2006 Experimental demonstration of a left-handed metamaterial operating at 100 GHz *Phys. Rev.* B **73** 193103

[6] Yen T J, Padilla W J, Fang N, Vier D C, Smith D R, Pendry J B, Basov D N and Zhang X 2004 Terahertz magnetic response from artificial materials *Science* **303** 1494–6

[7] Linden S, Enkrich C, Wegener M, Zhou J, Koschny T and Soukoulis C M 2004 Magnetic response of metamaterials at 100 terahertz *Science* **306** 1351–3

[8] Enkrich C, Wegener M, Linden S, Burger S, Zschiedrich L, Schmidt F, Zhou J F, Koschny T H and Soukoulis C M 2005 Magnetic metamaterials at telecommunication and visible frequencies *Phys. Rev. Lett.* **95** 203901
[9] Williams G P 2006 Filling the THz gap—high power sources and applications *Rep. Prog. Phys.* **69** 301–26
[10] Tonouchi M 2007 Cutting-edge terahertz technology *Nat. Photon.* **1** 97–105
[11] Zhang X C 2002 Terahertz wave imaging: horizons and hurdles *Phys. Med. Biol.* **47** 3667–77
[12] Crowe T W, Globus T, Woolard D L and Hesler J L 2004 Terahertz sources and detectors and their application to biological sensing *Phil. Trans. R. Soc.* A **362** 265–377
[13] Federici J F, Schulkin B, Huang F, Gray D, Barat R, Oliveira F and Zimdars D 2005 THz imaging and sensing for security applications—explosives, weapons, and drugs *Semicond. Sci. Technol.* **20** S226–80
[14] Liu H-B, Chen Y, Bastiaans G J and Zhang X-C 2003 Detection and identification of explosive RDX by THz diffuse reflection spectroscopy *Opt. Express* **11** 2549–54
[15] Kar S 2020 Terahertz technology—trends and application viewpoints *Terahertz Biomedical and Healthcare Technologies* ed A Banerjee *et al* (Amsterdam: Elsevier) ch 5 pp 89–111
[16] Zheng L, Sun X, Xu H, Lu Y, Lee Y, Rhee J and Song W 2015 Strain sensitivity of electric–magnetic coupling in flexible terahertz metamaterials *Plasmonics* **10** 1331–5
[17] Pendry J B, Holden J, Robbins D J and Stewart W J 1999 Magnetism from conductors and enhanced non linear phenomena *IEEE Trans. Microw. Theory Tech.* **47** 2075–84
[18] Moser H O, Casse B D F, Wilhelmi O and Saw B T 2005 Terahertz response of microfabricated rod-split-ring-resonator electromagnetic metamaterial *Phys. Rev. Lett.* **94** 063901
[19] Woolard D W, Loerop W R and Shur M S (ed) 2003 Terahertz sensing technology *Selected Topics in Electronics and Systems (2 Volumes)* vol 32 (Singapore: World Scientific)
[20] Hendrickson J, Guo J, Zhang B, Buchwald W and Soref R 2012 Wideband perfect light absorber at midwave infrared using multiplexed metal structures *Opt. Lett.* **37** 371–3
[21] Liu N, Mesch M, Weiss T, Hentschel M and Giessen H 2010 Infrared perfect absorber and its application as plasmonic sensor *Nano Lett.* **10** 2342–8
[22] Niesler F B P, Gansel J K, Fischbach S and Wegener M 2012 Metamaterial metal-based bolometers *Appl. Phys. Lett.* **100** 203508
[23] Wang B, Teo K H, Nishino T, Yerazunis W, Barnwell J and Zhang J 2011 Experiments on wireless power transfer with metamaterials *Appl. Phys. Lett.* **98** 254101
[24] Aydin K, Ferry V E, Briggs R M and Atwater H A 2011 Broadband polarization-independent resonant light absorption using ultrathin plasmonic super absorbers *Nat. Commun.* **2** 517–24
[25] Hedayati M, Faupel F and Elbahri M 2012 Tunable broadband plasmonic perfect absorber at visible frequency *Appl. Phys.* A **109** 769–73
[26] Ruck G T, Barrick D E, Stuart W D and Krichbaum A C K 1970 *Radar Cross Section Handbook* vol 2 (New York: Plenum)
[27] Padilla W J, Aronsson M T, Highstrete C, Lee M, Taylor A J and Averitt R D 2007 Electrically resonant terahertz metamaterials: theoretical and experimental investigations *Phys. Rev.* B **75** 041102
[28] YenT J, Padilla W J, Fang N, Vier D C, Smith D R, Pendry J B, Basov D N and Zhang X 2004 Terahertz magnetic response from artificial materials *Science* **303** 1494–6
[29] Landly N I, Sajuyigbe S, Mock J J, Smith D R and Padilla W J 2008 A perfect metamaterial absorber *Phys. Rev. Lett.* **100** 207402

[30] Tao H, Landy N I, Bingaham C M, Zhang X, Averitt R D and Padilla W J 2008 A metamaterial absorber for the terahertz regime: design, fabrication and characterization *Opt. Express* **16** 7181–8
[31] Lee Y P, Rhee J Y, Yoo Y J and Kim K W 2016 *Metamaterials for Perfect Absorption* (Springer Series in Material Science vol 236) vol 236 (Singapore: Springer Nature)
[32] Li M, Yang H L, Hou X W, Tian Y and Hou D Y 2010 Perfect metamaterial absorber with dual bands *Prog. Electromagn. Res.* **108** 37–49
[33] Shen X, Cui T J, Zhao J, Ma H F, Jiang W X and Li H 2011 Polarization-independent wide-angle triple-band metamaterial absorber *Opt. Express* **19** 9401–7
[34] Park J W, Tuong P V, Rhee J Y, Kim K W, Jang W H, Choi E H, Chen L Y and Lee Y 2013 Multi-band metamaterial absorber based on the arrangement of donut-type resonators *Opt. Express* **21** 9691–702
[35] Tuong P V, Park J W, Rhee J Y, Kim K W, Jang W H, Cheong H and Lee Y P 2013 Polarization-insensitive and polarization-controlled dual-band absorption in metamaterials *Appl. Phys. Lett.* **102** 081122
[36] Shen X, Yang Y, Zang Y, Gu J, Han J, Zhang W and Cui T J 2012 Triple-band terahertz metamaterial absorber: design, experiment, and physical interpretation *Appl. Phys. Lett.* **101** 154102
[37] Shan Y, Chen L, Shi C, Cheng Z, Zang X, Xu B and Zhu Y 2015 Ultrathin flexible dual band terahertz absorber *Opt. Commun.* **350** 63–70
[38] Zhao X, Fan K, Zhang J, Seren H R, Metcalfe G D, Wraback M, Averitt R D and Zhang X 2015 Optically tunable metamaterial perfect absorber on highly flexible substrate *Sensors Actuators* A **231** 74–80
[39] Wang W, Yan F and Tan S 2017 Zhou Hand Hou Y, Ultrasensitive terahertz metamaterial sensor based on vertical split-ring resonators *Photon. Res* **5** 571–7
[40] Alves F, Grbovic F, Kearney B, Lavrik N V and Karunasiri G 2013 Bi-material terahertz sensors using metamaterial structures *Opt. Express* **21** 13256–71
[41] Enkrich C, Wegener M, Linden S, Burgur S, Zschiedrich L, Schmidt F, Zhou J F, Koschny T and Soukoulis C M 2005 Magnetic metamaterials at telecommunication and visible frequencies *Phys. Rev. Lett.* **95** 203901
[42] Klein W M, Enkrich C, Wegener M, Soukoulis C M and Linden S 2006 Single-split split-ring resonators at optical frequencies: limits of size scaling *Opt. Lett.* **31** 1259–61
[43] Zhang S, Fan W, Panoiu N C, Malloy K J, Osgood R M and Brueck S R J 2005 Experimental demonstration of near-infrared negative-index metamaterials *Phys. Rev. Lett.* **95** 137404
[44] Eleftheriades G V and Balmain K G (ed) 2005 *Negative-Refraction Metamaterials—Fundamental Principles and Applications* (Piscataway, NJ: IEEE Press, Wiley Interscience)
[45] Valentine J, Zhang S, Zentgraf T, Ulin-Avila E, Genov D A, Bartal G and Zhang X 2008 Three-dimensional optical metamaterial with a negative refractive index *Nature* **455** 376–9
[46] Yao J, Liu Z, Liu Y, Wang Y, Sun S, Bartal G, Stacy A M and Zhang X 2008 Optical negative refraction in Bulk metamaterials of nanowires *Science* **321** 930
[47] Dolling G, Wegener M, Soukoulis C M and Linden S 2007 Negative index material at 780 nm wavelength *Opt. Lett.* **32** 53–5
[48] Valantine J, Zhang S, Zentgraf T and Zhang X 2011 Development of bulk optical negative index fishnet metamaterials: achieving a low-loss and broadband response through coupling *Proc. IEEE* **99** 1682–90

[49] Pendry J B, Martin-Moreno L and Garcia-Vidal F J 2004 Mimicking surface plasmons with structured surfaces *Science* **305** 847–8
[50] Sahalaev V M 2007 Optical negative-index metamaterials *Nat. Photon.* **1** 41–8
[51] Popov A K and Shalaev V M 2006 Compensating losses in negative-index metamaterials by optical parametric amplification *Opt. Lett.* **31** 2169–71
[52] Enkrich C, Perez-Williard F, Gerthsen D, Zhou J F, Koschny T, Soukoulis C M, Wegener M and Linden S 2005 Focussed-ion-beam nanofabrication of near-infrared magnetic metamaterials *Adv. Mater.* **17** 2547–9
[53] Bureck S R J 2005 Optical and interferometric lithography—nanotechnology enablers *Proc. IEEE* **93** 1704–21
[54] Ku Z and Bureck S R J 2007 Comparison of negative refractive index metamaterials with circular, elliptical and rectangular holes *Opt. Express* **15** 4515–22
[55] Boltasseva A and Shalaev V M 2008 Fabrication of optical negative-index metamaterials: Recent advances and outlook *Metamaterials* **2** 1–17
[56] Chou S Y, Krauss P R and Renstrom P J 1996 Nanoimprint lithography *J. Vac. Sci. Technol.* B **14** 4129
[57] Wu W *et al* 2007 Optical metamaterials at near and mid-IR range fabricated by nanoimprint lithography *Appl. Phys.* A **87** 143–50
[58] Wu W, Yu Z, Wang S-Y, Williams R S, Liu Z, Sun C, Zang X, Kim E, Shen Y R and Fang Y N 2007 Mid-infrared metamaterials fabricated by nanoimprint lithography *Appl. Phys. Lett.* **90** 063107
[59] Griffith S, Mondol M, Kong D S and Jacobson J M 2002 Nanostructure fabrication by direct electron-beam writing of nanoparticles *J. Vac. Sci. Technol.* B **20** 2768
[60] Morita T *et al* 2004 Nanomechanical switch fabrication by focused-ion-beam chemical vapour deposition *J. Vac. Sci. Technol.* B **22** 3137
[61] Takada K, Sun H-B and Kawata S 2005 Improved spatial resolution and surface roughness in photopolymerization-based laser nanowriting *Appl. Phys. Lett.* **86** 071122
[62] Formanek F, Takeyasu N, Tanaka T, Chiyoda K, Ishikawa A and Kawata S 2006 Three-dimensional fabrication of metallic nanostructures over large areas by two-photon polymerization *Opt. Express* **14** 800–9
[63] Kehagias N, Reboud V, Chansin G, Zelsmann M, Jeppesen C, Schuster C, Kubenz M, Reuther F, Gruetzner G and Torres C M S 2007 Reverse-contact UV nanoimprint lithography for multilayered structure fabrication *Nanotechnology* **18** 175303
[64] Pendry J B 2000 Negative refraction makes a perfect lens *Phys. Rev. Lett.* **85** 3966–9
[65] Grbic A and Eleftheriades G V 2004 Overcoming the diffraction limit with a planar left-handed transmission-line lens *Phys. Rev. Lett.* **92** 117403
[66] Fang N, Lee H, Sun C and Zhang X 2005 Sub-diffraction limited optical imaging with silver superlens *Science* **308** 534–7
[67] Liu Z, Lee H Y, Xiong Y, Sun C and Zhang X 2007 Far-field optical hyperlens magnifying sub-diffraction-limited objects *Science* **315** 1686
[68] Jacob Z, Alekseyev L V and Narimanov E 2006 Optical hyperlens: far-field imaging beyond the diffraction limit *Opt. Express* **14** 8247–56
[69] Xiong Y, Liu Z, Sun C and Zhang X 2007 Two-dimensional imaging by far-field superlens at visible wavelengths *Nano Lett.* **7** 3360–5
[70] Landly N I, Sajuyigbe S, Mock J J, Smith D R and Padilla W J 2008 A perfect metamaterial absorber *Phys. Rev. Lett.* **100** 207402

[71] Avitzour Y, Urzhumov Y A and Shvets G 2009 Wide-angle infrared absorber based on a negative-index plasmonic metamaterial *Phys. Rev.* B **79** 045131
[72] Jiang Z H, Yun S, Toor F, Werner D H and Mayer T S 2011 Conformal dual-band near-perfectly absorbing mid-infrared metamaterial coating *ACS Nano* **5** 4641–7
[73] Bossard J A, Lin L, Yun S, Liu L, Werner D H and Mayer T S 2014 Near-ideal optical metamaterial absorbers with super-octave bandwidth *ACS Nano* **8** 1517–24
[74] Aydin K, Ferry V E, Briggs R M and Atwater H A 2011 Broadband polarization-independent resonant light absorption using ultrathin plasmonic super absorbers *Nat. Commun.* **2** 517
[75] Cheng C-W, Abbas M N, Chiu C W, Lai K T, Shih M H and Chang Y C 2012 Wide-angle polarization independent infrared broadband absorbers based on metallic multi-sized disk arrays *Opt. Express* **20** 10376–81
[76] Pitchappa P, Ho C P, Kropelnicki P, Singh N, Kwong D L and Lee C 2014 Micro-electro-mechanically switchable near infrared complementary metamaterial absorber *Appl. Phys. Lett.* **104** 201114
[77] Zheludev N I 2010 The road ahead for metamaterials *Science* **328** 582–3
[78] Kar S 2021 Progress in metamaterial and metasurface technology and applications *Metamaterials: Technology and Applications* ed P K Choudhury (Boca Raton, FL: CRC Press of Taylor and Francis Group) ch 1 pp 1–29
[79] Plum E, Fedotov V A, Kuo P, Tsai D P and Zheludev N I 2009 Towards the lasing spaser : Controlling metamaterial optical response with semiconductor quantum dots *Opt. Express* **17** 8548–51
[80] Zheludev N I, Prosvirnin S L, Papasimakis N and Fedotov V A 2008 Lasing spaser *Nat. Photon.* **2** 351–4
[81] Bergman D J and Stockman M I 2003 Surface plasmon amplification by stimulated emission of radiation: quantum generation of coherent surface plasmons in nano-systems *Phys. Rev. Lett.* **90** 027402
[82] Fedotov V A, Rose M, Prosvirnin S L and Zheludev N I 2007 Sharp trapped-mode resonances in planar metamaterials with a broken structural symmetry *Phys. Rev. Lett.* **99** 147401
[83] Nikolaenko A E, De Angelis F, Boden S A, Papasimakis N, Ashburn P, Di Fabrizio E and Zheludev N I 2010 Carbon nanotubes in a photonic metamaterial *Phys. Rev. Lett.* **104** 153902
[84] Jakšić Z, Vuković S, Matovic J and Tanasković D 2011 Negative refractive index metasurfaces for enhanced biosensing *Materials* **4** 1–36
[85] Silalahi H, Chen Y-P, Chen Y-S, Hong S, Lin X-Y, Liu J-H and Huang C-Y 2021 Floating terahertz metamaterials with extremely large refractive index sensitivities *Photon. Res.* **9** 1970–8
[86] Chen T, Suyan L S and Sun H 2012 Metamaterials application in sensing *Sensors* **12** 2742–65
[87] Yang J J, Huang M, Tang H, Zeng J and Dong L 2013 Metamaterial sensors *Int. J. Antennas Propag.* **2013** 637270
[88] Chutinan A and John S 2008 Light trapping and absorption optimization in certain thin-film photonic crystal architectures *Phys. Rev.* A **78** 23825
[89] Akimov Y A, Koh W S and Ostrikov K 2009 Enhancement of optical absorption in thin-film solar cells through the excitation of higher-order nanoparticle plasmon modes *Opt. Express* **17** 10195–205
[90] Liu Y, Chen Y, Li J, Hung T-C and Li J 2012 Study of energy absorption on solar cell using metamaterials *Sol. Energy* **86** 1586–99

[91] El-Amassi D M, El-Khozondar H J and Mohammed M S 2015 Efficiency enhancement of solar cell using metamaterials *Int. J. Nano Stud. Technol.* **4** 84–7
[92] Jahani S and Jacob Z 2014 Transparent sub-diffraction optics: nanoscale light con-finement without metal *Optica* **1** 96–100
[93] Jahani S and Z. Jacob Z 2015 Breakthroughs in photonics 2014: relaxed total internal reflection *IEEE Photon. J.* **7** 1–5
[94] Jahani S and Jacob Z 2016 All-dielectric metamaterials *Nat. Nanotechnol.* **11** 23–6
[95] Shalaginov M Y, Vorobyov V V and Liu J *et al* 2015 Enhancement of single–photon emission from nitrogen-vacancy centres with N/(Al,Sc)N hyperbolic metamaterial *Laser Photon. Rev.* **9** 120–7
[96] Kar S 2017 Photonics and metamaterials (plenary lecture) *104th Indian Science Congress (Sri Venkateswara University, Tirupati, 3–7 January 2017)*
[97] Takeuchi S 2014 Recent progress in single-photon and entangled-photon generation and applications *Jpn. J. Appl. Phys.* **53** 030101

IOP Publishing

Metamaterials and Metasurfaces
Basics and trends
Subal Kar

Chapter 5

Modelling and characterization of metamaterials

The aim of life should be to become a role-model for others

5.1 Introduction

Metamaterials (MTMs) are artificially engineered composite structures, the simplest structure being a combination of metallic thin wires (TWs) and split-ring resonator (SRR) which are metallic inclusions embedded in dielectric substrate (see figure 2.4). Thus from a *microscopic* viewpoint, MTMs are intrinsically *inhomogeneous*. However, the period (lattice spacing) and the size of the metallic inclusions are much smaller than the wavelength of the incident electromagnetic wave. Thus from a *macroscopic* point of view an MTM is extrinsically a *homogeneous effective medium*. Hence MTM is characterized by bulk constitutive parameters like complex effective relative permittivity and effective relative permeability or complex effective refractive index using effective medium theory.

For modelling of metamaterials with effective-medium approach some type of averaging, also called homogenization [1], is performed on the electric and magnetic fields over a given period cell composing the metamaterial. It should be remembered that the so-called averaging is valid only when the wavelength (λ) is large enough with respect to the lattice constant (p) of the period cell, i.e. $\lambda > p$ (this is precisely the case with metamaterial). If used properly, the effective-medium theory approach can be a self-consistent and unique method for characterizing a metamaterial. Alhough the period-cell averaging for the field is correct for defining effective material properties, many researchers, prefer to use an approach based on reflection and transmission coefficients of a metamaterial sample of some defined thickness. The Nicolson–Ross approach [2], or a variant of it, is then used to obtain the effective material properties of the bulk metamaterial. The S-parameter-based retrieval method proposed by Smith *et al* [3] is also very popular. The parameter retrieval/extraction method proposed by Smith *et al* is based on Z, Y, or S-parameters and is used quite often in practice by microwave engineers.

doi:10.1088/978-0-7503-5532-2ch5

The analytical characterization of the bulk properties of metamaterial in order to evaluate the effective permittivity ($\varepsilon_{\text{reff}}$), effective permeability (μ_{reff}) and eventually the refractive index $n_{\text{reff}} = \sqrt{\varepsilon_{\text{reff}}\mu_{\text{reff}}}$ is traditionally done by Drude model (refer to equation (2.3), chapter 2) for TW characterization, while the Lorentzian model is used for the characterization of resonant structures like cut wire (CW), refer to equation (2.4), and SRR, refer to equation (2.6).

Since metamaterials are highly lossy and dispersive in nature, it becomes difficult for their parameter extraction, i.e. characterization via measurements—only a free-space characterization can be done to find the negative permittivity, permeability and refraction properties of fabricated metamaterial samples—some details of which will also be given in the later part of this chapter. Further, we will also discuss in brief the experimental technique for plane-slab 'superlens' characterization with near-field measurement technique [4].

5.2 Numerical techniques for modelling and characterization of metamaterials

5.2.1 Introduction

Various techniques have been developed from time to time for characterization of material parameters, in general, a brief discussion of which follows. Kadaba *et al* in 1984 [5] proposed a frequency domain technique to find out the complex permittivity and permeability of materials at high frequencies. However, this technique is valid for low-loss samples that are small in size, i.e. less than quarter guide wavelength. Ghodgaonkar *et al* [6] proposed a scheme to measure both complex permittivity and permeability using TRL (through reflect line) calibration, but still this technique has problems with lossy samples. In 2002, Tamyis *et al* [7] proposed an advanced algorithm-based method for magnetic materials using open and short circuit standards but it is still not applicable for highly lossy samples. Thus, all these conventional methods to obtain permittivity and permeability of material samples were found to be not useful for metamaterials, which are dispersive and lossy, and hence some new methods were developed subsequently.

Metamaterial modelling/characterization is mostly done by numerical techniques like homogenization by field-averaging technique, transfer matrix method (TMM), and finite difference time-domain (FDTD) method (see appendix E for the basics of the numerical techniques used in computational electromagnetics).

All the known analytical methods for characterizing the metamaterial on the basis of homogenization like Clausius–Mossotti and Maxwell–Garnet formulas are, however, valid under certain limitations and particular geometries or classes of structures. For metamaterials comprising conducting and in many situations resonant elements, and for which the periodicity is not necessarily negligible relative to the free-space wavelength, analytical homogenization techniques are unreliable or not applicable in many cases.

Given that Maxwell's equations can now be solved rapidly and efficiently—at least for domains relatively small with respect to the wavelength—with custom or commercial software running on modern personal computers, the local electric and magnetic fields, current distributions, and other quantities of interest can be found easily for virtually all types of structures. A purely numerical approach to the homogenization of artificial structures is thus feasible and avoids the limitations associated with the prior analytical homogenization models.

The numerical homogenization technique [1] is based on the averaging of the local fields in which a discrete set of averaged fields is defined at each unit cell, replacing the local fields that vary throughout the unit cell. In this sense the averaging scheme resembles a finite differencing of Maxwell's equations, with a resulting spatial discretization analogous to those used in FDTD and similar algorithms. It is observed that there is excellent agreement between the field-averaging of homogenization technique and the *S*-parameter retrieval method, the discrepancy, if any, is within the numerical error for these calculations.

The field-averaging process of homogenization technique provides useful means of characterizing the effective-medium parameters of an arbitrary metamaterial. It is particularly useful for complex structures in which the transmitted and reflected amplitudes are not easily interpreted; in such cases it may be difficult to choose the appropriate solution (out of the available set of solutions) that matches the expected effective-medium theory. In contrast the material parameters obtained by field-averaging method are not subject to the multiple branches that occur in the *S*-parameter method. When combined with dispersion curve calculations, field-averaging provides a wealth of information distinct and complementary to that obtained by *S*-parameter analysis and retrieval.

One approach to the assignment of electromagnetic parameters to a simulated structure is via *S-parameter-based retrieval method* proposed by Smith *et al* [3]. In the *S*-parameter retrieval method, the complex reflection and transmission coefficients are calculated for a finite thickness of metamaterial, usually one unit cell, and compared with analytical formulas for the reflection and transmission coefficients of a homogeneous slab with the same thickness. The retrieval method can be applied to simulated or measured *S*-parameter data, although thin samples (in the propagation direction) are the best. Despite its versatility and ease of use, *S*-parameter retrieval does not provide significant physical insight into the nature of the artificial material. The *S*-parameter observables—reflection and transmission coefficients—are not easily connected to analytical homogenization theories, which instead usually involves analysis of the local-field and current distributions.

Anyway, the *S*-parameter retrieval method proposed by Smith *et al* [3] used the TMM. The TMM represents a complete solution of Maxwell's equations in which the exact details of the scattering elements are taken into account. To determine the effective permittivity (ϵ) and permeability (μ) they analyzed the reflection and transmission coefficients calculated from transfer matrix simulation on finite lengths of electromagnetic metamaterials of wire, SRR and their combination. In their scheme, a plane wave is incident normally on a sample of finite thickness and scattering parameters are found as a function of frequency. Then this scattering parameter data is used to determine the refractive index n and wave impedance z of the sample. The permittivity and permeability of sample is found using refractive index and impedance of the sample from simple equations: $\varepsilon = n/z$ and $\mu = nz$. They found that the recovered frequency-dependent ε and μ are entirely consistent with analytic expressions predicted by effective medium arguments applied to metamaterials.

Other computational and purely simulation based methods are also available in literature for metamaterials. Deshpande *et al* [8] developed a robust procedure to extract electromagnetic properties of metamaterial from the knowledge of its reflection and transmission coefficients using the Nicholson–Ross method and genetic algorithms. In the method proposed by Nicholson and Ross [2] complex permittivity and permeability of linear materials are determined in the frequency domain by a single time-domain measurement. The technique involves placing an unknown sample in a microwave TEM-mode fixture and exciting the sample with a sub-nanosecond baseband pulse. The fixture is used to facilitate the measurement of the forward- and back-scattered energy, $S_{21}(t)$ and $S_{11}(t)$, respectively. The forward- and back-scattered time-domain 'signatures' are uniquely related to the intrinsic properties of the material viz. ε^* and μ^*. By appropriately interpreting $S_{21}(t)$ and $S_{11}(t)$, one is able to determine the real and imaginary parts of ε and μ as a function of frequency. Anyway, in this chapter we will discuss in detail the S-parameter-based retrieval method proposed by Smith *et al* [3] and their application to some metamaterial structures.

5.2.2 *S*-Parameter-based retrieval method for characterization of metamaterial inclusion structures

5.2.2.1 *The methodology*

This method was proposed by Smith *et al* [3] and used quite often by the metamaterial design engineers. The method is elaborated below [4].

ABCD matrix for a sample of finite thickness d is given by:

$$[ABCD] = \begin{bmatrix} \cos(nkd) & jZ\sin(nkd) \\ j\dfrac{\sin(nkd)}{Z} & \cos(nkd) \end{bmatrix} \tag{5.1}$$

where the *ABCD* parameters are related to the S-parameters by [9]

$$S_{11} = \frac{A + B/Z_0 - CZ_0 - D}{A + B/Z_0 + CZ_0 + D} \quad S_{21} = \frac{2}{A + B/Z_0 + CZ_0 + D} \tag{5.2}$$

Solving equations (5.1) and (5.2), S_{11} and S_{21} are given by:

$$S_{11} = \frac{j\sin(nkd)\left(Z_n - \dfrac{1}{Z_n}\right)}{2\cos(nkd) + j\sin(nkd)\left(Z_n - \dfrac{1}{Z_n}\right)}$$

$$S_{21} = \frac{2}{2\cos(nkd) + j\sin(nkd)\left(Z_n - \dfrac{1}{Z_n}\right)} \tag{5.3}$$

where Z_n is the impedance of sample. Using equation (5.3), Z_n and n are given by:

$$Z_n = \pm\sqrt{\frac{(1 + S_{11})^2 - S_{21}^2}{(1 - S_{11})^2 - S_{21}^2}} \cos(nkd) = \frac{1}{2S_{21}}[1 - S_{11}^2 + S_{21}^2]$$

$$\text{Im}(n) = \pm\text{Im}\left\{\frac{\cos^{-1}\left(\frac{1}{2S_{21}}[1 - S_{11}^2 + S_{21}^2]\right)}{kd}\right\}$$

$$\text{Re}(n) = \pm\text{Re}\left\{\frac{\cos^{-1}\left(\frac{1}{2S_{21}}[1 - S_{11}^2 + S_{21}^2]\right)}{kd}\right\} + \frac{2\pi m}{kd} \tag{5.4}$$

where m is an integer and S_{11} and S_{21} are the reflection and transmission coefficients of the sample, k is the free-space wave vector that depends upon frequency.

For passive materials, Re(z) and Im(n) must be greater than zero to satisfy casualty. These conditions will lead to unique solution of equation (5.4). When d is large, the branches due to integer m can lie arbitrarily close to one another, making the selection of the correct solution difficult in the case of dispersive materials like LHM. For this reason, best results are obtained for the smallest possible thickness of sample. Even with a small sample, more than one thickness must be measured to identify the correct branch(es) of the solution which yields consistently the same values for n. Note that the requirement that Im(n) > 0 uniquely determines the sign of Re(n), which is essential when the material may potentially have regions which are left-handed, i.e., materials in which Re(n) may be negative. Metamaterials do not have well-defined surfaces, and this makes reference plane selection difficult for this method. It was observed that the reference plane in the case of wire media is taken as first plane of wires and for SRR, the reference plane is not so critical. In combined structure, reference plane is taken as the first plane of wires due to dominance of wire in S-parameters.

The retrieval methodology discussed above might fail in some instances, such as when the thickness of the effective slab (exhibiting bulk properties) is not accurately estimated or when reflection (S_{11}) and transmission (S_{22}) data are very small in magnitude. Although these issues may be addressed to some extent, even then the retrieved results may be still unsatisfactory in some cases. In particular, the determination of the first boundary of the effective homogeneous slab, the selection of the sign of the roots when solving for the impedance z, the determination of the real part of refractive index n and so forth. Chen *et al* [10] suggested an improved and robust method to retrieve the effective parameters (index of refraction, impedance, permittivity and permeability) of metamaterials from transmission and reflection data. Improvements provided by this method when compared with the method proposed by Smith *et al* [3] include the determination of the first boundary and the thickness of the effective slab, the selection of the correct sign of the effective impedance, and a mathematical method to choose the correct branch of the real part of the refractive index.

Anyway, the above methods discussed for parameter retrieval of metamaterials are not applicable for inhomogeneous materials. For such materials, Smith *et al* have given an alternative approach [11] for parameter retrieval that is a modified version of their previous method [3]. For 1D metamaterials, unit cell can be chosen in such a way that structure becomes symmetric. But if we take 2D or 3D metamaterials, then we cannot avoid asymmetry and asymmetric analysis must be applied.

5.2.2.2 Simulated results for magnetic inclusion structures using S-**parameter retrieval method**

Standard parameter retrieval method [3] has been applied to characterize eight negative permeability magnetic inclusion structures [12] viz.: split-ring resonator (SRR), square split-ring resonator (SSRR), Two turn spiral resonator (TTSR), non-bi-anisotropic spiral resonator (NBSR), labyrinth resonator (LR), U-shaped split-ring resonator (USRR), broadside coupled split-ring resonator (BC-SRR), and interdigital capacitor loaded ring resonator (ICRR) shown in figure 5.1 (a brief discussion about the physical properties of these structures were discussed in sections 2.2.2 and 2.2.4 of chapter 2).

An finite element method (FEM)-based full-wave electromagnetic solver HFSS (high frequency structure simulator) from Ansoft (presently Ansys) [13] has been used for simulation of these negative permeability metamaterial structures. Periodic boundary conditions have been applied in x- and y-directions to intimate an infinite array, as shown in figure 5.2. Floquet ports are assigned in the z-direction which excites TM electromagnetic waves in z-direction and these ports are de-embedded up to unit cell substrate for determining exact phase of wave propagating through negative permeability structure. The solver typically returns the scattering parameters (i.e. S-parameters).

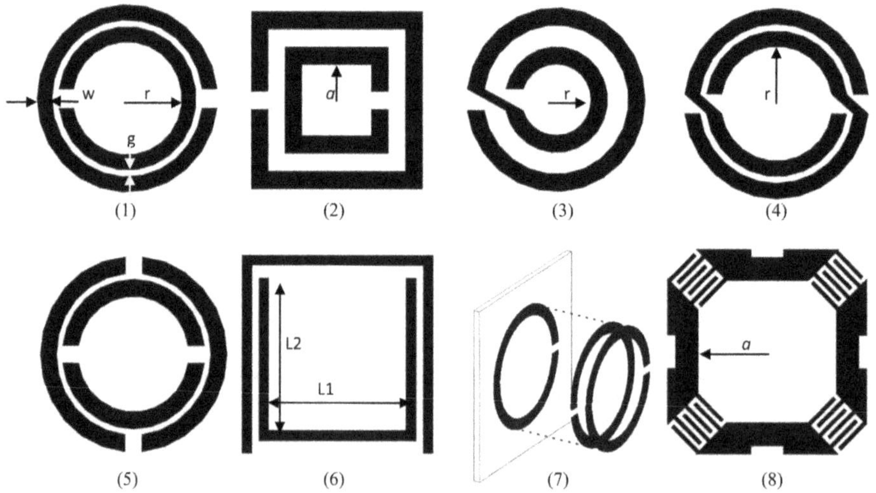

Figure 5.1. Schematic of unit cell of SRR (1), SSRR (2), TTSR (3), NBSR (4), LR (5), USRR (6), BC-SRR (7), and ICRR (8).

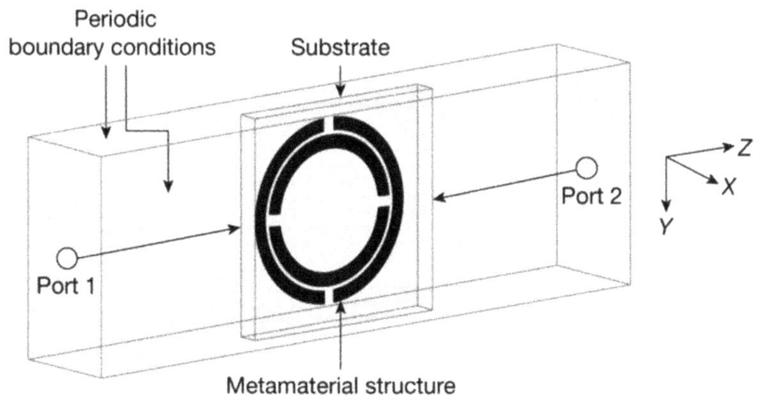

Figure 5.2. Simulation of a unit cell of negative permeability structure in HFSS.

Table 5.1. Dimension of negative permeability structures.

Negative permeability structure	Dimensions: Inner radius/inner dimension *r*/*a* (mm)	Width of metal strip/ring w (mm)	Gap *g* (mm)
SRR	0.9	0.25	0.1
SSRR	0.6	0.3	0.3
TTSR	**0.5**	0.25	0.3
NBSR	0.9	0.25	0.1
LR	1.4	0.3	0.1
USRR	L1 = 2 and L2 = 2.15	0.15	0.1
BC-SRR	1.25	0.25	0.508
ICRR	**1.722**	0.778	0.1

All negative permeability structures are designed and optimized at X band using the above- mentioned simulation technique and dimensions of these are summarized in table 5.1. Interdigital capacitor of ICRR consists of six metal strips of length 0.9 mm and width 0.1 mm separated by a gap of 0.1 mm. These structures are printed on a high-performance 20 mil Rogers RT/duroid 5880LZ dielectric substrate having relative permittivity 1.96 and loss tangent 0.0019 at 10 GHz. All structures have periodicity of 2.5 mm × 5 mm × 5 mm in *x*-, *y*- and *z*-directions, respectively, except ICRR which have periodicity of 4.6 mm × 5.5 mm × 5.5 mm.

The simulated results obtained for the effective negative permeability (real and imaginary parts) of all the eight magnetic inclusion structures using the *S*-parameter retrieval method is shown in figure 5.3.

5.2.2.3 Experimental and simulation results of some of the fabricated structures

Out of eight magnetic inclusion structures mentioned above, three magnetic inclusions structures SSRR, TTSR and NBSR have been fabricated and tested.

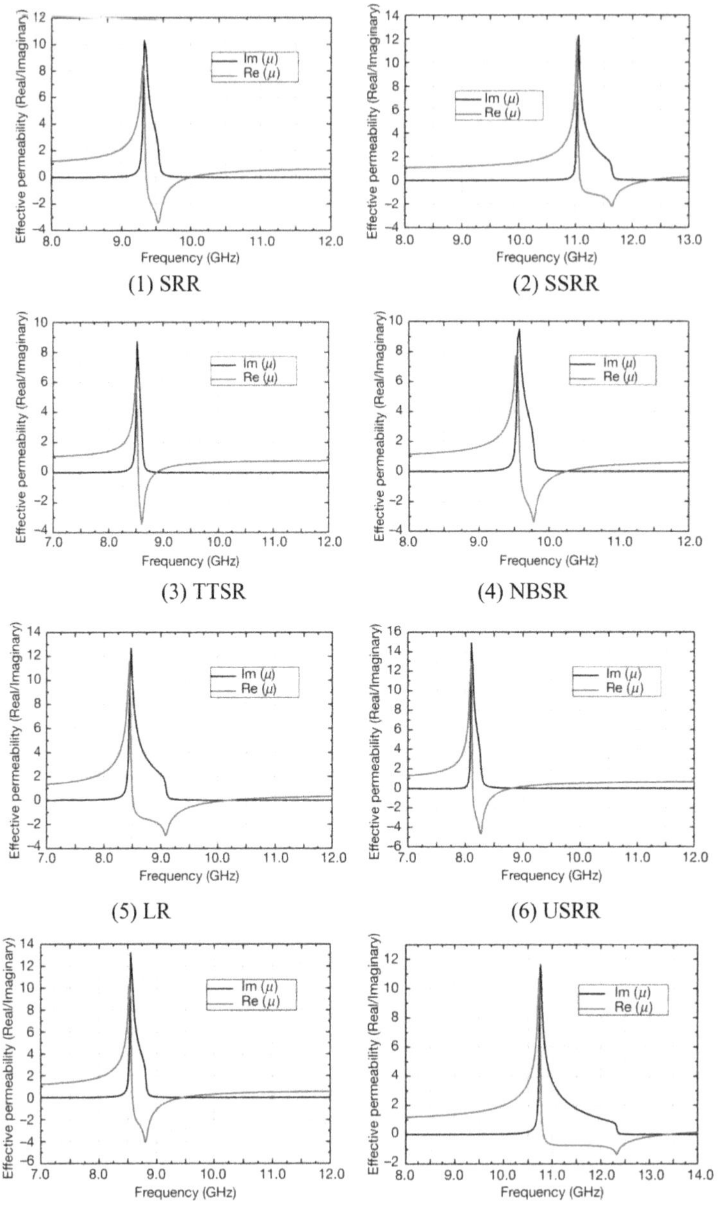

Figure 5.3. Real and imaginary parts of effective permeability of negative permeability structures plotted versus frequency.

Arrays of 10×10 unit cells have been fabricated on Rogers RT/Duroid 5880LZ substrate as shown in figure 5.4. These 10 substrates have been stacked using very low permittivity high-performance foam sheets to make a sample for testing.

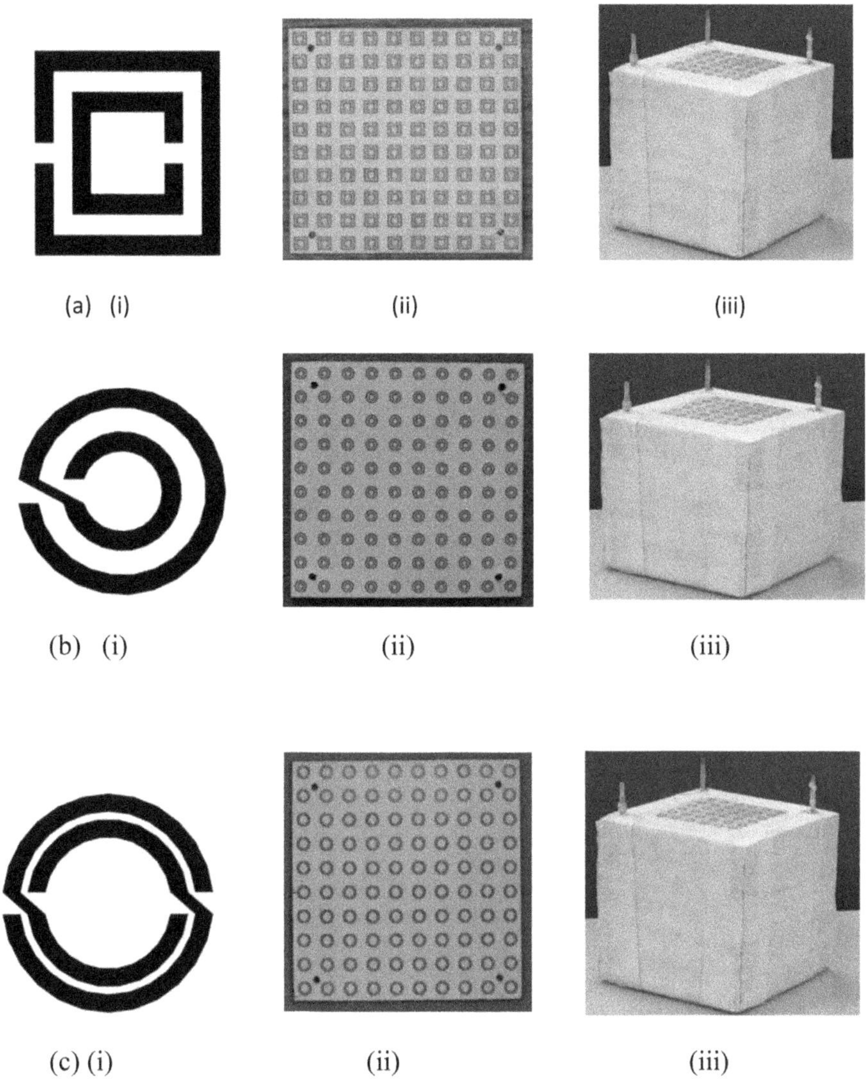

Figure 5.4. Three fabricated magnetic inclusion structures. (a) (i) SSRR unit cell, (ii) array of SSRR, (iii) stacked structure with SSRR arrays. (b) (i) TTSR unit cell, (ii) array of TTSR, (iii) stacked structure with TTSR arrays. (c) (i) NBSR unit cell, (ii) array of NBSR, (iii) stacked structure with NBSR arrays.

The fabricated samples of metamaterial are tested in a free-space focused beam LHM (i.e. metamaterial) characterization facility, a schematic of which is shown in figure 5.5.

The actual LHM characterization set-up at X band is shown in figure 5.6. It has a transmitter and a receiver section where each section consists of a standard X band pyramidal horn antenna and a double convex lens of diameter 175 mm. The convex lenses are made from a high-performance dielectric material Rexolite 1422. The test sample is placed between the transmitter and receiver section where the

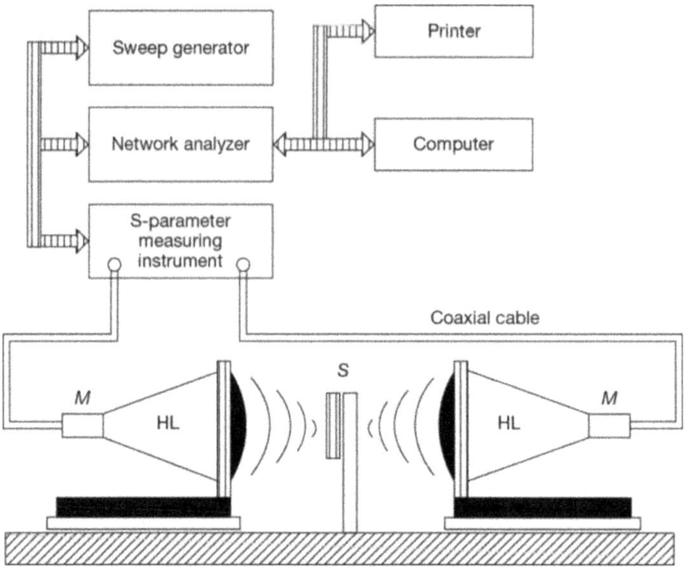

Figure 5.5. Schematic of experimental set-up for LHM sample characterization (S is the LHM sample, HL represents spot-focussing horn-lens antenna, and M stands for mode transition).

Figure 5.6. Free-space focused beam LHM characterization set-up at X band (courtesy: SAMEER Kolkata Centre).

electromagnetic wave is focused. Agilent's E8363B series VNA is used for measurement of scattering parameters. The set-up is calibrated for free-space path losses before testing of the samples. To reduce the possibility of picking up undue reflected signals, the whole set-up is surrounded by an anechoic environment of approximate size 4.5 m × 3 m × 2.5 m.

The simulated insertion loss through the array of 10 unit cells in the direction of propagation of wave for eight negative permeability structures is shown in figure 5.7. The measured insertion loss for the three structures SSRR, TTSR, and NBSR have

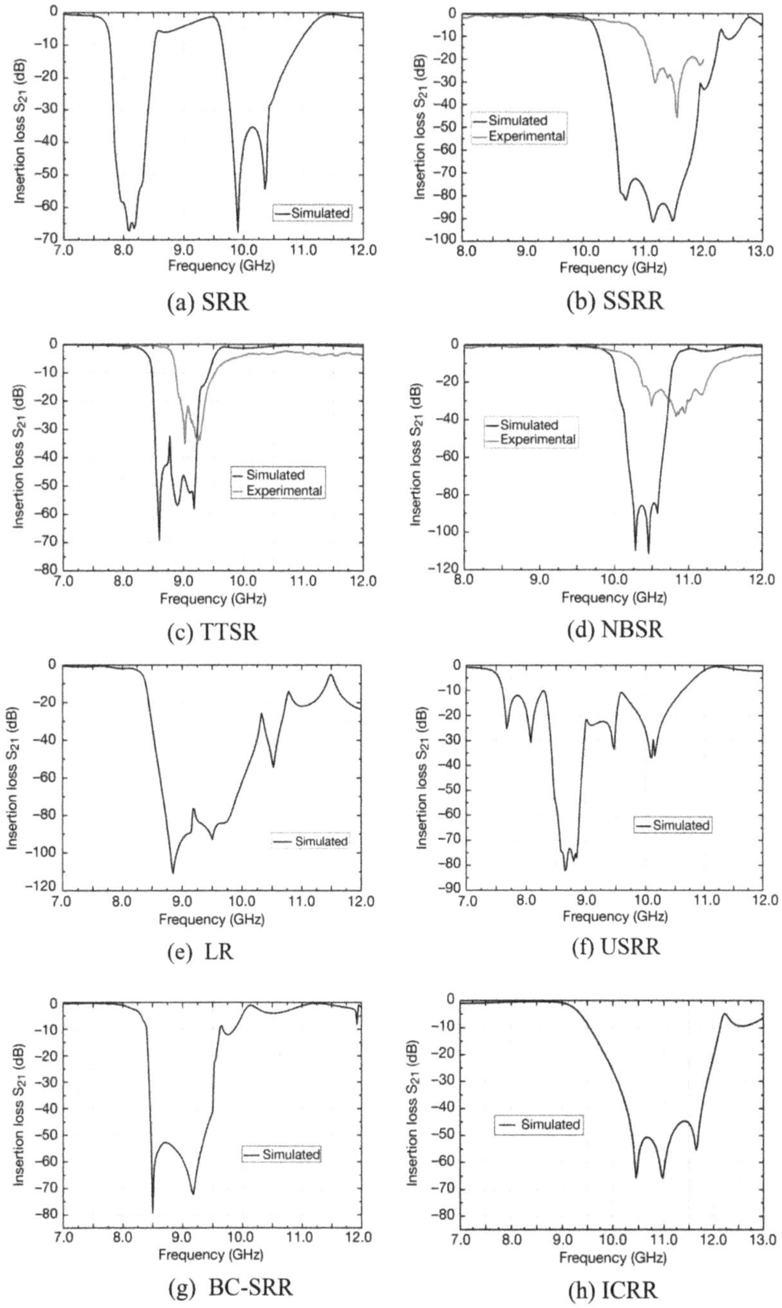

Figure 5.7. Insertion loss through negative permeability structures plotted versus frequency for all the eight magnetic inclusion structures (note: experimental result for fabricated structures: SSRR, TTSR, NBSR are also included).

also been shown along with their respective simulated insertion loss plots in the same lot of figures. SRR structure shows more than one stop-band which may be associated with coupling or extra electrical resonance or bi-anisotropy. It may be noted that the LR, USSR and ICRR structures shows wide stop-band which can be associated with coupling between adjacent unit cells.

Insertion loss obtained through simulation and experiments for each of the three fabricated structures, figure 5.7(b) SSRR (c) TTSR and (d) NBSR, are seen to be very close to each other. However, stop-band dip in simulation is lower than the experiments which may be associated with the finite size of fabricated sample or noise in the set-up. Also, there is slight frequency change in the case of TTSR and NBSR which may be due to fabrication tolerances.

Table 5.2 shows the comparison of magnetic resonance frequency f_{m0}, magnetic plasma frequency f_{mp}, negative permeability percentage bandwidth, maximum effective negative permeability and maximum losses for all the eight negative permeability structures. ICRR and LR structures have highest percentage bandwidth, while TTSR have the lowest percentage bandwidth and lowest losses. This behaviour is also observed in insertion loss (see figure 5.7) where ICRR and LR have wide stop-band while TTSR has narrow stop-band. USRR and BC-SRR have the highest realized negative permeability and highest losses while ICRR has lowest realized negative permeability (see figure 5.3). For the same resonant frequency, TTSR has the lowest dimension while ICRR has the highest dimension (see table 5.1).

Based upon the tabular data given in table 5.2, a designer may choose a suitable negative permeability metamaterial structure for desired application depending upon the priority which may be bandwidth or losses or realized negative permeability. For example, TTSR will provide highest selectivity while ICRR/LR will cover the wide range of RF/microwave frequencies.

Table 5.2. Performance comparison of eight negative permeability structures.

	Property					
Negative permeability structure	Magnetic resonance frequency f_{m0} (GHz)	Magnetic plasma frequency f_{mp} (GHz)	Bandwidth $\Delta f = f_{mp} - f_{m0}$ (GHz)	Percent (%) bandwidth	Maximum (μ_{reff})	($-j\mu_{reff}$)
SRR	9.36	10	0.64	6.61	−3.4	10.3
SSRR	11.09	12.32	1.23	10.51	−2.2	12.3
TTSR	8.54	8.88	0.34	**3.9**	−3.43	**8.75**
NBSR	9.59	10.25	0.66	6.65	−3.37	9.49
LR	8.51	10.2	1.69	**18.1**	−2.87	12.67
USRR	8.13	8.82	0.69	8.14	**−4.67**	**14.94**
BC-SRR	8.58	9.45	0.87	9.65	**−4.07**	**13.29**
ICRR	10.82	13.32	2.5	**20.71**	**−1.42**	11.63

5.3 Experimental characterization of metamaterial prism/wedge structure and plane-slab focussing structure

5.3.1 Introduction

In this section we will discuss how the free-space focused beam LHM characterization set-up of the type shown in figure 5.6 can be used for experimental characterization of a metamaterial sample (prism/wedge) and compare the experimental result thus obtained with analytical approach/simulation of the structure using an HFSS simulation tool [13]. Following this we will discuss how to characterize metamaterial plane-slab focussing which is known to act as a 'superlens'. The later characterization, however, uses a near-field characterization technique which will be clear from the details of the measurement technique to be discussed in section 5.3.3.

5.3.2 Characterization of metamaterial prism/wedge structure

The metamaterial prism was made out of the TW-LR structure that was shown in figure 2.1(a). The prism thus obtained is shown in figure 5.8(a), which is a 15° metamaterial prism. The experiment was carried out with a simple free-space measurement set-up with an improvised anechoic environment shown in figure 5.8(b). From the network analyzer output, see figure 5.8(c), with signal peak at 30.858 GHz a negative refractive index (NRI) of −1.89 GHz was obtained. The analytical result depicted in figure 5.8(d) shows a value of NRI to be −1.84 at 31.25 GHz [14].

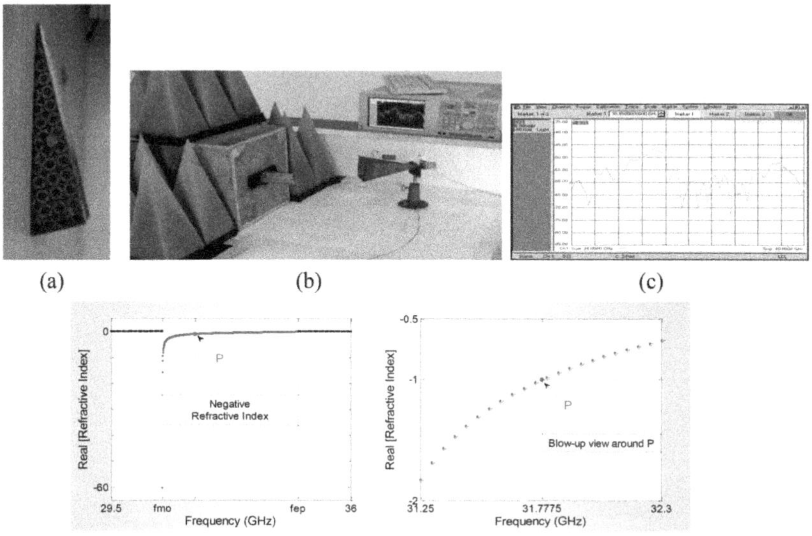

Figure 5.8. Characterization of TW-LR metamaterial prism. (a) The 15° prism. (b) Experimental set-up for measurement. (c) Network analyzer output showing signal peak at 30.858 GHz for angle of deviation 20° from normal. (d) Plot showing the analytical value of the real part of refractive index.

A 2D LR-CW structure; see figure 5.9(b), whose unit cell is shown in figure 5.9(a), was also used for experimental characterization and validated with simulation result [12, 15]. The LR and CW structures are printed on two separate dielectric substrates and are assembled in such a way that both substrates are back to back, having LR and CW structures on opposite faces. The wedge of combined LR and CW media was simulated in HFSS for which a refractive index of −1.68 was realized, while for the LR-CW LHM wedge when tested experimentally, a refractive index of −1.588 was obtained. The experimental set-up for characterization is shown in figure 5.9(c) and the network analyser display is shown in figure 5.9(d) in which peak signal was obtained at 9.899 GHz.

To find the amount of NRI in the LR-CW based LHM, a wedge has been simulated, as shown in figure 5.10, using 3D FEM solver of HFSS. The angle of wedge is 18.44° and it consists of a total of 35 unit cells. Only one unit cell of wedge is considered in x-direction and periodic boundary conditions are applied on yz-plane to intimate an infinite array. The wedge is finite in y- and z-directions. The whole structure is enclosed in a special box with radiation boundary conditions on its xy- and zx-planes. A positive z-direction propagating Gaussian beam is incident on left face of wedge at angle of incidence 0°. The positive or negative refraction will take place at right edge of wedge and the RHM and LHM regions are separated by

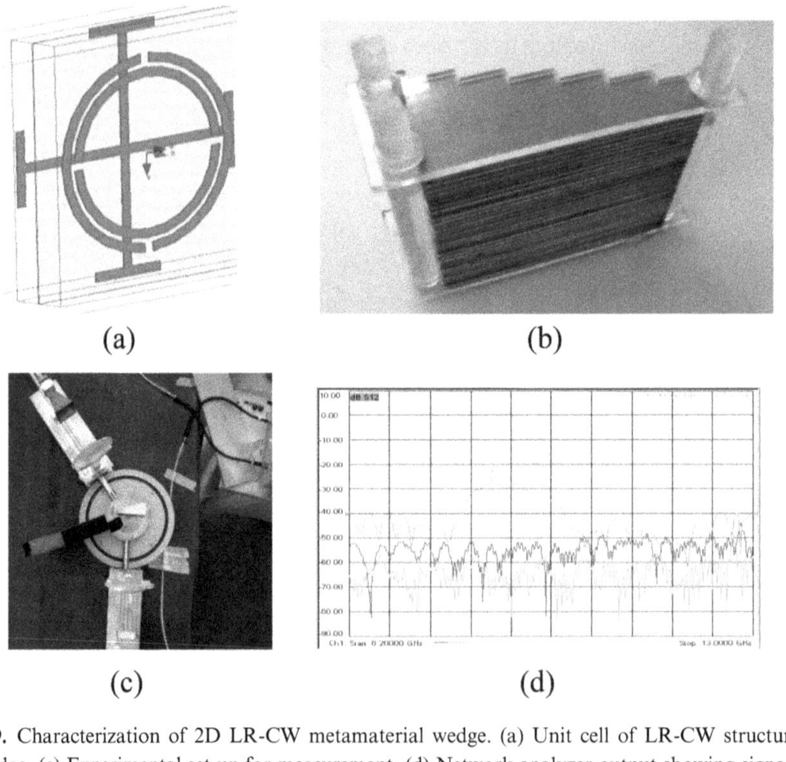

Figure 5.9. Characterization of 2D LR-CW metamaterial wedge. (a) Unit cell of LR-CW structure. (b) A 18.44° wedge. (c) Experimental set-up for measurement. (d) Network analyzer output showing signal peak at 9.899 GHz for angle of deviation 32° from normal.

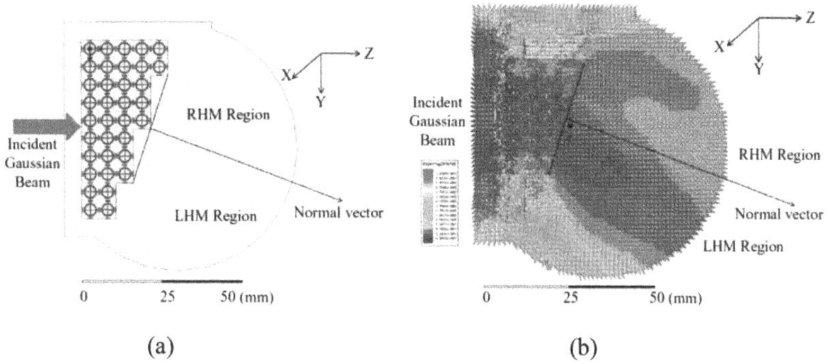

Figure 5.10. (a) Simulation methodology to find negative refraction in LR-CW metamaterial. (b) Simulated Poynting vector in LR-CW metamaterial wedge at 9.899 GHz.

normal vector. The Snell's law is used to find out the refractive index given by: $n = \sin(\delta)/\sin(A)$; where A is the angle of incidence which is 18.44° for the sample, a fixed value due to wedge shape and δ is the angle of refraction or deviation. The simulation results of the LR-CW metamaterial (LHM) at 9.899 GHz are shown in figure 5.10(b). A good amount of power is observed in the LHM region as compared to the RHM region which confirms the negative refraction in this LHM wedge. The angle of deviation from normal is 32° and hence the refractive index n is calculated to be −1.68.

In measurements, instead of single frequency, whole LHM pass-band is observed where the power received is sufficient after deviation. So, the NRI obtained in experiments is an averaged value. To observe the full profile of refractive index with frequency, one has to divide LHM pass-band into small frequency bands and then find out angle of refraction for each band separately. The final angle of deviation for LR-CW LHM wedge is 19.50°. The refractive index corresponding to this angle of deviation is −1.055. The received power at angle of deviation 19.50° is shown in figure 5.9(d). It can be observed that the power is maximum in LHM pass-band, i.e. from 10.45 GHz to 11.74 GHz. The received power is moderate below LHM pass-band because it is coming after diffraction in sample without any loss, whereas the power received in LHM pass-band region is coming after the losses in LHM structure.

5.3.3 Metamaterial plane-slab focussing characterization

It was mentioned in chapter 1 that focussing of electromagnetic signal with a plane-slab made of LH material is a near-field phenomena which is based on the so-called 'evanescent wave growth' by LH material providing very high resolution image of the object overcoming the diffraction limit of conventional optics. However, this focussing phenomenon is absolutely different from the far-field focussing we are familiar with in a convex lens made of RH or natural material. But it was also mentioned in section 4.3.4.1 of chapter 4 that using a hyperlens made of LH material in conjunction with an RH convex lens far-field focussing is also possible even at the optical frequency range [16]. But it may be noted that, owing to the use of curved surfaces in cylindrical hyperlens, such superlens may not be convenient for bio-imaging. Design of the planar hyperlens has been shown to be theoretically feasible via specific material dispersion based on transformation optics [17]. In such designs, the metamaterial properties are designed to bend light rays from sub-wavelength features in such a way as to form a resolvable image on a flat output plane.

Anyway, before we discuss the experimental characterization of focussing by plane metamaterial slab let us quickly recapitulate a few basic concepts related to focussing by lens in optics. An electromagnetic signal source has what are called the 'near-field', or evanescent wave components in addition to the 'far-field', or propagating wave components of the signal. The quasi-static near-field components decay exponentially with distance from the source (within approximately a quarter of a wavelength of the source signal frequency) which means that the final image with natural (i.e. RH) material lens always contain less information than is contained in the source (i.e. devoid of finer details or sub-wavelength details of the object carried by the evanescent waves). This 'diffraction limit'—which is associated with all positive-index optical components—means that the best resolution that is possible is limited to the extent of a wavelength of the incident signal (provided by the propagating wave) that is used to form the image of the object.

In 2000, John Pendry of Imperial College, UK, found that in addition to refocussing the far-field propagating components, a metamaterial plane-slab lens could also refocus the near-field components. However, this takes place due to an exponentially growing phenomenon of the evanescent wave, i.e. the near-field component of the incident wave as discussed in section 1.5.2 of chapter 1. In principle, such a lens would provide 'perfect image' reconstruction, prompting Pendry to dub the $n = -1$ plane-slab LHM lens as a 'perfect lens'. It may be noted that the imaging resolution of Pendry's perfect lens is not limited by diffraction limit, however, the losses in the metamaterial plane-slab diffuse the image formation, as discussed in section 1.7.2 of chapter 1.

Pendry's prediction proved unsettling for many, and sparked a vigorous debate. Several researchers drew attention to apparent conflicts with known physical limitations, such as energy conservation or the uncertainty principle. As remarkable as Pendry's prediction might seem, however, the prospects of a lens that can beat the diffraction limit have ultimately survived those challenges. Although limited to the near-field zone only, this kind of near-field superlens (NFSL) still enables many interesting applications including those of biomedical imaging and sub-wavelength photolithography. A slab of silver illuminated at its surface-plasmon resonance frequency is a good candidate for such an NFSL. Experiments with silver slabs have shown growth of evanescent waves [18] and imaging beyond the diffraction limit [19–21]. A near-field superlens operating in the mid-infrared part of the spectrum has also been demonstrated [22]. The NFSL based on bulk metal operates only at one frequency, i.e. narrow band. However, it has been shown that using metal–dielectric composites instead of bulk metals, one can develop an NFSL operating at any desired visible or near-infrared wavelength, with the frequency controlled by metal filling factor of the composite [23].

As mentioned earlier also, the prospect of having supermicroscope that can be built with plane metamaterial slab acting as perfect lens is enormous to which the scientists are inherently drawn that was viewed as a flat-out impossibility, some two decades back. Such a supermicroscope might one day make it possible to view vanishingly small objects such as viruses and individual strands of DNA. Such a tool could turbocharge biological research, advance computer chip manufacture, revolutionize education, and usher in an age of near-magical nanotechnology. In fact, we may quote Pendry: '*Fundamental physics sets no limit*' and he believes that the perfect lens he conceived will get steadily more powerful with applications that were never dreamt of. However, Pendry still remembers the day he realized that this new kind of lens would allow people to see invisibly small objects. In his own words: '*I was astonished and, frankly, frightened to publish the conclusion,*' he says, '*because I knew that everyone's reaction would be that it could not be true. I can tell you that I checked the result very many times!*'

One might ask what is so very special about a metamaterial plane-slab based supermicroscope? We have very sophisticated electron microscopes and x-ray diffraction—but these are like canes for blind people. To look at something with an electron microscope, for instance, one often have to coat the object to be observed with metal and place it in vacuum. Since this kills living things outright, it imposes serious limits on biological studies. We have no way to see living viruses infect cells, or to observe the interactions of proteins, or to watch DNA make a transcript of its biomolecular assembly instruction. Each year scientists are losing direct and dynamic visualization of such micro/nano phenomena—making the landscape of their studies go black. But a supermicroscope based on metamaterial plane-slab has the promise to aid in all these studies in their original dynamic platform of biological systems—this will definitely be a landmark feat for biological studies for the future.

With these few words about the promising possibility of LHM plane-slab let us now come back to our perspective discussion for the experimental characterization of focussing by LHM NRI plane slab. The schematic of experimental characterization is shown in figure 5.11 [4], which is similar to the one used by earlier researchers [24] for characterizing Veselago–Pendry type transmission-line (TL) metamaterial superlens.

The measurement apparatus consists of a source loop antenna and detector loop antenna connected to a network analyzer (not shown in the figure). The antennas may be constructed from semi-rigid 50 Ω microwave coaxial cable. It has a shielded topology that provides magnetic-dipole-type fields while simultaneously minimizing unwanted radiation from unbalanced currents on the coaxial feeding structure. The source (illuminating) antenna is placed at a distance of $t/2$ (t is the thickness of the NRI-TL lens) from the front face of the lens (on the left side in the figure), and the detector loop antenna, also placed $t/2$ from the back face of the plane-slab lens (on the right side in the figure), can be affixed to a computer-controlled *xyz*-translator and scanned behind the lens for measuring the field magnitude and phase distribution. The measurement apparatus is suggestive to be covered in an enclosure with microwave absorber. The noise in the measurement may be reduced by averaging at least 20–30 measurement data. The detector on receiver side is expected to have scanning capability

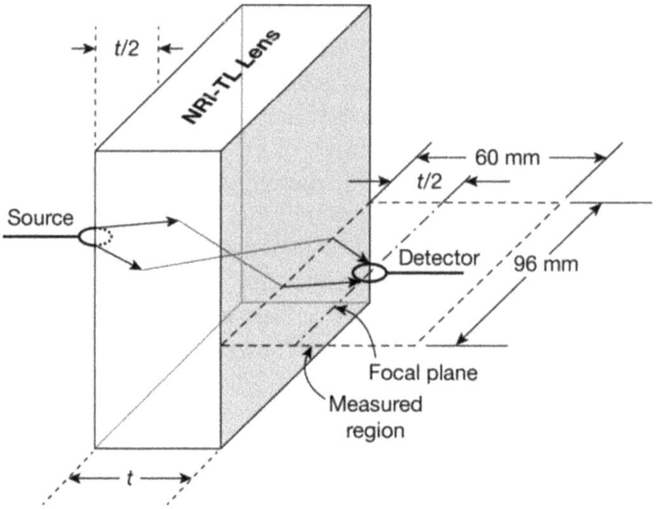

Figure 5.11. A scheme for experimental characterization of metamaterial plane-slab (NRI-TL lens) focussing.

in the area of approximately 60 mm by 96 mm to find out the focused image. However, the scanning area will depend on the overall sample size.

The researchers of the Toronto team [24], working with an NRI-TL plane-slab lens operating at 1.057 GHz, mapped the electromagnetic fields both within the slab and on the other side (image plane). They found that the recovered image does indeed have a resolution that is better than that which the diffraction limit implies—in agreement with Pendry's prediction. However, the image was found to be still broader than the source, which meant that it was not perfect. This was possibly due to the losses in the TL metamaterial, which place a limit on how well the near-fields can be recovered. Anyway, the focussing with NRI-TL metamaterial by Toronto group implied that the diffraction limit—which is the most fundamental limitation to image resolution in conventional optics —may, in fact, be circumvented by a plane-slab lens made of negative-index material.

References

[1] Smith D R and Pendry J B 2006 Homogenization of metamaterials by field averaging *J. Opt. Soc. Am.* B **23** 391–403

[2] Nicholson A M and Ross G F 1970 Measurement of the intrinsic properties of materials by time domain techniques *IEEE Trans. Instrumen. Meas.* **19** 377–82

[3] Smith D R, Schultz S, Markos P and Soukoulis C M 2002 Determination of effective permittivity and permeability of metamaterials from reflection and transmission coefficients *Phys. Rev.* B **65** 195104–8

[4] Majumder A, Das S, Kar S, (Guide-cum-Consultant) and Kumar A 2013 Technical Report on 'Left Handed Maxwell's Systems—Experimental Studies *Report number: Report on Project No.* 2009/34/37/BRNS/3176/ Dated-12-2-2010 https://researchgate.net/publication/324889314_Technical_Report_on_Left_Handed_Maxwell's_Systems_-Experimental_Studies/figures

[5] Kadaba P K 1984 Simultaneous measurement of complex permittivity and permeability in the millimeter region by a frequency-domain technique *IEEE Trans. Instrum. Meas.* **33** 336–40

[6] Ghodgaonkar D K, Varadan V V and Varadan V K 1990 Free-space measurement of complex permittivity and complex permeability of magnetic materials at microwave frequencies *IEEE Trans. Instrum. Meas.* **39** 387–94

[7] Tamyis N, Ramli A and Ghodgaonkar D K 2002 Free space measurement of complex permittivity and complex permeability of magnetic materials using open circuit and short circuit method at microwave frequencies Student Conf. on Research and Development (*SCOReD)* 394–8

[8] Deshpande M and Shin J 2005 Characterization of meta-materials using computational electromagnetic methods Proc. IEEE ACES Int. Conf. on Wireless Communications and Applied Computational Electromagnetics *(Honolulu, HI, 3–7 April 2005)* 421–4

[9] Kar S 2022 *Microwave Engineering—Fundamentals, Design, and Applications* 2nd edn (Hyderabad: Universities Press)

[10] Chen X T, Grzegorczyk T M, Wu B I, Pacheco J Jr and Kong J U 2004 Robust method to retrieve the constitutive effective parameters for metamaterials *Phys. Rev.* E **70** 0166081

[11] Smith D R, Vier D C, Koschny T and Soukoulis C M 2005 Electromagnetic parameter retrieval from inhomogeneous metamaterials *Phys. Rev.* E **71** 036617

[12] Kumar A, Majumder A, Das S and Kar S 2013 Simulation based characterization of negative permeability plasmonic structures at X band *Science and Information Conf. (London, October 7–9, 2013)* 675–9 https://ieeexplore.ieee.org/xpl/conhome/6653326/proceeding

[13] Ansoft Corporation High Frequency Structure Simulator (Pittsburgh, PA: Ansoft Corporation) www.hfss.com

[14] Kar S 2012 Advances in metamaterial research *99th Indian Science Congress (KIIT University, Bhubaneswar, 3–7 January 2012)*

[15] Kumar A, Majumder A and Kar S Simulation and experimental studies on novel cut-wire and LR based metamaterial (unpublished work, private communication)

[16] Liu Z, Lee H, Xiong Y, Sun C and Zhang X 2007 Far-field optical hyperlens magnifying sub-diffraction-limited objects *Science* **315** 1686

[17] Wang W, Xing H, Fang L, Liu Y, Ma J, Lin L, Wang C and Luo X 2008 Far-field imaging device: planar hyperlens with magnification using multi-layer metamaterial *Opt. Express* **16** 21142–8

[18] Liu Z, Fang N and Yen T-J 2003 Rapid growth of evanescent wave with a silver superlens *Appl. Phys. Lett.* **83** 5184–6

[19] Fang N, Lee H, Sun C and Zhang X 2005 Sub-diffraction limited optical imaging with silver superlens *Science* **308** 534–7

[20] Melville D O S, Blaikie R J and Wolf C R 2004 Submicron imaging with a planar silver lens *Appl.Phys. Lett.* **84,** 4403–5

[21] Melville D O S and Blaikie R J 2005 Super-resolution imaging through a planar silver layer *Opt. Express* **13** 2127–34

[22] Taubner T, Korobkin D, Urzhumov Y, Shvets G and Hillenbrand R 2006 Near-field microscopy through a SiC superlens *Science* **313** 1595

[23] Cai W, Genov D A and Shalaev V M 2005 Superlens based on metal–dielectric composites *Phys. Rev.* B **72** 193101

[24] Iyer A K and Elefteriades G V 2009 Free-space imaging beyond the diffraction limit using Veselago–Pendry transmission-line metamaterial superlens *IEEE Trans. Antennas Propag.* **57** 1720–7

Chapter 6

Metasurfaces—fundamentals, physics, modelling, and applications

The clam surface of an ocean is the signature of its depth

6.1 Introduction

As mentioned earlier, metamaterials are artificially engineered metal–dielectric composite materials in which a set of small scatterers (like SRR or TW) or apertures (like CSRR) are arranged in a regular (periodic) array, see figure 6.1(a), thus exhibiting some bulk electromagnetic behaviour that is not normally found in natural materials (like negative refractive index and so forth). It has also been observed in previous chapters that such composite media with sub-wavelength scale repetitive features (motifs) interact with incident electromagnetic waves and can be described by a set of homogenized material parameters. What differentiates metamaterials from ordinary composite media is that, with a careful design of motifs and their arrangement, new and tailored resonances are introduced to the system and the effective material properties such as electric permittivity and magnetic permeability can become extraordinary, attaining values that are very different from those of the constituent materials or any other material readily available in Nature. Over the last two decades, metamaterials have moved from being simply a theoretical concept to a thriving field in research and development leading to marketed applications in antennas, passive and active components and so forth (discussed in chapter 3).

Instead of being three-dimensional (3D) periodic structures like metamaterials, if the electrically small scatterers or holes are arranged into a planar, i.e. 2D pattern, in a geometrical sense, at a surface or interface—we get the *metasurface*, see figure 6.1(b). In fact, any geometrically 2D structure in which scatterers are periodically arranged and the thickness and periodicity of scatterers are small

doi:10.1088/978-0-7503-5532-2ch6

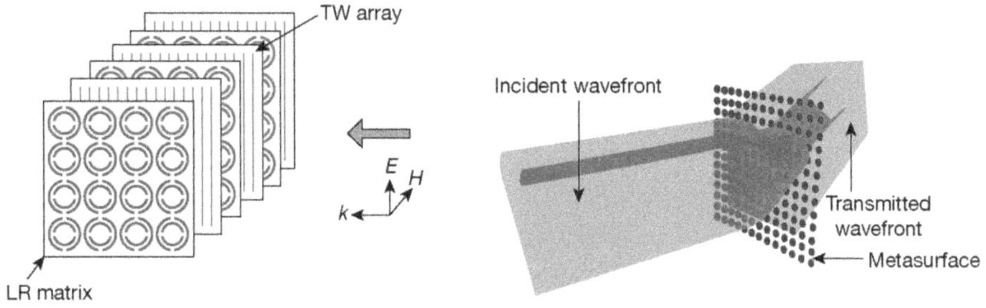

Figure 6.1. Metamaterial versus metasurface. (a) 1D periodic structure of metamaterial. (b) Metasurface showing wavefront transformation.

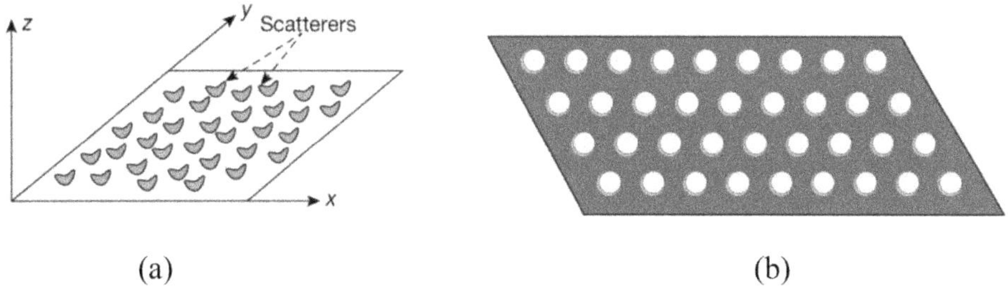

Figure 6.2. Schematic of two general class of metasurfaces: (a) metafilm and (b) metascreen.

compared to a wavelength in the surrounding media—we call it a metasurface. Within this general designation, we may identify two important subclasses of metasurface. Metasurfaces that have a 'cermet' topology—which refers to an array of isolated (non-touching) scatterers (figure 6.2(a))—are called *metafilms*, a term coined for such artificial surfaces. Metasurfaces with a fishnet structure, as shown in figure 6.2(b), will be called *metascreens*. These are characterized by periodically spaced apertures in an otherwise relatively impenetrable surface. The different types of metasurfaces that are used in practice lie somewhere between these two extremes. These 2D planar structures have the advantage of compactness, less loss and compatibility with planar fabrication technology. In recent times, the metasurfaces have had a wide range of potential applications in designing absorbers, controllable smart surfaces, novel wave-guiding structures, angular-independent surfaces, bio-medical devices, THz switches, and so forth.

6.2 Metasurface versus metamaterial

We are now quite familiar with metamaterials composed of periodic sub-wavelength metal/dielectric structures (i.e., meta-atoms) that resonantly couple to the electric or magnetic or both components of the incident electromagnetic fields, exhibiting frequency-dependent effective negative permittivity ($\varepsilon_{\text{reff}}$) and/or effective negative permeability (μ_{reff}). All the functionalities of metamaterials are controlled by the two negative constitutive parameters ($\varepsilon_{\text{reff}}$ and μ_{reff}) of the bulk material. This class of micro- and nano-structured artificial media, producing the desirable electromagnetic response and device functionalities by structural engineering, have received very active R&D efforts during the first decade of this century yielding groundbreaking electromagnetic and photonic phenomena with many practically usable and exotically possible applications. The initial overwhelming interest in metamaterials was centred on the realization of simultaneously negative electric and magnetic responses and, thereby, negative refractive index [1, 2], which can be used to accomplish super-resolution optical imaging [3, 4]. Scientists even speak of the possibility of super-lenses that would enable one to see objects down to nanometre dimension (a nanometre is a billionth of a metre) capable of visualizing even the individual strands of DNA. The capability of tailoring inhomogeneous and anisotropic refractive index resulted in electromagnetic invisibility cloak [5–7], another hallmark accomplishment using metamaterials.

These promising potential applications are, however, hindered in practice due to the high losses and strong dispersion associated with the resonant responses and the use of metallic structures in the design of metamaterials (dielectric metamaterials are also there with low loss), especially at optical frequencies. Another challenge in metamaterials is the difficult micro- and nanofabrication techniques required for the design of truly isotropic 3D-periodic structures as permittivity, permeability, and refractive index are essentially properties of bulk materials with which the metamaterial functionalities are primarily realized.

The so-called planar or 2D-geometrical (and not periodic) version of metamaterials, called metasurface, consisting of an ultrathin planar and periodic arrangement of sub-wavelength-size building blocks of metallic patches or dielectric etchings, however, can be readily fabricated using existing technologies such as lithography, nanoprinting and other nanofabrication techniques of today, driving many metamaterial researchers to focus on single-layer or few-layer stacks of planar structures (metasurfaces) that are more accessible, particularly in the optical regime. Such ultrathin structures have the unique ability to manipulate electromagnetic waves, with spatially arranged meta-atoms—fundamental building blocks of the metasurface—thereby blocking, absorbing, concentrating, dispersing or guiding the waves, from microwave through THz to optical frequencies. Like metamaterials, they are also capable of providing intriguing properties and varied applications, including far complex wave-manipulation properties. Unlike metamaterials, which need to be 3D-periodic isotropic structure, metasurfaces being planar in structure (2D-geometrical) can be easily fabricated using the planar fabrication tools. The planar fabrication process is cost effective, and being 2D-geometrical structure, the metasurfaces can be easily integrated into other devices (passive and active), which can make them a salient feature for nanophotonic circuits. Because the sub-wavelength thickness introduces minimal propagation phase, the effective permittivity, permeability and refractive index are of less interest in metasurfaces. In contrast, the significant importance is to be given to the surface or interface reflection and transmission resulting from the tailored surface impedance, including their amplitude, phase, and polarization states. Thus metamaterials are primarily understood to be constitutive property (permittivity/permeability) controlled while metasurface is boundary condition controlled. The ultrathin thickness in the wave propagation direction can greatly suppress the undesirable losses by using appropriately chosen materials and metasurface structures. Overall, metasurfaces can overcome the challenges encountered in bulk metamaterials while their interactions with the incident waves can be still sufficiently strong to obtain very useful functionalities. For this reason, metasurfaces have started dominating the general field of meta-research during the last decade or so, given their high potential in real world applications.

Metasurfaces can be engineered with spatially varying boundary conditions to convert a given incident electromagnetic field into a desired scattered waveform, resulting in wavefront transformations. Using novel configurations of meta-atoms, the incoming plane waves can be deflected to preferable directions, manipulate their polarization state, and generate special beams. It is important to note that a metasurface can efficiently manipulate the wavefront of an incident electromagnetic wave through just the sub-wavelength propagation distance compared to the traditional 3D metamaterial that does the same at a distance far larger than the wavelength. Therefore, this can largely alleviate the propagation loss. The controllable surface refractive index provided by metasurfaces can also be applied to lenses. In fact, they can be used to design 2D microwave/optical lenses like Luneburg and fish-eye lenses, which are applied in surface waveguides for antenna systems and planar microwave sources. Reconfigurable ultrathin surfaces provided by metasurface, resulting in wavefront transformations to the impinging waves to any imaginable degree, may form the basis for smart surfaces with a significant impact on nearly any electromagnetic and photonic application, from classical to quantum photonics, from radar to wireless technology. As metasurfaces comprise a rapidly growing field of research, there have been a few good review articles focussing on different aspects of it [8–13].

The merit of metasurfaces is the technological simplification: metasurfaces allow one to use a single-digital pattern (i.e., one lithographic mask) to create an arbitrary analogue optical phase response (e.g., amplitude, phase, polarization, and optical impedance). So far most of the metasurface structures are fabricated using electron-beam lithography, but manufacturing techniques for large-scale patterning of planar structures are available, such as nano-imprint lithography, soft lithography, and deep-UV lithography.

It was mentioned earlier too that the main advantages of metasurfaces with respect to the existing conventional technology of bulky metamaterial include their compact size, low cost, low level of absorption, and easy integration due to their thin profile. Because of these advantages, they are promising candidates for real-world solutions to overcome the challenges posed by the next generation of transmitters and receivers of future high-rate communication systems that require highly precise and efficient antennas, sensors, active components, filters, and integrated technologies. Although many of the early demonstrations of metasurfaces were in the microwave regime, due either to the reduced cost of manufacturing and testing or to satisfy the interest of the communications or aerospace industries, the real potential use of metasurfaces was subsequently found to be in the optical regime. In fact, the stupendous progress of photonic meta-research is now being felt to be due to the progress of the 2D or planar metamaterials, i.e. the metasurfaces with the advent of the advanced nanotechnology of today.

6.3 Frequency selective surface (FSS) versus metasurface

Metasurfaces resemble frequency selective surfaces (FSSs) in many respects, but they can also replace FSSs in many applications due to the sub-wavelength nature of 'meta-atoms' of which they are made. An FSS is a periodic structure made of composite material (or sometimes with only dielectric material) and designed to be transparent in some frequency bands while reflecting, absorbing or re-directing the incident signal [14]. FSSs typically have periodicity which is of the order of a wavelength (typically a half wavelength) of the resonant frequency. But the periodicity of individual elements of metasurface is of sub-wavelength order, so are the individual elements, i.e. the fundamental motif size of metasurface. Compared to FSSs, the sub-wavelength periodicity of the metasurface allows packing a large number of unit cells in a constrained space, which is highly useful for radome design and in many other applications demanding miniaturization due to space constraint. Further, it may be mentioned that an FSS is used, in general, for the specific purpose of frequency filtering. To perform this operation, the surface only needs to excite an electric *or* a magnetic response. Due to this, an FSS usually has limited control over EM wave propagation. But a metasurface can provide a more extensive control over EM wave propagation by exciting *both* electric *and* magnetic responses. A metasurface can provide controlled refraction, reflection, filtering, polarization control and so forth. These functions cannot be performed by an ordinary FSS. Thus one can think of a metasurface as a functional extension of a frequency selective surface.

To perform frequency selectivity (in simple terms the 'filtering') with FSS, see figure 6.3(a), power must be absorbed and/or redirected.

Devices are made absorptive by incorporating lossy materials. A 'Salisbury screen', see figure 6.3(b), was one of the first concepts for frequency selective surfaces and was used by the military to make military vehicles 'invisible' to radar. At the frequency in which the device is resonant, energy is absorbed in the lossy material.

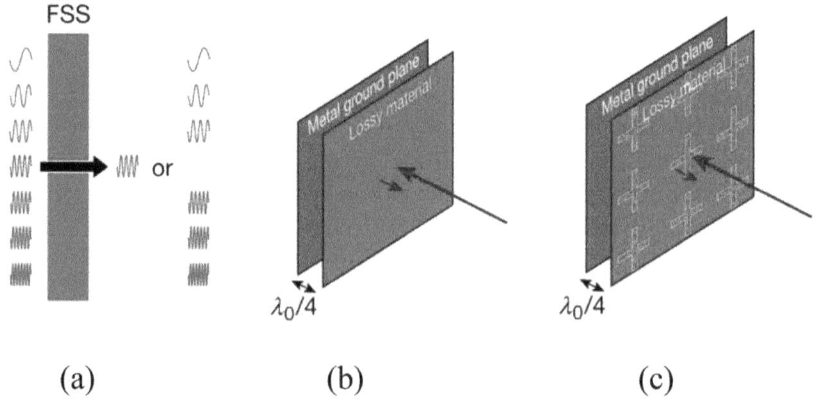

Figure 6.3. (a) FSS showing frequency selectivity i.e. filtering. (b) A Salisbury screen. (c) Salisbury screen with embedded resonant structure (circuit analogue absorber).

The absorption can also be amplified by incorporating some sort of resonant structures. A 'circuit analogue absorber', see figure 6.3(c), is like a Salisbury screen, but it incorporates periodic structures that amplify the absorption by enhancing the resonance.

Power can be redirected with FSS by using interference and diffraction. This can be simple reflection, diffraction from a grating or through a guided-mode resonance. In the technique shown in figure 6.4(a) (longitudinal resonance), the beam is incident from the top and partially reflects from each of the surfaces. It is the overall interference of the scattered waves that produces the frequency selectivity. In the technique in figure 6.4(b) (transverse resonance), the incident wave falls on a grating that scatters it into multiple directions. Frequency selectivity is produced by the inherent frequency dependence of scattering from a grating. In another technique, see figure 6.4(c) (diffractive), an external wave is coupled into a guided mode or surface wave. The guided-mode slowly leaks from the guide due to the grating. It is the interference between the applied wave and the 'leaked' wave that produces the frequency selectivity.

In recent times, all-dielectric FSS have also becoming popular for certain applications. This is because metals can be lossy, especially at optical frequencies. Dielectric FSSs can operate more safely in high-voltage environments and may be suited for high power applications. They can be easily adapted in monolithic designs too.

It may be recalled that both FSS and metasurface are 2D (geometrical and not periodic; metamaterial being 1D, 2D, 3D periodic) structures; however, the individual scattering elements in both FSS and metasurface are arranged with some periodicity—the periodicity being of the order of wavelength in the case of FSS while it is sub-wavelength (element size and periodicity) in the case of metasurface. Unlike FSS, whose functional nature is briefly mentioned above, the metasurface primarily functions with modified boundary conditions (unlike metamaterials that function with effective constitutive properties ε and μ). From an application point of view, the FSSs have been frequently employed to provide spectral filtering in signal communications. FSSs help in operating antennas in their desired frequency range by selectively transmitting or reflecting the electromagnetic waves by using the frequency selectivity property of FSS. FSSs are used in radomes where the FSS

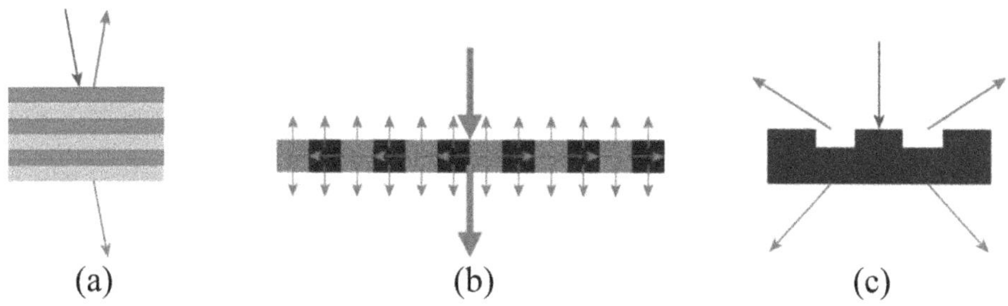

Figure 6.4. Power redirection with FSS: (a) longitudinal resonance, (b) transverse resonance, and (c) diffractive.

structure helps to reflect all the undesired frequencies and only pass the desired frequency. Another major application of FSS is in stealth technology, which is used to make aircraft or warships less visible to the radar signal by reducing the radar cross-sectional area of the aircraft. In common use, FSS can be found at the front door of a microwave oven for preventing the harmful microwaves from coming out of the oven but at the same time allowing visible light to go into the oven for visual purposes. On the other hand, metasurface has the special applications and varied use in transformation optics typically in tailorable wavefront shaping. In other words, it can arbitrarily control the following as a function of position: amplitude, angles (i.e. phase), frequency (nonlinear), reciprocity (nonlinear, anisotropic, etc) including polarization conversion and many more demanded by the emerging applications in photonic engineering of today.

6.4 Various physical aspects of metasurfaces related to application viewpoints of transformation optics

6.4.1 Introduction

Metasurfaces have emerged in recent years as a powerful tool to realize newer features of transformation optics that can control the fabric of 'electromagnetic space' (and thus light propagation) which are not feasible with conventional homogenous optical metamaterial. They act as a platform to design sub-wavelength-thick optical components ('flat optics'), which can be used to implement any optical function (beam deflection, focussing, wave plates, and so forth). These flat optical components can be fabricated using a single lithographic step. Metasurfaces make it possible to manipulate all properties of light (amplitude, phase, and polarization), which enable us to build a large variety of flat optical components, including planar lenses, quarter-wave plates, holograms for vector beam generation, including ultrathin perfect absorbers and so forth.

Here we will try to provide, from the viewpoints of diverse theoretical backgrounds, the physics behind the extraordinary properties exhibited by metasurfaces. This would shed light on the insight regarding various tailored optical phenomena that has brought phenomenal changes in the landscape of conventional optics leading to the possibilities of developing new optical functionalities useful for many applications of today that was not dreamt of even a decade ago. Though such phenomena, in principle, are applicable at lower electromagnetic frequencies like microwaves and THz too but optical applications are more demanding—thus our discussion, in general, will be centred around optical frequency regime.

Conventional optical components are based on reflection, refraction, absorption, and/or diffraction of light, and light manipulation is achieved via propagation through media of given refractive indices, which can be engineered to control the optical path of light beams. In this way phase, amplitude, and polarization changes are accumulated through propagation in optical components based on refraction and reflection, such as lenses, wave plates, and optical modulators. Ancient people even used ice lenses to focus sunlight and start fires, one example of harnessing the light propagation.

For incidence of light in the interface between two media, the reflection and transmission coefficients and their directions, in conventional optics, are determined by the continuity of field components at the boundary, and are given by the Fresnel equations and Snell's law, respectively. If we add to the interface a periodic array of sub-wavelength resonators of negligible thickness forming a metasurface, the reflection and transmission coefficients will be then dramatically changed because the boundary conditions are modified by the resonant excitation of an effective current within the metasurface. Thus with the use of metasurface, one can effectively control the direction of wave propagation, at the designer's will, and the shape of wavefront including the state of polarization. Likewise, various other tailored optical phenomena have been realized with the use of metasurface that have been significantly enriching, every new morning, the photonic functionalities and device developments which we are going to discuss in the following.

6.4.2 Anomalous reflection and refraction and tailorable wavefront shaping with metasurfaces

6.4.2.1 Introduction

It has already been known to us that metamaterials can exhibit interesting transformation optics phenomena like bending of light in unusual ways leading to negative refraction, sub-wavelength focussing with superlens, and this excites our imagination to the possibility of realizing a 'magic device' like cloaking. Metasurfaces, or the so-called 2D metamaterials, have gone a step further resulting in the possibility of realising anomalous reflection and refraction phenomena with tailorable wavefront shaping. The shaping of the wavefront of light with conventional optical components such as lenses and prisms, as well as diffractive elements such as gratings and holograms relies on gradual phase changes accumulated along the optical path. But the phase discontinuities or the 'phase jumps' provided by metasurface have great flexibility in the design of light beams and hence wavefront shaping, which is totally different from conventional optics and whose possibility of applications is enormous [15]. We can directly engineer the phase, amplitude and polarization along the surface using optical resonators (which the individual scatterers of metasurface are) so that we can make a new class of metasurface-based optical components. In this sense, the refractive index is not a particularly useful quantity to characterize metasurfaces—as we are concerned here with the issue of 'surface property', i.e. with boundary conditions and not the bulk property of the bulk medium which we do with metamaterials.

6.4.2.2 Analytical background

The anomalous reflection and refraction and tailorable wavefront shaping with metasurface may be understood analytically [10, 15] with the Fermat's principle of optics. Fermat's principle states that the trajectory taken between two points A and B by a ray of light is that of the least optical path, $\int_A^B n(\vec{r})dr$ where $n(\vec{r})$ is the local index of refraction, and readily gives the laws of reflection and refraction between two media. In its most general form, Fermat's principle can be stated as the principle of stationary phase; that is, the derivative of the phase $\int_A^B d\phi(\vec{r})$ accumulated along the actual light path will be zero with respect to infinitesimal variations of the path. Now, the metasurface placed at the interface between the two media introduces a spatial distribution of abrupt phase shift $\phi(\vec{r}_s)$ or 'phase jumps' in the optical path between the two media over the scale of the wavelength; $\phi(\vec{r}_s)$ depends on the coordinate $\vec{r}_s$ along the interface, due to electromagnetic scattering at its constitutive scatterers. Thus, the actual optical path in the presence of these phase jumps should be stationary in the total accumulated optical phase: $\phi(\vec{r}_s) + \int_A^B \vec{k}.\, d\vec{r}$; $\vec{k}$ is the wave vector of the propagating light. This law of stationary phase ensures that wavelets starting from a source point with the same initial phase will arrive at the point of destination in-phase after reflecting from or transmitting through the metasurface, and thus constructively interfere, which makes the optical path a physical path of optical power. A set of generalized/modified/anomalous laws of reflection and refraction can be derived from the law of stationary phase as [15]:

The generalized law of reflection:

$$\begin{cases} \sin(\theta_r) - \sin(\theta_i) = \dfrac{1}{n_i k_0}\dfrac{d\Phi}{dx} \\ \cos(\theta_r)\sin(\phi_r) = \dfrac{1}{n_r k_0}\dfrac{d\Phi}{dy} \end{cases} \tag{6.1a}$$

The generalized Snell's law of refraction:

$$\begin{cases} n_t \sin(\theta_t) - n_i \sin(\theta_i) = \dfrac{1}{k_0}\dfrac{d\Phi}{dx} \\ \cos(\theta_t)\sin(\phi_t) = \dfrac{1}{n_t k_0}\dfrac{d\Phi}{dy} \end{cases} \tag{6.1b}$$

In the above expressions $d\Phi/dx$ and $d\Phi/dy$ are, respectively, the components of the phase gradient parallel and perpendicular to the plane of incidence. The definitions of the angles are shown in figure 6.5.

The generalized/anomalous laws of reflection and refraction given by equations (6.1a) and (6.1b) can be derived by considering the conservation of wave vector along the interface as indicated in figures 6.6(a)–(f). Figures 6.6(a) and (b) show the standard law of reflection and refraction, respectively. Figure 6.6(c) shows how the phase accumulates across surface while figure 6.6(d) shows what happens if the surface affects the phase. Figure 6.6(e) and figure 6.6(f), respectively, shows the modified (anomalous) law of reflection and modified law of refraction due to the presence of metasurface at the interface that causes the so-called 'phase jump'.

These generalized laws indicate that the reflected and transmitted (i.e. refracted) light beams can be bent in arbitrary directions, depending on the direction and magnitude of the interfacial phase gradient ($d\Phi/dx$), introduced with the presence of

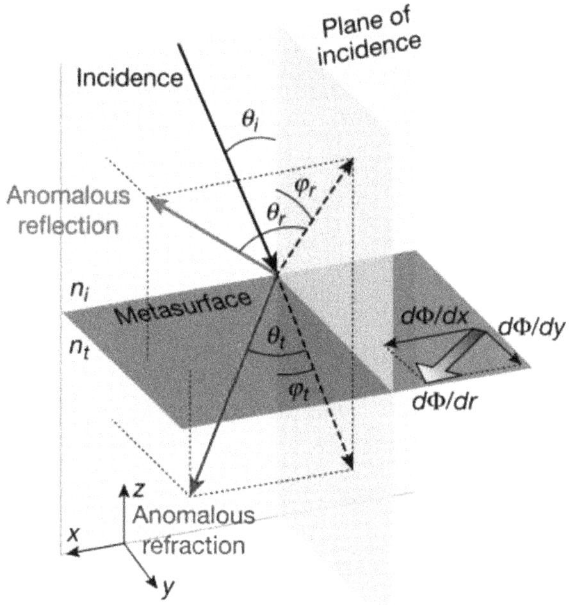

Figure 6.5. The schematic shows that an interfacial phase jump $d\Phi/dr$ provides an effective wave vector along the interface that can bent transmitted and reflected light into arbitrary directions resulting in anomalous refraction and reflection. (Reproduced/adapted with permission from [10], copyright IOP Publishing. All rights reserved.)

metasurface, as well as the refractive indices of the surrounding media. This is possible because in the modified reflection and refraction laws (see equations (6.1a) and (6.1b)) there is a nonlinear relation between θ_r and θ_i and also between θ_t and θ_i which is markedly different from what we observe in reflection and refraction phenomena of conventional optics.

In the above derivation, we have assumed that Φ is a continuous function of the position along the interface; thus, all the incident energy is transferred into the anomalous reflection and refraction. However, because experimentally we use an array of optically thin resonators with sub-wavelength separation to achieve the phase change along the interface, this discreteness implies that there are also regularly reflected and refracted beams, which follow conventional laws of reflection and refraction ($(d\Phi/dx) = 0$ in equations (6.1a) and (6.1b)). The separation between the resonators controls the amount of energy in the anomalously reflected and refracted beams. We have also assumed that the amplitudes of the scattered radiation by each resonator are identical, so that the reflected and refracted beams are plane waves.

It may be noted here that there is a fundamental difference between the anomalous refraction phenomena caused by phase discontinuities (introduced by metasurface) and those found in bulk designer metamaterials, which are caused by either negative dielectric permittivity and negative magnetic permeability or anisotropic dielectric permittivity with different signs of permittivity tensor components along and transverse to the surface [16].

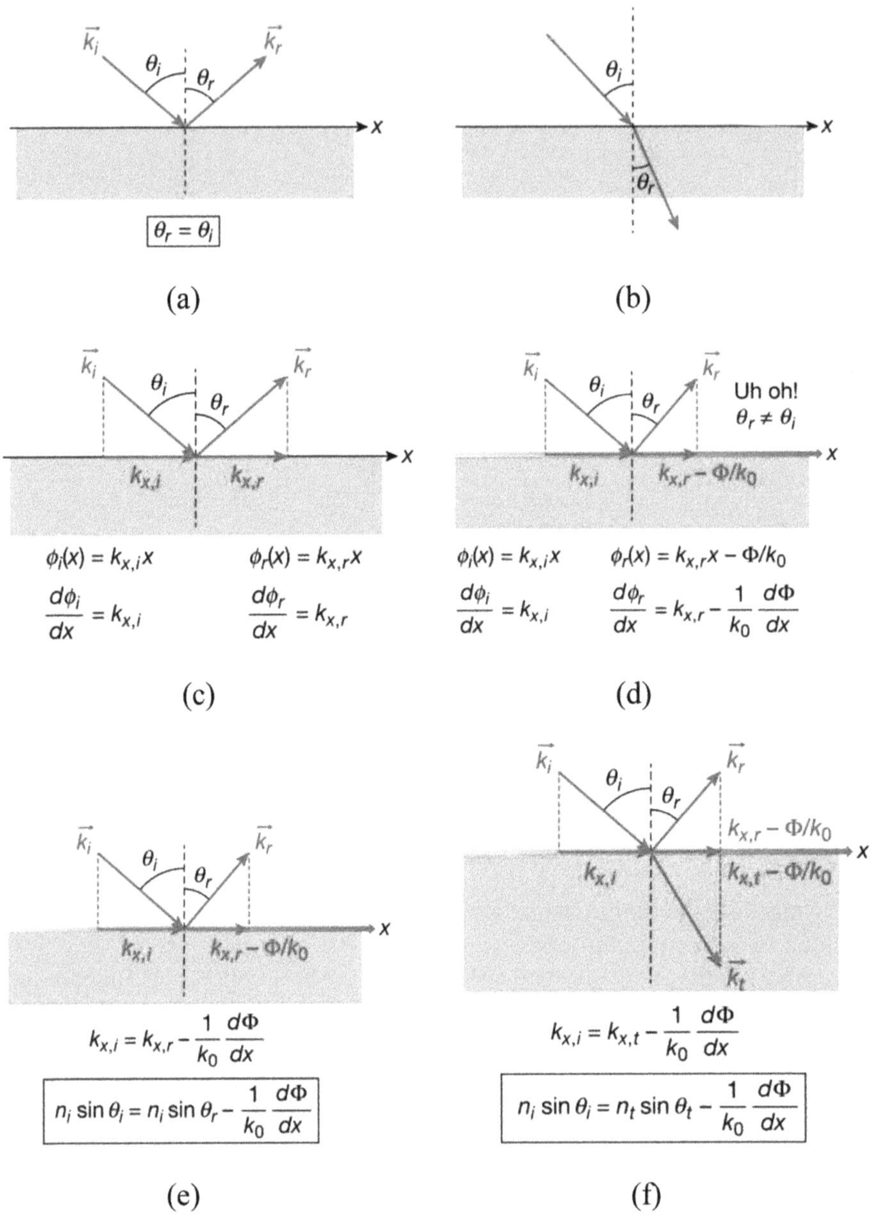

Figure 6.6. (a) Standard law of reflection. (b) Standard law of refraction. (c) Phase accumulation across the surface. (d) What happens when the surface affects the phase. (e) Modified (anomalous) law of reflection. (f) Modified law of refraction.

6.4.2.3 Practical realization of the metasurface to verify anomalous refraction and reflection phenomena

We have many choices to create abrupt optical phase shifts or the so-called 'phase jumps' that lead to anomalous refraction and reflection phenomena with metasurfaces. Any type of optical resonators can be used, including plasmonic antennas, dielectric resonators, quantum dots, and nanocrystals. The plasmonic antennas are commonly used because of their widely tailorable optical properties and the ease of fabricating planar antennas of nanoscale thickness. However, the optical resonators, that are commonly used, have to satisfy the following requirements: (i) they should have sub-wavelength thickness; (ii) they must have small footprints so that they can be packed together with sub-wavelength separations; (iii) their phase response should cover the entire 0 to 2π range so as to provide full control of the wavefront; and (iv) the scattering amplitude must be uniform and large across the array.

The phase shift between the emitted and the incident radiation of an optical resonator changes appreciably across a resonance. By spatially tailoring the geometry of the resonators in an array and hence their frequency response, one can design the phase discontinuity along the interface and mould the wavefront of the reflected and refracted beams practically in any arbitrary way.

The metasurface design is based on V-shaped optical antenna array as shown in figure 6.7(a) [15], where the upper panel depicts one unit cell of the fabricated structure and the lower panel reveals a schematic of the metasurface (Λ is the grating period). The V-shaped antenna consists of two arms of equal length h connected at one end at an angle Δ, and it has double-resonance property. V-antennas support 'symmetric' and 'antisymmetric' modes which are excited by electric-field components along $\hat{s}$ (unit vector along the symmetry axis of the antenna) and $\hat{a}$ (unit vector perpendicular to $\hat{s}$) axes, respectively, see figure 6.7(b). In figure 6.7(b) the V-antenna supporting symmetric and antisymmetric modes has been shown. The schematic current distribution is represented by colours on the antenna (blue for symmetric and red for antisymmetric mode), with brighter colour representing larger currents. The direction of current flow is indicated by arrows with colour gradient. In the

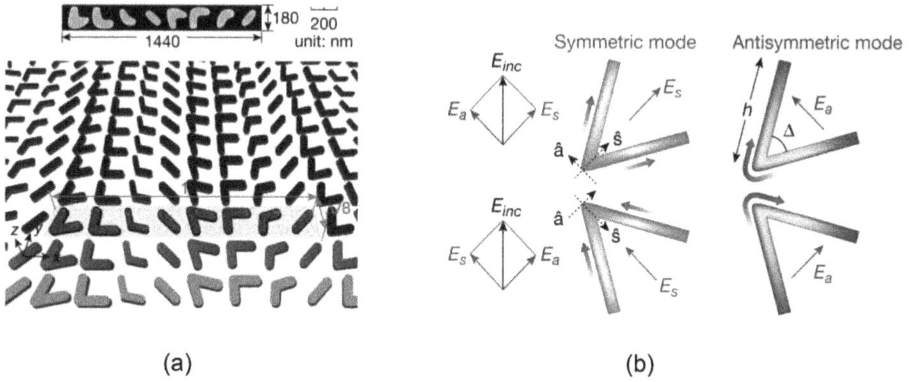

Figure 6.7. (a) A metasurface based on V-shaped antenna used to demonstrate generalized laws of reflection and refraction in the near-infrared. (b) A V-antenna supporting symmetric and antisymmetric modes. (Reproduced/adapted from [15], with permission from The American Association for Advancement of Science.)

symmetric mode, the current distribution in each arm approximates that of an individual straight antenna of length h and therefore the first-order antenna resonance occurs at $h \approx \lambda_{eff}/2$, where λ_{eff} is the effective wavelength. In the *antisymmetric mode*, the current distribution in each arm approximates that of one half of a straight antenna of length $2h$ and the condition for the first-order resonance of this mode is $2h \approx \lambda_{eff}/2$. The polarization of the scattered radiation is the same as that of the incident light when the latter is polarized along $\hat{s}$ or $\hat{a}$. For an arbitrary incident polarization, both antenna modes are excited but with substantially different amplitude and phase because of their distinctive resonance conditions. As a result, the scattered light can have a polarization different from that of the incident light. These modal properties of the V-antennas allow one to design the amplitude, phase, and polarization state of the scattered light. In the particular design, the incident polarization has been chosen to be at 45° with respect to $\hat{s}$ and $\hat{a}$ so that both the symmetric and antisymmetric modes can be excited and the scattered light has a substantial component polarized orthogonal to that of the incident light. Experimentally, this allows using a polarizer to decouple the scattered light from the excitation.

Generalized optical laws discussed analytically in the previous sub-section were first demonstrated using V-shaped optical antennas in the mid-infrared spectral range [15] and later confirmed in the near-infrared [17]. The antenna spacing in the array, see figure 6.7(a), is between one tenth and one fifth of the free-space wavelength. The metasurface creates anomalously refracted and reflected beams satisfying the generalized laws over a wide wavelength range, with negligible spurious beams. The broadband performance is due to the fact that the two eigenmodes supported by the V-antennas form a broad effective resonance over which the scattering efficiency is nearly constant and the phase response is approximately linear. The scalability of metasurface allows the extension of this concept to other frequency ranges, e.g., broadband anomalous refraction was also demonstrated at THz frequencies using C-shaped metallic resonators [18].

6.4.3 Polarization conversion

6.4.3.1 Introduction

Polarization state is an intrinsic property of electromagnetic waves, and the conversion between polarization states is very often highly desirable (or even necessary) for many modern electromagnetic and photonic applications. For instance, in advanced communication and sensing, converting linear polarization to circular polarization makes a beam resistant to environmental variation, scattering and diffraction. During recent years, conversion among polarization states using metasurfaces has attracted increasing interest due to their design flexibility and compactness. The accompanied capability of tuning a phase delay spanning the entire 2π range over a broad bandwidth and with a deep sub-wavelength resolution could potentially address some critical issues in the development of flat optics.

Highly symmetric simple meta-atoms can be advantageous in maintaining polarization states. Breaking the symmetry can, however, provide additional degrees of freedom to achieve customized functionality that enables the manipulation of polarization states. Through tailoring the two eigenmodes corresponding to orthogonal linear polarizations, it is possible to have equal transmission magnitude but a relative phase delay $\Delta\varphi$ at a specific frequency. Narrowband polarization conversions between linear and circular polarization states ($\Delta\varphi = \pi/2$, quarter-wave plates), or linear polarization rotation ($\Delta\varphi = \pi$, half-wave plates) have been realized using metasurfaces [19]. The efficiency is limited, in general, up to 50%; however, the low level of polarization conversion efficiency can be addressed by the implementation of few-layer metasurfaces.

6.4.3.2 Linear-to-circular polarization conversion

An antenna array backed with a ground plane has been widely exploited at microwave frequencies to enhance the radiation efficiency and beam directionality. This configuration also enhances the polarization conversion in reflection for anisotropic sub-wavelength metallic resonator arrays. Early work at microwave frequencies demonstrated that narrowband conversion to various polarization states, including linear-to-circular polarization and linear polarization completely perpendicular to the incident one, is possible depending on the structural parameters, incident angle, and frequency [20].

New device functionalities could be realized by controlling spatial distribution of polarization response using metasurfaces. Figure 6.8 shows a metasurface-based quarter-wave plate [21] that generates high-quality circularly polarized light (>0.97) over a broad wavelength range (λ = 5–12 μm). The unit cell of the metasurface comprises two subunits (coloured pink and green in the figure 6.8). Upon excitation by linearly polarized incident light, the subunits generate two co-propagating waves with equal amplitudes, orthogonal linear polarizations, and a $\pi/2$ phase difference (when offset $d = \Gamma/4$), which produce a circularly polarized anomalously refracted beam that bends away from the surface normal.

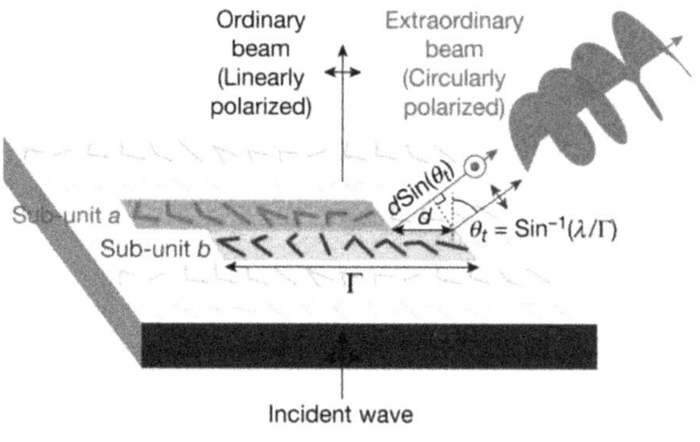

Figure 6.8. Schematic of a metasurface quarter-wave plate, with the unit cell of the metasurface comprising two subunits (pink and green). Each subunit contains eight V-antennas. (Reproduced/adapted with permission from [10], copyright IOP Publishing. All rights reserved.)

6.4.4 Dielectric metasurfaces

The majority of metasurface research has focused on using sub-wavelength metallic structures, where ohmic losses pose a severe problem, particularly in the optical frequency range, limiting the performance of arguably any desirable functions. Low-loss, high-refractive-index dielectric materials have received much attention during recent years in order to address the efficiency issue in metallic metasurfaces. On the other hand, the capability of tuning the magnetic and electric resonances through tailoring the geometric dimensions and spacing of dielectric resonators also facilitates metasurface functionalities beyond metallic metasurfaces.

Dielectric resonators (DRs) are well known to microwave and millimetre-wave engineers that are used for designing low phase-noise oscillators and compact antennas at millimetre-wave frequency regime [23]. DR consists of a high dielectric constant ceramic material which may be of various shapes—hemispherical, cylindrical, and rectangular; preferably cylindrical. Increasing the dielectric constant (ε_r) can significantly reduce the required size d of the resonators, which is related to the free-space resonant wavelength λ_0 by $d \approx \lambda_0 / \sqrt{\varepsilon_r}$. However, increasing the dielectric constant also reduces the radiation efficiency (when acting as an antenna) and narrows the operational bandwidth, which is inversely related to the dielectric constant. Typical values of the dielectric constant used range from 8 to 100 in order to balance the compactness, radiation efficiency and bandwidth requirements. Very often dielectric resonators are mounted on top of a metal ground plane, which improves the radiation efficiency and acts as electrical symmetry plane to improve the compactness.

In the optical regime, low-loss dielectric particles support strong electric and magnetic scattering known as Mie resonances. Multiple modes thus generated herewith are determined by the particle size and structural properties, in contrast to metallic particles where the resonance scattering is dominated by the electric resonances. In most dielectric resonators of regular shapes such as spheres, cubes, cylindrical disks and rods, the lowest resonant mode is the magnetic dipole resonance and the second lowest mode is the electric dipole resonance. The magnetic resonance mode originates from the excitation of circulating displacement currents, resulting in the strongest magnetic polarization at the centre, similar to the case of magnetic resonant response in metallic SRRs. The contribution from other higher order modes can be safely ignored as the coefficients of these modes are orders of magnitude lower. Sub-wavelength dielectric resonators can be used as the basic building blocks of metamaterials and metasurfaces. O'Brien and Pendry used such resonators to obtain magnetic activity in dielectric composites [24]. A class of Mie resonance-based dielectric metamaterials have been consequently demonstrated, with some early work reviewed in [25], where high dielectric constant materials are used to enable sub-wavelength resonators for the realization of negative electric and magnetic responses from microwave to optical frequencies with different dielectric materials for the DR [26].

The loss reduction enabled by dielectric metasurfaces becomes clear when functioning as a linear polarization rotator, as shown in figure 6.9, where an array of anisotropic (rectangular) silicon resonators is separated from a metal ground plane by a thin layer of PMMA (*Poly Methyl Methacrylate—a* synthetic resin

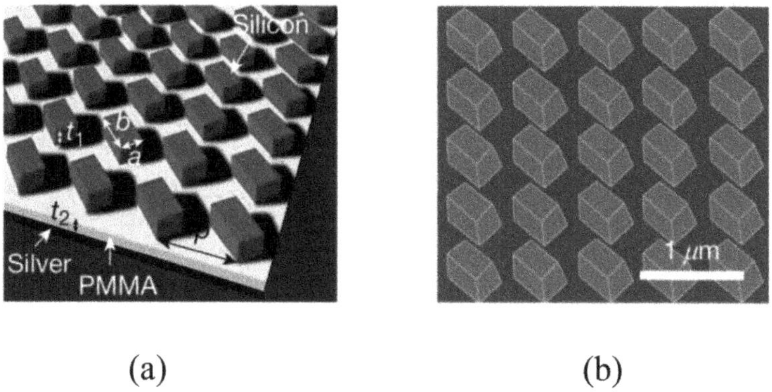

Figure 6.9. Dielectric metasurface for broadband polarization conversion in reflection. (a) Schematic and (b) SEM (Scanning Electron Microscope) image of the dielectric metasurface structure. (Reproduced/adapted with permission from [10], copyright IOP Publishing. All rights reserved.)

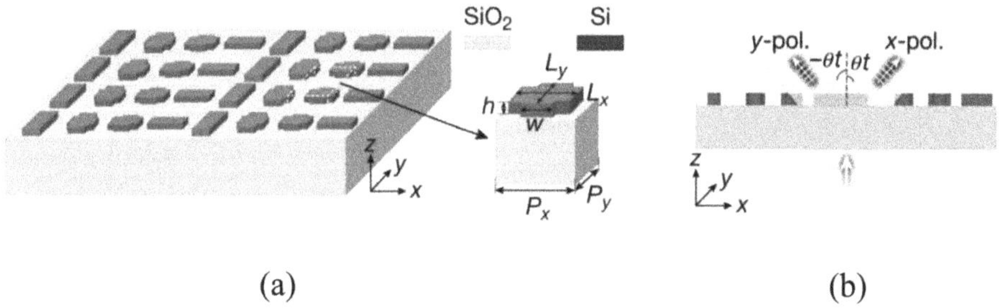

Figure 6.10. (a) The metasurface structure, and (b) polarization splitting. (Reproduced/adapted from [28], copyright (2019) with permission of Springer Nature.)

produced from the polymerization of methyl methacrylate which is a transparent and rigid plastic). In experiments, linear polarization conversion with more than 98% conversion efficiency was demonstrated over a 200 nm bandwidth in the short wavelength infrared band [27]. This result exemplifies the significant loss reduction enabled by the use of dielectric metasurfaces instead of metallic resonators, particularly in the infrared and visible frequency ranges.

Similar to metasurfaces consisting of plasmonic metal resonators, wavefront control and beam forming can be accomplished by using dielectric metasurfaces too. By varying the dimensions of the rectangular silicon resonators shown in figure 6.9, a phase variation can span over the entire 2π range. Through varying the geometric dimensions and coupling strength between dielectric resonators, it is possible to create the required phase profiles to simultaneously control the wavefront at multiple wavelengths. This approach was exploited in the demonstration of an achromatic dielectric metasurface lens operating near telecommunication wavelengths [27].

Let us also discuss here the realization of a polarization split in the visible region [28], see figure 6.10(a), that uses an all-dielectric gradient metasurface, composed of periodic arrangement of differently sized cross-shaped silicon nanoblocks resting on the fused silica substrate. The cross-shaped silicon block arrays can induce two opposite transmission-phase gradients along the x-direction for the linear x-polarization and along the y-direction for y-polarization. With proper design, the metasurface can separate the linearly polarized light into x- and y-polarized ones, which propagate at the same angle along the left and right sides of the normal incidence in the x–z-plane, as shown in figure 6.10(b). The polarization beam splitter is expected to play an important role for future free-space optical devices.

6.4.5 Active, tunable/reconfigurable and nonlinear metasurfaces

6.4.5.1 Introduction

During its initial stage of developments, the term 'metasurface' was predominantly used in relation to phenomena associated with wavefront tailoring/modifications of electromagnetic radiation by diffraction on a planar metamaterial. The field of metasurfaces gained considerable prominence and massive follow-up after the demonstration of anomalous refraction on gradient metasurfaces [15] and the development of high-throughput metasurfaces for focussing of light.

With the maturing of metasurface research, it is felt that in addition to various wavefront manipulation techniques, an important feature required in many applications is agile tunability/reconfigurability. While most realization of metasurfaces thus far was strictly passive and static, efforts were then made to include active elements in the metasurface unit cells at microwave, infrared and optical frequencies, opening exciting new opportunities to broaden the impact of this technology for various practical goals and applications. Active elements not only provided ways to reconfigure the response in real time, but also helped to compensate loss and overcome fundamental challenges of passive metasurface technology. In fact, by the insertion of active elements, metasurfaces can break the fundamental limitations of passive and static systems. The dynamic metasurfaces are now the promising new platforms for 5G/6G communications, remote sensing and radar and many other exciting and emerging applications.

Active control of metamaterials and metasurfaces extends their exotic passive properties by allowing fine resonance tuning to adapt the operational conditions, and enabled a switchable resonant response, for instance, resulting in signal modulation for communication and imaging. As compared to bulk metamaterials, the planar nature of metasurfaces facilitates the integration of active functional materials and adoption of active device architectures in a more effective way. There are a variety of functional materials and structures that have been successfully implemented to enable active metasurfaces through the application of an external stimulus that includes voltage bias, optical pump, magnetic field, thermal excitation, and mechanical deformation. Semiconductors and graphene have become the materials of choice in the arena of the development of active metasurfaces due to the developed device physics and the availability of mature integration and fabrication technology.

Anyway, despite various technological challenges still to be addressed, the steady progress in theoretical concepts, material science, modulation schemes, and nanofabrication techniques, offers a rosy prospect on the future of active, tunable, reconfigurable and nonlinear metasurfaces. It is envisioned that all these developments of metasurfaces will revolutionize the next generation of sensing, storing, computing and imaging devices based on light.

6.4.5.2 Semiconductor-hybrid and graphene-hybrid active metasurfaces

The conductivity in semiconductors can be increased by orders of magnitude through doping, enabling metal-like behaviour. Furthermore, active tuning of the conductivity of the semiconductor material can be realized by carrier injection and depletion through bias voltage and photo-excitation. Such a unique capability makes semiconductors ideal materials for integration into metamaterial/metasurface structures to accomplish active and dynamic functionalities, particularly in the microwave and THz frequency range. Varactor diodes have been widely used to realize frequency tunable and nonlinear response [29] in microwave metasurfaces. At THz frequencies, SRR arrays can be directly fabricated on top of semiconducting substrates such as intrinsic silicon and gallium arsenide, and the resonant response is switched through photo-excitation of free charge carriers at the substrate surface resulting in an ultrafast switching speed [30]. On the other hand, semiconductors can be used as part of the resonant structure. In this case, photo-excitation dynamically modifies the structural geometry of the resonator, enabling switchable or frequency tunable response. As shown in figure 6.11(a), pair of silicon bars form a part of the capacitive gap in an electric SRR unit cell. Under photo-excitation with near-infrared light, the silicon bars become metallic thereby increasing the SRR capacitance. Therefore, the frequency of the SRR *LC* resonance is tuned to a lower frequency with the tuning range about 20% [31]. Optically modifying the metasurface geometric structure enables the transition between different types of resonances. In figure 6.11(b) silicon is integrated at the gaps of a metal patch array which exhibits a dipolar resonance without photo-excitation and allows high transmission below the resonance frequency. Under photo-excitation, the metallic silicon connects the metal patches, effectively forming a metal wire grating that blocks the low frequency THz waves resulting in ultra broadband THz modulation [32].

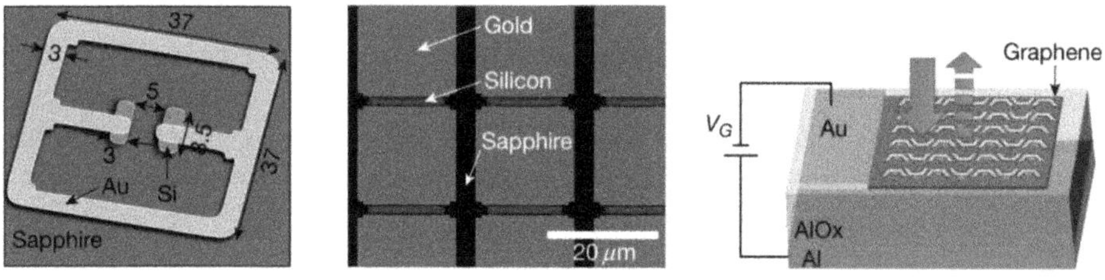

(a) (b) (c)

Figure 6.11. Semiconductor-hybrid and graphene-hybrid metasurfaces. (a) A pair of silicon bars form a part of the capacitive gap in an electric SRR unit cell. (b) Silicon is integrated at the gaps of a metal patch array. (c) Schematic of an ultrathin mid-infrared modulator based on a tunable graphene-based metasurface absorber. (Reproduced/adapted with permission from [10], copyright IOP Publishing. All rights reserved.)

Semiconducting hybrid metasurfaces also feature electrical tuning of resonances via the application of a voltage bias, which is more convenient and practical for real-world applications. The most prominent examples are the integration of varactor diodes for microwaves and Schottky junctions for THz frequencies. The first demonstration of an electrically switchable THz metasurface [33] featured an unprecedented 50% modulation depth, which was further improved to 80% through structure optimization by the same group.

Excellent mechanical properties (stronger than steel and harder than diamond, yet offers high flexibility) and the tunable carrier density of graphene make it an excellent material to enable active metasurfaces. Graphene, a 2D allotrope of carbon where carbon atoms are arranged in a honeycomb lattice, has largely tunable optical conductivity in the mid-infrared and THz frequency ranges. The doping of graphene can be adjusted through changing the bias voltage by a factor of 10 at room temperature, which will lead to a large change in its sheet conductivity σ and therefore the in-plane electric permittivity $\varepsilon_{\parallel} = 1 + j\sigma/\varepsilon_0\omega t$, where $t = 0.33$ nm is the thickness of graphene. For graphene, the resonant response of metasurfaces is of particular importance to enhance interactions with mid-infrared and THz radiation because of its atomic thickness and weak absorption. Metallic plasmonic antennas are able to capture light from free space and concentrate optical energy into sub-wavelength spots. The electric field at these spots can be two to three orders of magnitude larger than the incident field. By placing graphene in the hot spots created by metallic plasmonic antennas and by tuning the optical conductivity of graphene, one can switch the resonance or tune the resonance frequency of the composite over a wide range. In the THz frequency range, graphene by itself has been used to enable broadband electrical modulation and patterned graphene structures also exhibit resonant plasmonic response. Integrating graphene into metallic resonators has enabled the demonstration of THz hybrid metasurface electrical modulators [34].

Mid-infrared metasurfaces with electrically tunable spectral properties have been experimentally demonstrated by controlling the carrier density of graphene. Optimizing optical antenna designs has improved both the frequency tuning range and the modulation depth. In figure 6.11(c), the upper metasurface layer is separated from a back aluminium mirror by a thin aluminium oxide film. Such a reflect-array structure exhibits nearly perfect absorption [35]. That is, at the OFF state the reflection is nearly zero, and the frequency at which near-zero reflection occurs can be tuned by applying a voltage bias that modifies the dispersion of the top graphene-antenna array. This approach provides modulation speed in the tens of MHz range and an optical modulation depth close to 100%.

6.4.5.3 Tunable/Reconfigurable metasurfaces

Tunable metasurfaces are typically realized by combining passive meta-systems with materials whose electromagnetic properties can be modulated by external stimuli like temperature, voltage, electric or optical fields. Many schemes were proposed to realize tunable/reconfigurable metasurfaces at different frequency domains [36–39] since the inception of active metasurfaces.

Electrically tunable meta-atoms (ETMs) have been realized at frequencies ranging from microwave to optics. A microwave ETM can be designed by integrating a meta-atom with a varactor/PIN diode. Through varying the bias voltage applied on the dipode, one can dynamically switch the functionality of the ETM. However, although microwave ETMs have proven to be powerful systems to actively control EM waves, the realized devices still suffer from certain issues, such as limited working bandwidths and difficulties in transmission mode realizations. Integrating doped semiconductors with meta-atoms is another intriguing way to get ETMs with high performance (i.e. broad bandwidth and high modulation speed). A broadband terahertz (THz) phase modulator was experimentally demonstrated based on such a scheme [36]. The Schottky diode formed at the metal–semiconductor interface allows the efficient control of carrier density in doped GaAs by external bias voltage, resulting in real-time active amplitude (60%) and phase ($\pi/6$) modulation on the transmitted THz wave with a speed up to 2 MHz.

Thermal stimuli can dramatically modify the EM properties of many materials. A promising approach to achieve active metasurfaces is to hybridize metasurface with thermally responsive materials, such as phase change materials (PSMs), liquid crystals (LCs), and superconductors. For instance, vanadium dioxide, a typical PSM sensitive to temperature changes, has been combined with passive metasurfaces to achieve the 'ON' and 'OFF' states in controlling the transmitted light. LCs with both voltage and temperature-tuned refractive index can help realize tunable metasurfaces at different frequencies, such as tunable absorbers and spatial light modulators [37].

Optically tunable metasurfaces are constructed through combining photoconductive semiconductors and passive metasurface. By controlling the conductivity of the involved semiconductor through carrier photo-excitation, one can dynamically modulate the EM responses of the entire metasurface [31], leading to exotic effects such as tunable chirality and negative refractive index. Such meta-systems can offer ultrafast optical control on light but with relatively weak tunability.

Micro-electro-mechanical-system (MEMS) technologies have been adopted to realize reconfigurable metasurfaces [38], which can dynamically modulate EM waves ranging from microwave to infrared, with deformable cantilevers or thermal bimorphs frequently used to control the device configurations. There are other materials/technologies to realize tunable metasurfaces, such as incorporating stretchable dielectric substrate or ferrite into the metasurface designs, or using microfluidic technology to dynamically control the device configurations.

A few words may also be mentioned in regard to reconfigurable metasurfaces that have opened up new application areas. Chen *et al* reported a sub-wavelength reconfigurable Huygens' metasurface realized by loading it with controllable active elements [39]. The proposed design provides a unified solution to the aforementioned challenges of real-time local reconfigurability of efficient Huygens metasurfaces. As an example it may be mentioned that a reconfigurable metalens at the microwave frequencies is experimentally realized, which demonstrates that multiple and complex focal spots can be controlled simultaneously at distinct spatial positions and reprogrammable in any desired fashion, with fast response time and high efficiency. Such an active Huygens metalens may offer unprecedented potential for real-time, fast, and sophisticated electromagnetic wave manipulation such as dynamic holography, focussing, beam shaping/steering, imaging, and active emission control. Many other studies have also been reported on reconfigurable metasurfaces and their variants like metagratings [40] and so forth.

Overall, considerable progress has been made in achieving tunable/reconfigurable metasurfaces based on different mechanisms, indicating the promising future of active metadevices for diversified applications. The past decade has witnessed a fast development on tunable/reconfigurable metasurfaces while grand challenges still exist on physics/technology/material aspects. Anyway, these challenges may be considered to be the driving forces to push the field forward, eventually making tunable/reconfigurable metasurfaces a promising platform for realizing functional devices which will be able to address versatile application requirements.

6.4.5.4 Nonlinear metasurfaces

Nonlinear metasurfaces represent an important research direction that is expected to expand the metasurface functionalities and applications in a great way. Thus at the end of this section it will be worthwhile to mention a few words about the nonlinear metasurface applications. Conventionally, the phase matching condition in nonlinear processes, such as second-harmonic generation (SHG), has to be satisfied in bulk nonlinear crystals to achieve efficient nonlinear optical generation. In the case of phase matching, nonlinearly generated optical signals constructively build up, and optical power will be continuously transferred from the pump(s) to the nonlinear optical signal. Metasurfaces greatly relax the requirement for phase matching as nonlinear processes occur within metasurfaces that have sub-wavelength thicknesses. Significant second-harmonic response has been experimentally demonstrated in plasmonic metasurfaces that were integrated with nonlinear media, see figure 6.12(a) [41].

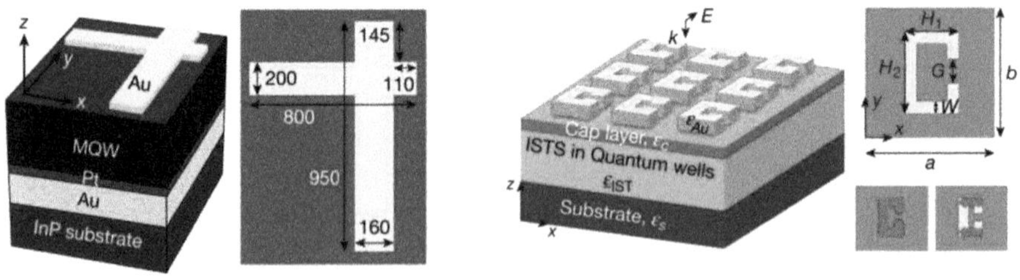

(a) (b)

Figure 6.12. (a) A unit cell of the SHG metasurface; the dimensions of the gold nanocross, on the right, are given in nm, and the unit cell has a dimension of 1000 nm × 1300 nm. (b) Left panel: schematic of a metasurface consisting of SRRs on top of a stack of MQWs. Upper right panel: top view of one SRR. Lower right panel: schematic showing the two main resonant modes of the SRR at the pump and SH frequency, respectively. (Reprinted/adapted with permission from [10], copyright IOP Publishing. All rights reserved.)

Specifically, InGaAs/AlInAs multiple quantum wells (MQWs) were used as the nonlinear media, which exhibit such SHG and electrically tunable nonlinear coefficients in the mid-infrared. The nonlinear metasurfaces achieved a nonlinear conversion efficiency of $\sim 2 \times 10^{-6}$ using a pump intensity of only 15 kW cm^{-2} [41], corresponding to an effective second-order nonlinear coefficient of $\chi^{(2)} \sim 30$ nm V^{-1}, about three orders of magnitude larger than that of $LiNbO_3$. Even larger $\chi^{(2)} \sim 250$ nm V^{-1} has been experimentally demonstrated in a nonlinear metasurface consisting of an array of SRRs and InGaAs/AlInAs multiple quantum wells, see figure 6.12(b) [42]. The two plasmonic resonances of the SRRs enhance, respectively, the pump and SH signal being generated.

6.5 Metasurface modelling and characterization

6.5.1 Introduction

Metasurfaces or the 2D/planar metamaterials find an important precedent in the planar periodic or quasi-periodic arrays of printed patches (or the complementary structure of apertures in a metallic screen) commonly used in microwaves and antenna engineering; for instance, frequency selective surfaces (FSSs) [14] and reflectarrays/transmitarrays [43]. A common characteristic of the scatterers of these periodic structures is that their size is of the order of the wavelength at the operation frequencies. This feature makes it that the rigorous electromagnetic analysis of the structures has to resort to full-wave computational methods demanding high computational resources. However, an alternative approach is also tried for that which requires much lower computational resources but still provides sufficiently accurate results, along with a good qualitative physical insight of the electromagnetic scattering phenomena—which is the equivalent circuit approach (ECA) [44]. The closed-form ECA may also be an efficient tool to study metasurfaces, provided the frequency band of interest lies inside the limit of validity of the approach. This validity frequency range extends up to and close to the second resonance of the individual scatterer of the periodic metasurface, which covers many practical situations of interest.

Metasurface inclusions/scatterers are usually small in comparison with the wavelength, which is sub-wavelength in nature, opening up the possibility to use homogenization theories for their analytical modelling and design. Through averaging the tangential fields and induced polarizations, homogenization theories allow description of the electromagnetic properties of metasurfaces by effective parameters, like susceptibilities (relating averaged fields and surface polarization densities), or collective polarizabilities (relating local fields and surface polarization densities) and sheet impedances. Homogenization models are powerful tools for the design of functional metasurfaces, allowing us to predict and optimize their electromagnetic response. Each homogenization model has different advantages. For example, the polarizability-based model allows calculation of the required polarizabilities of the individual particles. On the other hand, the values of susceptibilities immediately tell if the metasurface is lossy or active. Finally, the impedance matrix representation is a powerful tool for the analysis of a cascade of metasurfaces using the transmission-line approach. Detailed descriptions of these homogenization methods can be found in a standard textbook on this subject [45].

The concept of metasurfaces has been extended to non-homogenizable metasurfaces too. For instance, metasurfaces where the distances between inclusions are sub-wavelength but the structural periodicity allows the excitation of higher-order Floquet modes require novel approaches to modelling. The gradient metasurface is one such case that has led to many studies focused on phase-controlling metasurfaces [15, 46–50], bi-anisotropic metasurfaces [51] or strongly non-local metasurfaces for wavefront manipulations [46–49]. For the analysis and practical realizations of such metasurfaces, it is important to distinguish between the macroscopic and mesoscopic responses of the metasurface. The macroscopic behaviour refers to the scattering properties of the whole non-uniform structure, for example, the energy distribution between Floquet modes. The mesoscopic properties are studied considering each unit cell of a gradient metasurface in a periodic lattice of identical cells (the locally-periodic approximation).

The research and development of metasurface devices follows the line of using more and more complex structures of the unit cells, including non-reciprocal, time-dependent, and even active elements. On the other hand, researchers play with complex gradient metasurfaces which have resonant response of unit cells and at the same time space-modulation periods comparable with the wavelength. Some features of such complex metasurfaces can be revealed using numerical simulations, but it is clear that we need more advanced analytical models to interpret, design, and fully exploit the properties of emerging metasurfaces.

6.5.2 Modelling techniques for metasurfaces

Initial attempts to model and characterize metasurface were tried with the effective-medium bulk-parameter analysis of metamatrials but have not been a successful approach [52]. In such studies the surface/film was modelled as a single-layer metamaterial in which effective bulk properties of the metasurface were obtained by arbitrarily introducing a non-zero thickness parameter into the analysis. However, such a bulk property characterization of metasurface is incorrect at a fundamental level—as the thickness in such analysis is then a physically artificial parameter. The effective bulk properties extracted in this way was found to display non-physical behaviour too—violating causality and the passive material was found to exhibit a bulk permittivity or permeability with gain and so forth.

A metasurface is a complex electromagnetic structure, which is typically an electrically very thin ($\delta \lll \lambda$), electrically relatively large in size (L_x, $L_y > \lambda$), homogenizable (p_x, $p_y \ll \lambda$) which is composed of sub-wavelength scattering particles with extremely small features, see figure 6.13(a). It is therefore challenging to model and typically requires a holistic design approach of the type described in figure 6.13(b), which includes both synthesis and analysis operations [53].

The synthesis operation consists in determining the physical (geometrical and electromagnetic) parameters of the metasurface structure, such as the metasurface size and scattering particle geometries, to achieve a specified wave transformation. It

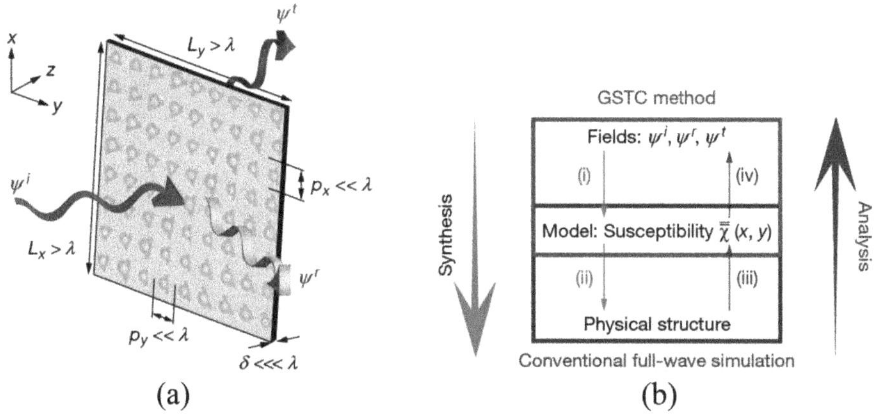

Figure 6.13. (a) A generalized metasurface where ψ^i, ψ^r and ψ^t, respectively, represent the incident, reflected and transmitted waves. (b) A holistic design methodology for metasurface.

may be decomposed in the following two operations: the determination of a continuous tensorial susceptibility function, $\overline{\overline{\chi}}(x, y)$, characterizing the metasurface from the specified fields ((i) in figure 6.13(b)) and, based on this function, the determination of the aforementioned physical parameters ((ii) in figure 6.13(b)). The analysis operation is the inverse of the synthesis. It consists in determining the scattered fields of a given physical metasurface, and may be itself decomposed in the following two operations: the extraction of the susceptibility function $\overline{\overline{\chi}}(x, y)$ characterizing the metasurface structure ((iii) in figure 6.13(b)) and the subsequent computation of the scattered fields ((iv) in figure 6.13(b)). Being typically much thinner than the operating wavelength, a metasurface is most efficiently modelled as a sheet of zero thickness ($\delta = 0$). Such a complex zero-thickness sheet cannot be modelled by conventional boundary conditions, and no currently existing commercial software can efficiently simulate it.

Because the sub-wavelength unit cells of metasurfaces can be homogenized with electric and magnetic surface susceptibilities or equivalent impedances and admittances, unlike effective-medium description of metamaterial, the metasurfaces can be best modelled by generalized sheet-transition conditions (GSTCs) [54, 55]. GSTCs relate the difference between tangential electric and magnetic fields on both sides of the metasurface to electric and magnetic surface polarization densities, themselves linearly related to the sum of fields on both sides of the metasurface. The details of the surface geometry are incorporated here into the boundary condition through the surface susceptibilities or porosities. The GSTCs allow a metasurface to be modelled by a boundary condition that is applied across an infinitely thin *equivalent* surface. In fact, physically speaking, the metasurface acts like an infinitesimal sheet: one that causes a phase shift (and possibly a change in amplitude too). The coefficients appearing in the GSTCs for any given metasurface are all that are required to model the metasurface's interaction with the electromagnetic field. The fields that appear in the generalized sheet-transition conditions are 'macroscopic' fields, in the sense that they do not exhibit variations on a length scale comparable to scatterer or aperture dimensions or the lattice spacing, but only on larger scales of the wavelength in the surrounding medium. The surface susceptibility and porosity dyadics that appear explicitly in the GSTCs are uniquely defined (unlike in a bulk effective-parameter model), and, as such, represent the physical quantities that uniquely characterize the metasurface—and are the most appropriate manner of characterizing metafilms/metasurfaces [52].

For a metafilm, the GSTC relates to the electromagnetic fields on both sides of the metasurface [54]

$$\begin{aligned} \vec{a}_z \times \vec{H}\,\Big|_{z=0^-}^{0^+} &= j\omega\overleftrightarrow{\chi}_{\mathrm{ES}} \cdot \vec{E}_{\mathrm{t,\,av}}\,\Big|_{z=0} - \vec{a}_z \times \nabla_{\mathrm{t}}\left[\chi_{\mathrm{MS}}^{zz} H_{z,\,\mathrm{av}}\right]_{z=0} \\ \vec{E}\,\Big|_{z=0^-}^{0^+} \times \vec{a}_z &= j\omega\overleftrightarrow{\chi}_{\mathrm{MS}} \cdot \vec{H}_{\mathrm{t,\,av}}\,\Big|_{z=0} + \vec{a}_z \times \nabla_{\mathrm{t}}\left[\chi_{\mathrm{ES}}^{zz} E_{z,\,\mathrm{av}}\right]_{z=0} \end{aligned} \tag{6.2}$$

where the subscript 'av' represents the average of the field on either side of the metasurface, the subscript 't' refers to components transverse to z, and $\vec{a}_z$ is the unit vector. The parameters $\overleftrightarrow{\chi}_{\mathrm{ES}}$ and $\overleftrightarrow{\chi}_{\mathrm{MS}}$ are the dyadic surface electric and magnetic susceptibilities. These have units of length and are related to the electric and magnetic polarizability densities of the scatterers per unit area. These dyadics vanish when the scatterers are absent and equation (6.2) reduces to the ordinary conditions of continuity of the tangential components of $\vec{E}$ and $\vec{H}$ fields.

Metasurfaces with other structures will require a different form of GSTCs. For example, metascreens consisting of periodic isolated apertures in a zero-thickness perfect conductor obey the Kontorovich form of GSTC [56], which can be written in the form

$$\vec{E}\,\Big|_{z=0} \times \vec{a}_z = j\omega\mu\overleftrightarrow{\pi}_{\mathrm{MS}} \cdot \vec{H}\,\Big|_{z=0^-}^{0^+} + \frac{1}{\varepsilon}\vec{a}_z \times \nabla_t[\pi_{\mathrm{ES}}^{zz} D_z]_{z=0^-}^{0^+} \tag{6.3}$$

where the tangential $\vec{E}$ field and the normal $\vec{B}$ fields are continuous across the metascreen. Here the dyadics $\overleftrightarrow{\pi}_{\mathrm{ES}}$ and $\overleftrightarrow{\pi}_{\mathrm{MS}}$ are what we call the electric and magnetic porosities of the metascreen, and, like surface susceptibilities of the metafilm, these have dimensions of length. When the apertures are absent and we have only the perfectly conducting plane, the electric and magnetic porosity dyadics vanish, and equation (6.3) reduces to the vanishing of tangential $\vec{E}$ field, as expected. The most general form of metascreen boundary conditions has not yet been determined, and its development is the subject of ongoing work. The same is true for metasurfaces such as wire gratings, which exhibit features of both metafilms and metascreens.

The GSTC may also be cast in the form of impedance-type boundary conditions [56]. Impedance boundary conditions (IBCs) [57] may be viewed as a special case of the GSTCs. Considering the plane wave fields of the form: $e^{-j\vec{k}.\vec{r}_t}$(where $\vec{k} = k_x\vec{a}_x + k_y\vec{a}_y$ and $\vec{r}_t = x\vec{a}_x + y\vec{a}_y$), the variation of which is parallel to the surface of a metasurface, and using Maxwell's equations we can write the GSTCs of equation (6.2) as [56]

$$\begin{aligned} \vec{a}_z \times \vec{H}\,\Big|_{z=0^-}^{0^+} &= \overleftrightarrow{Y}_{ES} \cdot \vec{E}_{t,\,av}\,\Big|_{z=0} \\ \vec{E}\,\Big|_{z=0} \times \vec{a}_z &= \overleftrightarrow{Z}_{MS} \cdot \vec{H}_{t,\,av}\,\Big|_{z=0} \end{aligned} \tag{6.4}$$

In which the spatially dispersive ($\vec{k}$-dependent) surface transfer admittance and transfer impedance are given by:

$$\begin{aligned}\overleftrightarrow{Y}_{\text{ES}} &= j\omega\overleftrightarrow{\chi}_{\text{ES}} + \frac{j\chi_{\text{MS}}^{zz}}{\omega\mu}(\vec{a}_z \times \vec{k})(\vec{a}_z \times \vec{k}) \\ \overleftrightarrow{Z}_{\text{MS}} &= j\omega\overleftrightarrow{\chi}_{\text{MS}} + \frac{j\chi_{\text{ES}}^{zz}}{\omega\varepsilon}(\vec{a}_z \times \vec{k})(\vec{a}_z \times \vec{k})\end{aligned} \tag{6.5}$$

Such boundary conditions can also be interpreted as equivalent transmission-line circuits [58]

Metafilms can also be used as a means of extracting bulk properties of metamaterials [59]. In this technique, plane wave reflection and transmission coefficients from metalfilm are used to extract the surface susceptibilities. The polarizabilities of the individual scatterers in the metafilm are then obtained [54]. These are finally inserted into the Clausius–Mossotti relations to obtain the effective permittivity and permeability of a metamaterial realized by stacking these metafilms in a third dimension. This technique is a useful alternative to standard parameter-extraction methods when transmission through a sample of many layers is very small.

6.5.3 Characterization of metasurfaces

Unlike metamaterials, the bulk effective permittivity and permeability (as well as the refractive index) of a metasurface are not uniquely defined parameters; rather, the GSTC along with Maxwell's equation are all that is required to analyze the interaction of electromagnetic field with the metasurface. The surface susceptibility and porosity dyadics that appears explicitly in the GSTC are uniquely defined (unlike in a bulk-effective-parameter model), and, as such, represent the physical quantities that uniquely characterize the metasurface. The retrieval method for uniquely characterizing the metasurface is based on an inversion of its reflection and transmission coefficients to obtain the susceptibilities [52, 60] that will be discussed in the following.

The reflection (R) and transmission (T) coefficients of a metafilm for both TE and TM polarized plane waves, see figure 6.14, were derived in [61].

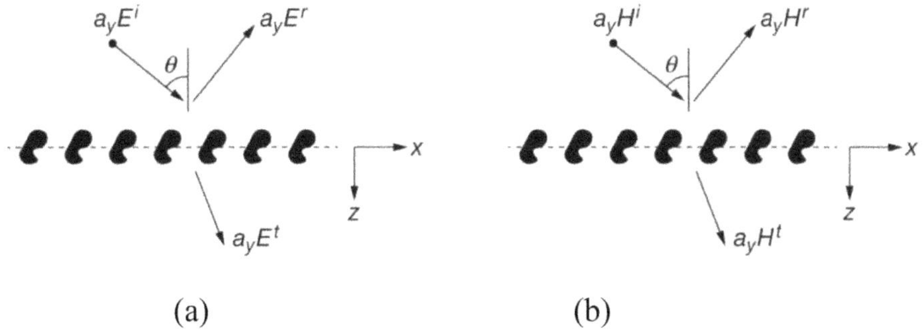

Figure 6.14. A plane wave incident onto a metafilm: (a) TE polarization and (b) TM polarization.

The reflection coefficient, R, and transmission coefficient, T, for TE polarized plane wave are given by [52, 60]:

$$R_{\mathrm{TE}}(\theta) = \frac{-j\frac{k_0}{2\cos\theta}\left(\chi_{\mathrm{ES}}^{yy} + \chi_{\mathrm{MS}}^{zz}\sin^2\theta - \chi_{\mathrm{MS}}^{xx}\cos^2\theta\right)}{D_1}$$

$$T_{\mathrm{TE}}(\theta) = \frac{1 + \left(\frac{k_0}{2}\right)^2 \chi_{\mathrm{MS}}^{xx}\left(\chi_{\mathrm{ES}}^{yy} + \chi_{\mathrm{MS}}^{zz}\sin^2\theta\right)}{D_1} \tag{6.6}$$

$$D_1 \equiv 1 - \left(\frac{k_0}{2}\right)^2 \chi_{\mathrm{MS}}^{xx}\left(\chi_{\mathrm{ES}}^{yy} + \chi_{\mathrm{MS}}^{zz}\sin^2\theta\right) + j\frac{k_0}{2\cos\theta}\left(\chi_{\mathrm{ES}}^{yy} + \chi_{\mathrm{MS}}^{xx}\cos^2\theta + \chi_{\mathrm{MS}}^{zz}\sin^2\theta\right)$$

and for TM polarized plane wave is given by:

$$R_{\mathrm{TM}}(\theta) = \frac{-j\frac{k_0}{2\cos\theta}\left(\chi_{\mathrm{ES}}^{xx}\cos^2\theta - \chi_{\mathrm{MS}}^{yy} - \chi_{\mathrm{ES}}^{zz}\sin^2\theta\right)}{D_2}$$

$$T_{\mathrm{TM}}(\theta) = \frac{1 + \left(\frac{k_0}{2}\right)^2 \chi_{\mathrm{ES}}^{xx}\left(\chi_{\mathrm{MS}}^{yy} + \chi_{\mathrm{ES}}^{zz}\sin^2\theta\right)}{D_2} \tag{6.7}$$

$$D_2 \equiv 1 - \left(\frac{k_0}{2}\right)^2 \chi_{\mathrm{ES}}^{xx}\left(\chi_{\mathrm{MS}}^{yy} + \chi_{\mathrm{ES}}^{zz}\sin^2\theta\right) + j\frac{k_0}{2\cos\theta}\left(\chi_{\mathrm{MS}}^{yy} + \chi_{\mathrm{ES}}^{xx}\cos^2\theta + \chi_{\mathrm{ES}}^{zz}\sin^2\theta\right)$$

where k_0 is the free-space wavenumber. These reflection and transmission coefficients apply to a metafilm, the constituent scatterers of which have sufficient symmetry and that the surface susceptibility dyadics are diagonal. The more general case of non-symmetric, bi-anisotropic surface susceptibility dyadics could also be handled in a similar way, as well as that of a metascreen or more general metasurface.

Once the reflection and transmission coefficients are obtained (either from measurements or from numerical simulations), the surface susceptibilities can be determined. For TE polarized wave, the three unknown surface susceptibilities are determined from the following:

$$\chi_{\mathrm{MS}}^{xx} = \frac{2j}{k_0}\frac{R_{\mathrm{TE}}(0) - \mathrm{TE}(0) + 1}{R_{\mathrm{TE}}(0) - \mathrm{TE}(0) - 1}$$

$$\chi_{\mathrm{ES}}^{yy} = \frac{2j}{k_0}\frac{R_{\mathrm{TE}}(0) + \mathrm{TE}(0) - 1}{R_{\mathrm{TE}}(0) + \mathrm{TE}(0) + 1} \tag{6.8}$$

and

$$\chi_{\mathrm{MS}}^{zz} = -\frac{\chi_{\mathrm{ES}}^{yy}}{\sin^2(\theta)} + \frac{2j}{k_0}\frac{\cos(\theta)R_{\mathrm{TE}}(\theta) + \mathrm{TE}(\theta) - 1}{\sin^2(\theta)R_{\mathrm{TE}}(\theta) + \mathrm{TE}(0) + 1} \tag{6.9}$$

where $R(0)$ and $T(0)$ are the reflection and transmission coefficients for normal incidence, and $R(\theta)$ and $T(\theta)$ are the reflection and transmission coefficients at some oblique incidence angle θ.

For TM polarized wave, the three unknown surface susceptibilities are determined by:

$$\begin{aligned}\chi_{\mathrm{ES}}^{xx} &= \frac{2j}{k_0}\frac{R_{\mathrm{TM}}(0) + \mathrm{TM}(0) - 1}{R_{\mathrm{TM}}(0) + \mathrm{TE}(0) + 1}\\ \chi_{\mathrm{MS}}^{yy} &= \frac{2j}{k_0}\frac{R_{\mathrm{TM}}(0) - \mathrm{TE}(0) + 1}{R_{\mathrm{TM}}(0) - \mathrm{TE}(0) - 1}\end{aligned} \tag{6.10}$$

and

$$\chi_{\mathrm{ES}}^{zz} = -\frac{\chi_{\mathrm{MS}}^{yy}}{\sin^2(\theta)} + \frac{2j}{k_0}\frac{\cos(\theta)\mathrm{T}_{\mathrm{TM}}(\theta) - 1 - R_{\mathrm{TM}}(\theta)}{\sin^2(\theta)\mathrm{T}_{\mathrm{TM}}(\theta) + 1 - R_{\mathrm{TM}}(\theta)} \tag{6.11}$$

This approach can be applied equally well either to numerically or experimentally determined values of R and T. However, because of the difficulty of separating the incident and reflected components at normal incidence in measurement, it may be more beneficial to rewrite these equations for two arbitrary incidence angles both of which differ from zero, as discussed in [52]. The interesting thing to notice is that this retrieval approach for characterizing a metafilm by its surface susceptibilities does not pose the problem of choosing the appropriate sign of a square root, nor require the assumption of an arbitrary layer thickness, as is the case when trying to infer effective properties for a bulk metamaterial model, as was discussed in section 5.2.2 of chapter 5.

These expressions for retrieving the surface susceptibilities of a metafilm were validated by taking some examples [8, 52] and it was observed that the surface susceptibilities obtained from the retrieval approach were virtually identical to those obtained from the approximate analytical expressions. Also, it may be commented that the surface susceptibilities are unique properties of a metafilm, and as such serve as the most appropriate way to characterize them.

6.6 Application potential of metasurfaces

6.6.1 Introduction

In the past few years, metasurfaces have achieved groundbreaking progress, providing unparalleled control of light, including the construction of arbitrary wavefronts and realizing active and nonlinear optical effects—opening up new and exciting applications at THz and optical frequencies, including those in nano-photonics. It was mentioned earlier too that metamaterial, being 3D, is difficult to fabricate while metasurface being planar is easy to fabricate and can be easily integrated into other devices, which can make it a salient feature for nanophotonic circuits. Absorbers, that are required in many applications like anechoic chamber, solar cells, photodetectors and so forth starting from microwave through THz to optical frequencies, have to be compact and low-loss for modern design. But the conventional absorbers, including those made of metamaterial, are usually composed of multilayer structures that are lossy and bulky. Metasurface-based absorbers are thus a good catch for their low profile, light weight and simplicity of construction with simple metallic structures.

The transformation optics that deals with the control of electromagnetic waves leading to designer demanded tailoring of its wavefront thereby manipulating the wave dynamics, has undergone unprecedented progress with the use of metasurfaces. The wavefront shaping with conventional techniques like lens, hologram etc can be realized over a distance that is larger than the wavelength of operation. The same is the case with metamaterials or double negative materials in which the control of the wavefront of an electromagnetic signal is done by accumulating the phase through propagating over a distance larger than a wavelength. But with metasurfaces, wavefront shaping and focussing of energy can be done over sub-wavelength distances and thus alleviating the propagation loss. This makes metasurfaces potentially superior for this purpose and they are fast replacing the bulky metamaterials in many applications. Many of the issues of transformation optics physics and applications with metasurface have already been discussed in section 6.4.

Other applications of metasurface include the design of very thin lenses that use the non-uniform impedance surfaces possible to realize using metasurfaces. Metasurfaces have already revolutionized antenna design, especially the leaky-wave antennas, from microwave to THz and higher frequencies. Metasurface leaky-wave antennas have advantages in that they are low profile and have simple feed structures. They also have frequency-dependent beam-scanning properties (beam squint). Further, metasurface can be used in the design of a very promising cloaking device at optical frequency. Compared to the metamaterial-based cloak, the skin cloak based on metasurface is miniaturized in size and scalable for hiding macroscopic objects too (as was discussed in section 1.6.2 of chapter 1).

Metasurface finds interesting applications in the design of plasmonic laser or lasing spaser, whose bandwidth can be controlled by tailoring the metamolecules of metasurface (as discussed in section 4.3.5.2 of chapter 4). Since metasurfaces have high bio-tissue sensitivity, they can be used in biosensors for inside-body examination and bio-tissue discrimination, including cancer disease diagnosis. Apart from those mentioned above, metasurface is opening up many emerging applications too that will be discussed in the following.

6.6.2 Metasurface-based absorbers

It has already been mentioned that absorbers are widely used from microwave through visible frequencies in applications including but not limited to scattering control, anechoic chambers, microbolometers, photodetectors, and solar cells.

Details of metamaterial-based absorbers at microwave through THz and optical frequencies have been discussed in detail in chapter 4. But with modern systems requiring more compact designs, absorbers with miniature structures are needed which can be met with metasurface absorbers. Metasurface absorbers are broadly studied for their low profile and simplicity of fabrication.

A metasurface-based passive absorber is shown in figure 6.15(a) [11, 62] that consists of an array of gold material patches on MgF_2-substrate that are capable of absorbing energy independent of polarization and with a wide-angle range of up to ±80°. The proposed structure is designed to be polarization-independent in the x- and y-directions at normal incidence, thus yielding a polarization-insensitive property. However, the all-dielectric metasurface absorber is capable of eliminating the ohmic losses and thermal effects of metallic unit cells, as shown in figure 6.15(b) [11, 63]. Here the THz all-dielectric metasurface absorber is based on hybrid dielectric waveguide resonances. The metasurface geometry is tuned in order to overlap electric and magnetic dipole resonances at the same frequency, thus achieving an experimental absorption of 97.5%. The concept of an all-dielectric metasurface absorber offers a new route for control of the emission and absorption of electromagnetic radiation from surfaces with potential applications in energy harvesting, imaging, and sensing.

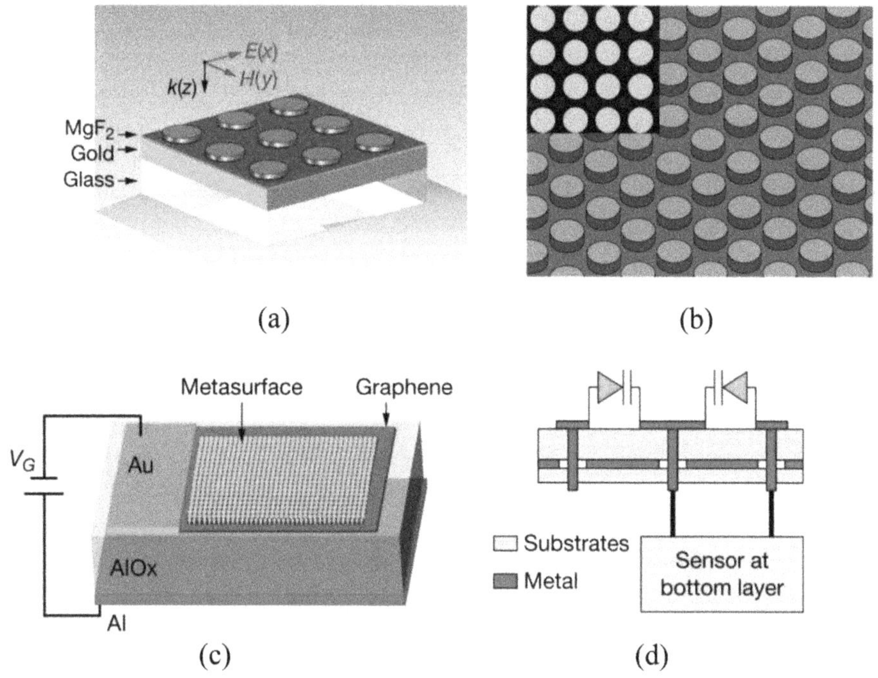

Figure 6.15. Metasurface absorbers; (a) passive absorber at infrared frequency, (b) THz all-dielectric metasurface absorber, (c) active optical tunable metasurface absorber, and (d) electrically tunable active metasurface absorber. (Reproduced/adapted from [11], with permission from De Gruyter, copyright (2018) Li *et al.*)

The passive metasurface-based absorbers, discussed above, though have advantages of low profile and light weight with simple structures; however, such passive absorbers have limitations in terms of their absorption bandwidth, absorption rate, and air breakdown at high power. To overcome the intrinsic limitations of passive metasurfaces, active electronics play an important role in advanced metasurfaces by enabling switchable absorption, tunable resonant frequency, and tunable nonlinear response.

A tunable metasurface absorber, which can be used from tens of GHz to near infrared due to the broadband optical response of graphene, is shown in figure 6.15(c) [11, 64]. Figure 6.15(d) [11, 65] shows a metasurface that includes microwave power sensors at each unit cell. Feedback circuits change the surface impedance in response to the local microwave power level by tuning varactor diodes on the top surface. This nonlinear metasurface can enable self-focussing of surface waves, analogous to the self-focussing effect that occurs in some nonlinear optical materials. It can also behave as a power-dependent surface wave absorber.

Anyway, the details of many more metamaterial/metasurface-based absorbers are also available in suitable text [66].

6.6.3 Metasurface in antenna design

Recent developments in metasurfaces have opened new opportunities in antenna design leading to the development of low cost, light weight, and compact antennas capable of producing arbitrary aperture fields. Traditionally, the tailoring of aperture fields in phase, amplitude and polarization for microwave and millimetre-wave antennas is done with reflectors/lenses excited by feed antennas. This conventional approach results in large and heavy structures, due to the bulky reflectors/lenses, gimbals and so forth. Alternatively, apertures can be composed of discrete antenna elements, and electronically controlled, as in the case of phased arrays. Phased arrays, however, are costly and exhibit high feed network losses that grow with aperture size and frequency.

Metasurface antennas can transform fields of a source excitation into an arbitrary radiating aperture field in various ways of which leaky-wave technique is quite popular. Metasurface leaky-wave antennas have advantages in that they are low-profile and have simple feed structures. The principle of leaky-wave antenna in brief is the following. When a leaky-wave antenna is fed by a point source, the field of the surface can be represented as: $\Psi(\rho, z) = \Psi_0 \exp(-j\gamma\rho - ik_z z)$, where $\gamma = (\alpha + j\beta)$ represents the propagation constant on the surface of the leaky-wave antenna and $k_z = \sqrt{k_0^2 - \beta^2}$ is the propagation constant normal to the antenna surface. In the fast wave regime, when $\beta < k_0$, k_z is purely real, indicating that energy is coupled into free space as radiation. The direction of the radiation can be controlled by controlling the value of β.

Figure 6.16(a) [11, 67] depicts a sinusoidally modulated graphene leaky-wave antenna with electronic beam-scanning capability. By applying bias voltages to the different grating pads beneath the graphene substrate, the graphene surface reactance can be modulated, resulting in a versatile beam-scanning capability at THz frequencies. Another 2D leaky-wave antenna operating at THz frequencies

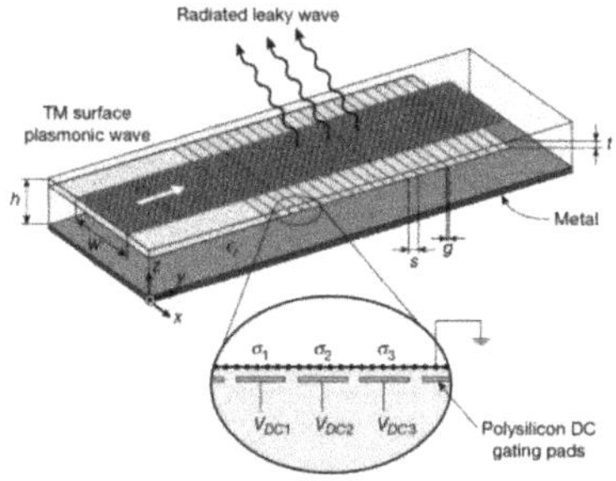

(a)

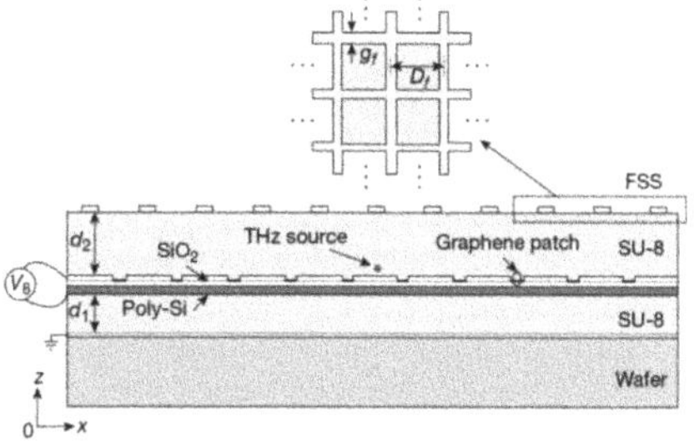

(b)

Figure 6.16. Leaky-wave metasurface antennas. (a) Sinusoidally modulated graphene leaky-wave antenna with electronic beam-scanning capability. (b) 2D leaky-wave antenna operating at THz frequencies with tunable frequency and beam angle. (Reproduced/adapted from [11], with permission from De Gruyter, copyright (2018) Li *et al.*)

with tunable frequency and beam angle is shown in figure 6.16(b) [11, 68]. It is based on tuning the conductivity of the graphene.

Metasurfaces have also been successfully used to design and manufacture high-gain holographic antennas for which the surface wave is the major incident wave. Holographic antennas borrow their design concept from optical holography, which is used for recording and recreating a complex optical wavefront. In the microwave region, this technique has been applied for planar antenna designs. Sievenpiper and his colleagues [69] used the concept of holography to design impedance surfaces that convert a given surface wave into a freely propagating wave with desired far-field radiation pattern and polarization in the microwave spectral range. The impedance surface is essentially a hologram, which is the interference pattern between a reference beam and an object beam, and carries information of the phase, amplitude and polarization of the desired object beam. The object beam is reconstructed when the reference beam impinges on the hologram. In Sievenpiper's implementation, a source antenna produces the reference beam in the form of a surface wave, E_{surf}, and the object beam is the desired wave, E_{rad}, propagating in the half space above the surface, figure 6.17(a) [11]. Microwave holograms are created according to the interference pattern produced by the two waves. Both scalar and tensor forms of the impedance surfaces were experimentally demonstrated.

A surface covered with an array of sub-wavelength metallic patches on a grounded dielectric substrate provides variable surface impedance controlled by the geometry of the patches. A closer view of the pattern is shown in figure 6.17(b). By varying the shape of the cells, anisotropic impedance values can also be achieved, enabling polarization control.

Metasurfaces can also improve the performance of horn antennas. A metasurface designed using a genetic algorithm has been used as an inner surface for a conical horn, and the cross-polar and side lobe levels have been improved over the entire Ku-band [70]. A similar approach was applied to improve the performance of a hybrid mode square horn antenna using metasurfaces [71].

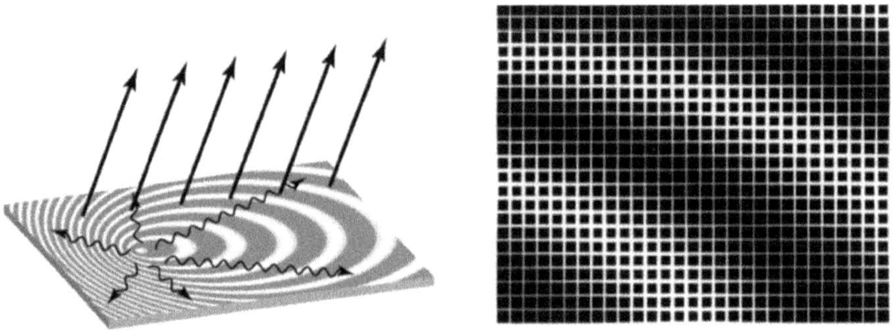

(a) (b)

Figure 6.17. Metasurface-based holographic antenna. (a) Schematic of holographic leaky-wave antenna concept showing the surface waves (undulating arrows) excited on an artificial impedance surface, which are being scattered by variations in the surface impedance to produce the desired radiation (straight arrows). (b) A close view of the isotropic holographic metasurface. (Reproduced/adapted from [11], with permission from De Gruyter, copyright (2018) Li *et al.*)

Metasurface-based lenses that are directly integrated with feed antennas have demonstrated an order of magnitude reduction in thickness compared to traditional lens-based antennas. An experimental lens-antenna system operating at 9.9 GHz with approximate radiation efficiency of 50%, aperture efficiency of 70%, half-power gain bandwidth around 8%, and impedance bandwidth of around 4% was reported [72].

Although metasurface-based antennas show great promise for dramatically reducing the size and complexity of antenna systems, certain challenges lie ahead for their mass deployment. Metasurfaces consist of sub-wavelength resonators with electromagnetic properties that are often times frequency dispersive. In other words, the resonant nature of a metasurface's constitutive elements can restrict the usable bandwidth of metasurface-based antennas. Single-layer metasurface antennas with operational bandwidths of up to 8% have been reported [73]. In comparison, a commercially available Ku-band SATCOM antenna covers global receive and transmit bands nearing a total bandwidth of 23%. Furthermore, if simultaneous operation in multiple satellite communication bands is desired with a single antenna, even larger bandwidths or multiband performance are needed. Thus performance and synthesis challenges, as well as tuning/reconfiguration mechanisms must be addressed before their use in antenna design becomes widespread in application areas ranging from terrestrial/satellite communication systems and aerial platforms, to radar and surveillance systems.

6.6.4 Metalens with metamaterial/metasurface

It was discussed in chapter 4 that metamaterial-based hyperlens has gained much enthusiasm to the metamaterial community since its inception in 2007. However, all of the hyperlenses, although they perform as super-resolution lens for far-field imaging, by converting near-field image to the far-field, they lack the ability to focus plane waves. For conventional optical lenses, refraction and focussing of light rely on curved interfaces, fundamental for Fourier transforms and imaging. But hyperlenses cannot realize the Fourier transform function due to their lack of a phase compensation mechanism. A *metamaterial-based metalens* [73] can not only provide the advantages of super-resolving capabilities of hyperlens, in addition, but also has the Fourier transform function, making it an exceptional combination of super-resolution with desirable functions of conventional lenses. Anyway, it may be mentioned that metamaterial-based hyperlens and the metalens share the same material requirement as far as super-resolution is concerned, but for phase matching extra care needs to be taken in metamaterial-based metalens. Various types of phase compensation mechanism have been proposed for focussing plane waves in such metalenses. The underlying design principle is to satisfy the phase matching condition to constructively bring plane waves from air into deep sub-wavelength focus inside the metamaterial. This can be accomplished by either geometric variations, such as plasmonic waveguide couplers [74] or material refractive-index variations like gradient-index metamaterials [75].

Luneburg lens is a spherical multi-layered lens which is capable of focussing a wave from any direction. By placing a series of feeds on the opposite side of the desired direction of the signal, the Luneburg lens is capable of receiving/transmitting simultaneously in any direction. Using Luneburg lens technology it is possible to construct a multi-beam antenna capable of simultaneous multi-satellite tracking or capable of receiving/transmitting several signals in any direction (360°) simultaneously.

A traditional Luneburg lens becomes increasingly heavy and difficult to manufacture with increased diameter, thus restricting its size (approximately 200 kg for a 1.0 m traditionally built Luneburg lens). But a Luneburg lens made of metasurface weighs approximately eight times less for the same size Luneburg lens made of conventional material or RHM.

Introduction of non-uniform impedance surfaces in lenses with the help of metasurface has resulted in very thin and ready-to-manufacture lenses.

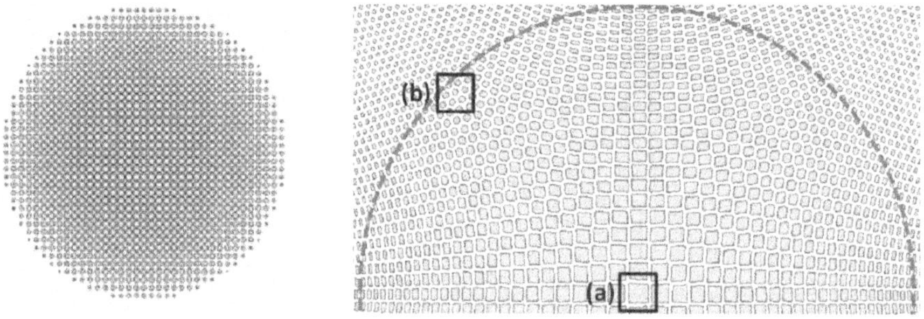

(a) (b)

Figure 6.18. Luneburg lens (a) with metasurface having gradually decreasing radii of unit cells from center to the rim. (b) Metasurface elements having smoothly varying, i.e., asymmetric polygons. (Reproduced/adapted from [11], with permission from De Gruyter, copyright (2018) Li *et al.*)

Figure 6.18(a) shows one such metasurface lens in which the radii of the patches, i.e., the unit cells that make up the metasurface, are gradually decreased as one moves from the centre of the lens [11, 76]. On such a surface, the travelling electromagnetic wave encounters gradually varying surface impedance and the corresponding change in phase velocity. The impedance profile is obtained by combining the Luneburg lens design with TM surface wave dispersion relation. In another design, the metasurface elements have smoothly varying, i.e., asymmetric polygons, as shown in figure 6.15(b). Both the methods are useful to design surfaces with spatially varying refractive indices.

Further, specially designed 2D arrays of nanometer-scale metasurface antennas may allow bulky optical components to be shrunk down to a planar device structure. Khorasaninejad *et al* [77] show that arrays of nanoscale fins of TiO can function as high-end optical lenses. At just a fraction of the size of optical objectives, such planar devices could turn our phone camera or our contact lens into a compound microscope.

6.6.5 Metasurface-based waveguide and wave coupling

In addition to the application of metasurfaces in free-space and surface wave manipulation including radiation of electromagnetic waves, they can also be used in the design of novel waveguides and also for wave coupling.

Electromagnetic energy can be trapped and guided in the region between two metasurfaces, see figure 6.19, since metasurface can be designed to have total reflection of an incident wave.

Assuming a waveguide mode propagating along z-direction, with propagation constant $\beta = k_0 \sin\theta$, the normalized propagation constant for TE mode propagation, for total internal reflection to occur, may be written as

$$n_e = \frac{\beta}{k_0}\sqrt{\frac{\chi_{\rm MS}^{zz}\chi_{\rm ES}^{yy} + \dfrac{4}{k_0^2}}{-\chi_{\rm MS}^{zz}\chi_{\rm ES}^{xx}}} \tag{6.12}$$

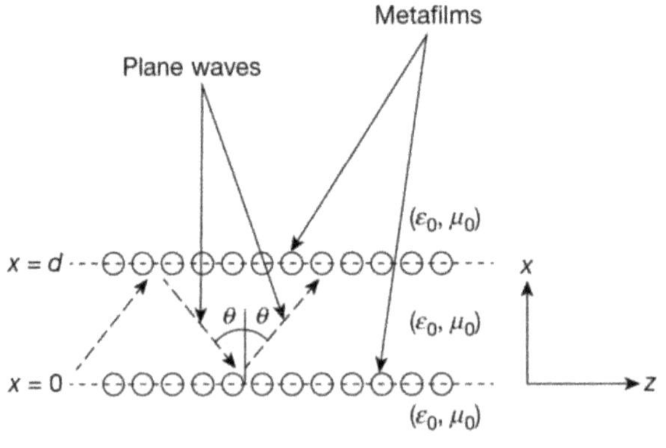

Figure 6.19. An illustration of metasurface-based waveguide.

Since for TE mode the electric and magnetic susceptibilities must satisfy the condition for total reflection as [8, 61]

$$k_0^2\chi_{\mathrm{MS}}^{xx}\left(\chi_{\mathrm{ES}}^{yy} + \chi_{\mathrm{MS}}^{zz}\sin^2\theta\right) = -4 \tag{6.13}$$

Once the scatterers that compose the metasurface are chosen to meet the above criterion and β is determined, the transverse wavenumber in the x-direction is given by

$$k_x = \sqrt{k_0^2 - \beta^2} = k_0\sqrt{1 - n_e^2} \tag{6.14}$$

and the separation distance, d, between the two metasurfaces is given by

$$d = \frac{\left(n + \dfrac{1}{2}\right)\pi + 2\tan^{-1}\left(\dfrac{k_x\chi_{\mathrm{MS}}^{zz}}{2}\right)}{k_x} \tag{6.15}$$

where $n = 1,2,3, \ldots$.

It is desirable to find conditions for which the imaginary part of n_e is as small as possible, and we need to ensure that $Im(n_e) < 0$ and $0 < Re(n_e) < 1$ (the latter conditions because otherwise the mode will basically be a surface wave localized near the two metasurfaces, and will likely suffer increased attenuation as a result). Similar sets of expressions may also be derived for TM modes [78].

The metasurface-based waveguide promises to be compact, with low material and radiation losses. If the metasurfaces are constructed in polymer type material, then it should be possible to develop a flexible wave-guiding structure. If the scatterers composing the metafilm are then chosen properly, a flexible low-loss waveguide could be developed, which could have potential applications at THz frequencies. It is also possible to control the properties of the scatterers. This could be used to design controllable or smart frequency-agile wave-guiding structures.

Other practical metasurface-waveguide structures may also consist of a narrow strip of high impedance bordered by two low impedance sheets [79]. In other words, the guide is made using an anisotropic impedance region, where the transverse impedance matches the surrounding surfaces. Initially it was observed that below a certain radius of curvature, it was impossible to bend the propagating wave without suffering from significant power leakage. However, ultimately it was found that bending to arbitrary angles is possible by controlling the impedance profile of the guide [80].

In relation to the concept of photonic topological insulators, it is possible to build unidirectional, polarization-locked waveguides where light is guided along an arbitrary path at the interface between two complementary impedance surfaces, such as a capacitive surface in contact with an inductive surface. Referred to as 'line waves', these structures support a composite TE/TM mode at the interface. The mode has a singularity at the interface and is therefore highly confined. Its polarization-locked propagation ensures reflection-free, quasi non-reciprocal behaviour. By controlling the impedance on each side of the guide, it is possible to curve

EM waves around bends smoothly. It is also possible to build a variety of conventional microwave and optical devices, such as the magic-T structure, smooth bends and so forth [11].

In addition to the controlled guidance of electromagnetic waves, metasurfaces can also be used for coupling between guided waves and waves propagating in free space. By designing the in-plane effective wave vector using metasurface structures, we are able to realize conversion between two different guided modes or between a guided mode and a mode propagating in the free space. It may be mentioned that there are a number of major differences between mode conversion using metasurfaces and using conventional grating-based mode convertors [10].

6.6.6 Miniaturized resonators based on metasurface

The classical resonator size is known to be $n\lambda/2$ (n = 1,2,3,...) with minimum size being $\lambda/2$ at the operating wavelength. This classical lower bound for resonator can be reduced if a cavity is partially filled with a negative-index material, i.e. metamaterial [81]. But if one uses metasurface instead of metamaterial the resonator size can be further miniaturized as 2D metasurface requires less physical space compared to 3D metamaterial. The phase matching conditions required at resonance for the separation distance, d, between two metal plates with a metasurface placed in the centre is given by [82]

$$
\begin{aligned}
d &= \frac{\lambda}{2}\left[1 + n\pi - \frac{2}{\pi}\tan^{-1}\left(\frac{\pi}{\lambda}\chi_{\mathrm{E}}\right)\right] \\
d &= \frac{\lambda}{2}\left[2n\pi - \frac{2}{\pi}\tan^{-1}\left(\frac{\pi}{\lambda}\chi_{\mathrm{M}}\right)\right]
\end{aligned}
\tag{6.16}
$$

for n = 0,1,2,3, ...

Where $n = 0$ is not allowed if $\chi_{\mathrm{M}} \geqslant 0$. From these expressions it may be observed that by judiciously choosing the metasurface, it is possible to have resonators that significantly overcome the $\lambda/2$ size limit; various cases for different conditions on the scatterers have been investigated by researchers [82]. A square-patch metafilm placed at the center of a resonator was found to reduce the size by as much as 56% [8]. If metasurfaces are redesigned with scatterers having more elaborate polarizability characteristics (e.g. judiciously chosen resonant behaviour), it should be possible to achieve even greater reduction in size. In fact, by controlling the properties of the metasurface, a frequency-agile resonator could be realized.

6.6.7 RF to electricity—energy harvesting with metasurface

Energy harvesting or energy scavenging is the process of harvesting readily available energy from the environment or from surrounding systems and converting it to usable electrical energy. In 1968, Glaser introduced the concept of harvesting power from the Sun known as space-based solar power (SBSP). SBSP collects the solar energy (power) by using arrays of solar cells and then converting the solar power to microwave power. By using highly directive antennas, this power is transmitted to the Earth where large rectenna (rectifying antenna) arrays placed at the receiver site receive the microwave power and convert it back into DC [83]. Radio frequency (RF) energy harvesting [84]—that seems to sound like something out of a science fiction novel—is expected to be a very promising solution for the future, replacing batteries and supplying power to autonomous electronic devices. RF harvesters are also the focus of much attention in recent times because urban areas are strongly backed with RF waves at different frequency bands, produced by the ubiquitous presence of television stations, FM radios, bluetooth devices, cell phones and other communication devices, WiFi stations and much more. With such a huge explosion of radio-based technologies, from which the emitted RF energy is otherwise getting wasted (also polluting the environment with radio frequency signal) this needs to be collected/absorbed and harnessed for fruitful purposes. With the advancements of the metamaterial and metasurface technologies of today the time is ripe to look for the new devices and techniques that could benefit from collecting this wasted RF energy and putting it to good use for mankind.

A research team from the university of south Florida recently reported (March 2022) [85] that an energy efficient metasurface antenna, with high absorption capability of radiowaves, is able to transform low-intensity radio wave from cell phone tower to electrical power. Their test done with radio source between 0.7 and 2.0 GHz, found that the metamaterial/metasurface antenna could harvest 100 microwatt of electrical power from the radiowaves of a cell phone tower 100 m away from the antenna. This electrical power may be useful for powering smart home sensors, such as those used for temperature, lighting and motion, or sensors used to monitor the structure of buildings or bridges, where replacing a battery might be difficult or impossible. Further, by eliminating wired connections and batteries, these antennas could help reduce costs, improve reliability and make some electrical systems more efficient. This metamaterial/metasurface antenna can also be built to be more transportable, allowing better electrical harvesting while on the go or in the background with smart devices. The present metamaterial/metasurface antenna for RF to electricity conversion is somewhat larger in size and with low efficiency of conversion in collecting the RF energy from cell phone environment/cell phone tower. However, research is ongoing to make a version that is able to collect RF energy from multiple radio wave sources simultaneously with larger electrical power and with better RF-to-electrical power conversion efficiency.

6.7 A brief scenario of the developments and challenges/new possibilities ahead for meta-research

6.7.1 Introduction

The phenomenal and continued interest in metamaterials and metasurfaces over the past 20 years has been driven by the idea that any phenomenon, technique, or device based on the properties of its constituent materials may be open to reexamination, for it may possess a 'meta' counterpart with paradigm shifting implications, useful new attributes, or improved performance. As a result of the attention garnered by metamaterials/metasurfaces in the academic community, there has also been a growing recognition of their potential commercial value in industry. Thus, meta-research has arrived at a crossroads, at which multiple academic disciplines and industries have subsequently been brought together to initiate a mass transition from proof-of-concept to practical applications in real-world problems, so as to encourage their adoption as commercially viable alternatives to existing mainstream technologies. In this context, several exciting trends have emerged in the recent metamaterial/metasurface research, opening important opportunities for a wide range of technologies.

At present, there are several hundreds of research groups actively pursuing 'meta' research programs worldwide, and several institutional networks and funding mechanisms have been established solely for the purpose of advancing this science and technology. Several tens of thousands of articles have been published, during the last two decades, at a truly astonishing rate in the leading journals in several research communities. With its continuous maturity with time, the field is becoming significantly more interdisciplinary. Beginning with the principles of traditional solid-state physics, microwave engineering to plasmonics of optics, and materials science it has subsequently added to its list acoustics, thermal engineering, quantum concepts and so forth. Hundreds of novel metamaterial-/metasurface-infused devices have been proposed in fields ranging from biomedicine to aerospace/defence to telecommunications and many more. This trend has necessitated concurrent advances in computing and fabrication technology, which have enabled investigations that may have previously been considered out of reach; as such, artificial intelligence, machine learning, and micro- and nanotechnologies are expected to be integral facets of future metamaterial/metasurface research. A number of review papers are available [86, 87] and [8–13] that provide glimpses of the developments made so far and challenges/possibilities ahead for metamaterials and especially for metasurfaces research in future. Reports are also available that project the global market for metamaterials and metasurfaces looking towards 2033 [88]. It is forecasted that the metamaterial/metasurface market, by the next decade, will become a multi-billion dollar market which is at present just >100 million dollar market with more than 50 metamaterials/metasurface commercial product developers worldwide in recent times. Thus, in the concluding part of this textbook let us summarize in brief all the relevant issues related to meta-research, especially the challenges and new possibilities ahead for it.

6.7.2 The scenario of the developments of metamaterials and metasurfaces for various applications

With the first demonstration of negative refraction in 2001 with metamaterial at microwave frequency (on the basis of TW and SRR proposed by J B Pendry in 1996 and 1999, respectively) a flurry of research activity was ignited in the field of metamaterials. As metamaterial science and technology demands knowledge from solid-state physics to microwaves and plasmonics of optics, it has attracted immediate and sustained interest from the microwave and optical communities, each emboldened by the essentially limitless possibilities of discovering one or another new property or application thereof. This frenetic pace was periodically energized by the continued contributions from hundreds of researchers around the world. It was again J B Pendry in 2000 who first proposed that a properly designed negative-index plane slab could offer, at least theoretically, infinite resolution, breaking the classical ‘diffraction limit’. Pendry further suggested in 2006 that metamaterials could be used to alter electromagnetic space (i.e., bend light) in just the right way as to render objects ‘invisible’. Both the revolutionary concepts of the so-called ‘superlens’ by plane metamaterial slab and ‘invisibility cloak’ were ultimately proved in experiments. With continued innovations in the field of metamaterials, several metamaterial implementations beyond the TW/SRR medium were also developed like: planar transmission-line metamaterials, complementary-SRR-based metamaterials, truly plasmonic metamaterials at optical frequency, all-dielectric metamaterials, to name just a few. All these led to many practical applications benefiting from or inspired by metamaterial properties, such as leaky-wave antennas capable of back-fire to end-fire radiation through broadside, ultra-miniaturized efficient resonators and antennas, and compact phase shifters, power dividers, including other microwave passive and active components and devices. Developments of metamaterials at terahertz and optical frequency range have opened up a new vista of applications with functionalities that was not possible with natural materials—and now we have the scope for some such metamaterial-based applications that could not be dreamt of two decades back. Chapters 1–4 of this book deal with all these developments in details, and chapter 5 is devoted to metamaterial modelling and characterization techniques.

Although the history of development of metasurfaces can be traced back to a work of 2003 by Sievenpiper *et al* [89], the pioneering works of Capasso and Shalaev in the area of gradient metasurfaces [15, 17], based on the manipulation of the amplitude and phase of the impinging wavefront with engineered nanostructures, jumpstarted a remarkable interest in the control of the optical wavefront with the engineered ultrathin 2D metamaterial, i.e. the metasurface. This has henceforth opened up a new avenue of applications of metasurfaces, especially in terahertz and optical frequency domain. Active, tunable/reconfigurable and nonlinear metasurfaces are now available with many potential applications. Metasurface-based absorbers, metasurface in a variety of antenna design, metasurface-based metalens developments and so forth are all enriching the RF and photonic applications every day—details of all these developments have been covered in this chapter, chapter 6, of this book.

6.7.3 Challenges and possibilities ahead for meta-research

Although the physics of metamaterial (MTM) is fairly well established theoretically, numerically and experimentally with the realization of many exciting phenomena like negative refractive index and near-field focussing with plane slab have been abundantly demonstrated in literature including cloaking and other microwave through THz to optical domain applications; however, many challenges and possibilities are still ahead for this artificial material that is believed to mimic the natural material but with counter-intuitive characteristics.

The average lattice constant p in current metamaterial is still electromagnetically too large for high-quality refraction. Current metamaterials are typically characterized by $p/\lambda = 1/5 \ldots 1/20$. Although refraction phenomena dominate in such metamaterial structures; diffraction/scattering affects the purity of refracting effects, thereby diffusing the focussing spot and increasing the transmission losses. To obtain pure refractive phenomena, it is necessary to homogenize the metamaterial in true sense, i.e., to decrease the structural feature p/λ by one order of magnitude or more. If this challenge is met, metamaterial will behave as real artificial materials, i.e. the unit cell of metamaterial would be so small that they would really behave like atoms in natural materials and would be capable of producing truly homogeneous macroscopic response, thereby faithfully mimicking the microscopic response of natural material. Just for a comparison, in natural media (such as air or Teflon) the atomic lattice constant is of the order of angstrom ($p \approx 10^{-10}$ m); therefore, ideally at microwave frequency perfect homogenization needs p/λ to be of the order of 10^{-9}. Anyway, such a huge ratio between λ and p is not required in practice to obtain good refraction. A decrease of p/λ from ~0.1 to 0.001 would be quite sufficient, since scattering would have then become completely negligible making refraction the only dominant phenomena—as desired in metamaterial to mimic natural material. But the question is whether it is at all possible or realistic to realize/fabricate structures with such dimension at THz and optical frequencies or even at microwave frequencies with presently existing fabrication technologies.

In the case of resonant-type metamaterial, a true homogenization even in practical sense is difficult to be solved. At the same time, bandwidth increase without prohibitive increase of losses is impossible in resonant-type metamaterial. However, with transmission-line (TL) metamaterial the problem of homogenization may be met to some extent at microwave/millimetre-wave frequencies [90]. The requirement of RH parameters L_{Rh} and C_{Rh} that demands a decrease in their value by a factor of, say ζ, can be realizable as the inductance and capacitance of a structure decreases when the size of the structure decreases. But the LH parameters demand an increase of L_{Lh} and C_{Lh} to be increased by the factor ζ. This seems to represent a real challenge as much more electric and magnetic energy needs to be stored in a much smaller volume! This challenge requires new architectures and new technologies. Development of novel 2D configurations, including high-permittivity ceramic or ferroelectric metal–insulator–metal (MIM) capacitors as well as high permeability ferrimagnetic spiral inductors is expected to address these challenges. Capacitors will not be of much problem as that will need very small spacing or very high permittivity in MIM configuration. But due to the absence of magnetic charges, it is difficult to realize inductors. One possible way to synthesize high-value inductance in a very small volume could be to embed inductor wires within nano-structured (much smaller pitch than the unit cell of the metamaterial) materials made of nano-ferrite particles that can achieve both high permeability and low loss. Nano-ferrites are composite ferromagnetic materials made of clusters of magnetic dipoles isolated from each other. Further, for 1D CRLH TL structure with a given unit cell size p the *increase* of the product $L_{\mathrm{Lh}}C_{\mathrm{Lh}}$ and *decrease* of the product L_{R} C_{Rh} would lead to increase in bandwidth too.

Superconducting and quantum metamaterials offer an incredibly fertile arena for research [91, 92]. Superconducting metamaterials will provide a dramatic reduction of losses, accompanied by access to extreme sensitivity of the superconducting state to external stimuli and exceptional nonlinearity of superconductors enabling low energy switching at sub-attojoule level. Negative dielectric constant and dominant kinetic resistance also make superconductors an intriguing plasmonic media. Even more intriguing is the fact that the classical object of metamaterial research—the ubiquitous split-ring metamolecule—has much in common with the fundamental unit of superconductivity, the Josephson junction ring. In fact, an array of superconducting Josephson rings could be a truly quantum metamaterial, where each metamolecule is a multilevel quantum system supporting phase qubits. However, applications of superconducting metamaterials will be limited to the microwave domain for niobium-based metamaterials, and to the THz spectral domain if high-temperature superconductors are used. This is because higher frequencies destroy the superconducting phase. In natural solids, optical response is determined by the quantum energy-level structure of the constituent atoms or molecules. By contrast, the electromagnetic properties of metamaterials are derived from the resonant characteristics of the sub-wavelength plasmonic resonators (SRR etc) from which they are constructed. Thus 'quantum metamaterials' would provide a much closer analogy to natural crystals which the metamaterial mimics. Anyway, the manufacture of large metamaterial arrays of Josephson rings is a highly sophisticated process that is not yet widely available. It may, however, be noted that the cryo-cooling requirements for superconductors are no longer a serious technological limitation as compact cryodevices are now widely deployed in telecommunications and sensing equipment.

However, no progress in metamaterial research will be possible without further developments in fabrication technology. New techniques will have to achieve perfection of nanostructures at close to the molecular level, and at low cost. We need to go beyond electron-beam lithography, focussed ion beam milling, and nano-imprint. The new techniques must be able to build metamaterials to almost any blueprint. In addition, the real challenge is to create truly volume (i.e. 3D) metamaterials, isotropic and homogeneous ones; a great deal of innovative efforts have been made on that and we can expect marvelous results in years to come.

Although it has been the objective of most researchers to develop isotropic metamaterials, anisotropic metamaterials may also be useful in specific applications, such as quasi-optical beam forming or reflector system. An 'orthogonally anisotropic' RH/LH TL metamaterial has been reported [93] and also a 'resonance cone' spatial spectrum analyzer metamaterial (constituted of a planar wire-grid network with inductors along one direction and capacitors in the other orthogonal direction) have been reported [94] in literature. The idea of anisotropy can be pushed further to non-uniform or graded-index metamaterials, which have been used in beam-tuning CRLH antenna [95].

Nonlinear metamaterials may be another possibility for optical applications. The first nonlinear metamaterial was reported by Zharov *et al* [96]. This paper theoretically considered two different types of nonlinearity in the SRR-TW structure: nonlinearity due to the presence of a nonlinear dielectric (Kerr effect) embedding the TWs and SRRs and nonlinearity due to dependence of μ_r on the local field intensity in the gap of the rings. A similar analysis for SRR array alone and with parameters appropriate to optical frequencies was reported [97]. Zharov *et al* further investigated the possibility of soliton phenomena in nonlinear metamaterials [98]. Solitons are a particular type of solitary waves, which have the specific property of being orthogonal, in the sense that when two of these waves cross one another in the medium their intensity profile is not altered—only phase shifts are imparted as a result of the interaction. To be more technical, solitons are pulse-like stationary waves, which may propagate without ever altering in shape as they propagate in a dispersive and nonlinear medium as a consequence of an exact balancing between the phenomena of group velocity dispersion and self-phase modulation [99].

It may be commented herewith that exponentially growing research has been going on for more than one decade or so on 'metasurfaces'—the so-called 2D metamaterials that has opened up a new horizon with new capabilities and possibilities for advancing the meta-research of the future, especially in THz and optical regime, which is expected to change the whole landscape of meta-research of the future as it evolves into a new dimension with undreamt of applications in the coming days.

Anyway, a few more glimpses of the developments and new possibilities and challenges ahead for meta-research, reported in recent times, may be summarized below, which will show that the wealth of potential concepts and applications of metamaterials/metasurfaces is limited only by our imagination.

Although the chiral metamaterials have greatly impacted the field of optical sensing over the past decade [100] metasurfaces have attracted significant attention in recent times because of their superior ability to enhance the chiroptical response (such as optical rotator dispersion and circular dichroism, CD, between left and right circularly polarized light) by manipulating amplitude, phase and polarization of electromagnetic fields in a novel way [101]. The field of chiral plasmonics has met huge progress with machine learning-mediated metamaterial prototyping too. Ashalley *et al* [102] presented an end-to-end functional bidirectional deep-learning model for 3D chiral metamaterial design and optimization. This model efficiently explored the sensing of biomolecular enantiomers and showed a promising potential to other applications including photodetectors, polarization-resolved imaging and CD spectroscopy.

Bi-anisotropic metasurfaces which exhibit magneto-electric coupling, i.e. which can acquire magnetic (electric) polarization when excited by electric (magnetic) external field have become important from an application point of view [51]. Non-reciprocal bi-anisotropic metasurfaces can be used to design, for instance, various types of isolators with simultaneous control of amplitude and polarization of transmitted waves [103].

An increasing interest in bringing novel functionalities enabled by flat photonics to the realm of quantum optics has recently been noticed. Quantum optical technologies require sources of single photons, entangled photons and other types of non-classical light, as well as novel methods of detection. The quantum states could be based on different degrees of freedom of light polarization, direction and orbital angular momentum. Metasurfaces have a real potential for the realization of each of these states. Rapid progress in the development of metaphotonics has allowed bulky optical assemblies to be replaced with metasurfaces, opening a broad range of novel and superior applications of flat optics to the generation, manipulation and detection of classical light [104, 105].

Active liquid crystals (LCs)-based metamaterials have been attracting an increasing amount of attention in the past few years because of their controlled characteristics leading to emerging applications, including modulators, switches and filters. The use of LCs allows for the achievement of metadevices with dynamic functions induced by electro-, magneto-, photo- and temperature-sensitive properties. Kowerdziej *et al* [106] reported on the shortening of switching times of various soft-matter-based tunable metamaterials to improve the functionality of modern active devices. As a matter of fact, the frequency convertible dielectric anisotropy of the dual-frequency mixture has shown to provide the opportunity to create a fast-response in-plane switching metasurface at the nanoscale, which could be tuned by an electrical signal with different frequencies [107].

In recent years, there has been tremendous progress in the theory and experimental implementations of non-Hermitian photonics, including parity-time (PT) symmetric systems [108, 109]. PT symmetry is a fascinating concept to make sense of non-Hermitian Hamiltonians. In particular, non-Hermitian concepts can be translated to electromagnetic structures by means of spatial modulation of loss and gain, which is becoming technologically viable in artificial materials and metamaterials [110]. The intrinsic capability of photonics of creating and superposing non-Hermitian eigenstates through optical gain and loss is ideal for exploring various non-Hermitian paradigms. PT metamaterials and metasurfaces may operate in microwave frequencies and higher, suppress losses and provide tunability through the natural appearance of a transition to a PT-broken phase. These systems enable new pathways for metasurface design using phase, symmetry and topology as powerful tools [111]. The research area of non-Hermitian photonics is relatively new and it is still largely unexplored. Novel theoretical ideas combined with new experimental schemes are expected to produce more surprising results, leading to altogether different, previously unknown, means for controlling light–matter interactions both in the classical and the quantum regimes.

During the recent pandemic of COVID-19, the THz metamaterial researchers were not sitting idle. In the diagnosis of severe contagious diseases, including COVID-19, there is an urgent need for protein sensors with large refractive index sensitivities. THz radiation in the electromagnetic spectrum with its non-ionizing property has the ability to sense materials with an extremely high accuracy. THz devices have potential applications in diverse areas such as imaging, high-resolution spectroscopy and biomedical analytics. However, the THz metamaterials developed till 2020 could not be used to develop such highly sensitive protein sensors due to their low refractive index sensitivities. Silalahi *et al* [112] proposed an efficient method based on a patterned photoresist to float the SRRs of a terahertz metamaterial at a height of 30 μm from its substrate that is deposited with complementary SRRs, and is compatible with all geometrical structures of THz metamaterials to increase their refractive index sensitivities. This floating THz metamaterial has demonstrated an extremely large refractive index sensitivity of 532 GHz/RIU (RIU stands for 'refractive index unit') because its near field is not distributed over the substrate and also because the complementary SRRs confine the field above the substrate. The floating THz metamaterial was able to sense bovine serum albumin (BSA) and the protein binding of BSA and anti-BSA when BSA, and anti-BSA solutions with low concentrations that are smaller than 0.150 μmol L^{-1} were sequentially dropped onto it.

Gradient metasurfaces [15] and their various applications from optics (metalenses, meta-holograms) to radiowaves have been driving the meta-research community to more applied directions. Commercial interest in these efforts has been growing fast and this field of technology promises tremendous opportunities for real-life impact on the near future. One missing component in this growing field is the possibility of real-time reconfigurability of the functionality encoded in metasurfaces. Reconfigurable ultrathin surfaces providing ad hoc wavefront transformations to the impinging waves [39, 40, 113], ideally responding in real time to changes in the environment, may form the basis for smart surfaces with significant impact on nearly any electromagnetic and photonic application, from classical to quantum photonics, from radar to wireless technology.

Further, it may be predicted that the currently maturing nanotechnologies with novel metamaterials/metasurfaces will give birth to intelligent novel metamaterials/metasurfaces with unique performance, such as programmable or automatic reconfigurability for instance. One report [114] already shows the way to the so-called digital metamaterial/metasurface and even field programmable metamaterial/metasurface which can be used to control electromagnetic waves in real time. Such metamaterial/metasurface, it is expected, can lead to software-defined metamaterials/metasurfaces and cognitive metamaterials/metasurfaces. In the future, metamaterials/metasurfaces may become part of complex radio-frequency integrated circuits (RFICs) and monolithic integrated circuits (MMICs) technology implementation. Anyway, it may finally be concluded that in addition to pursuing the fundamental issues of meta-research there is also a large push toward application-driven research, with the end goal of dispelling the hyperbole accumulated over the years and engendering real confidence in their practical application as viable alternatives to established, more mainstream technologies. Thus the future of human society is expected to greatly benefit with the ripening fruits in various gardens of meta-research, whose seeds have been sown today with its humble beginning in the dawn of this century.

References

[1] Shelby R, Smith D R and Schultz S 2001 Experimental verification of a negative index of refraction *Science* **292** 77–9

[2] Das B 2009 India joins the metamaterials club *Nat. India* **August 2009** (https://www.nature.com/articles/nindia.2009.273)

[3] Pendry J B 2000 Negative refraction makes a perfect lens *Phys. Rev. Lett.* **85** 3966–9

[4] Iyer A K and Elefteriades G V 2009 Free-space imaging beyond the diffraction limit using Veselago–Pendry transmission-line metamaterial superlens *IEEE Trans. Antennas Propag.* **57** 1720–7

[5] Schurig D, Mock J J, Justice B J, Cummer S A, Pendry J B, Starr A F and Smith D R 2006 Metamaterial electromagnetic cloak at microwave frequencies *Science* **314** 977–80

[6] Gharghi M, Gladden C, Zentgraf T, Liu Y, Yin X, Vlentine J and Zhang X 2011 A carpet cloak for visible light *Nano Lett.* **11** 2825–8

[7] Ni X, Wong Z J, Mrejen M, Wang Y and Zhang X 2015 An ultrathin invisibility skin cloak for visible light *Science* **349** 1310–4

[8] Holloway C L, Kuester E F, Gordon J A, OHara J, Booth J and Smith D R 2012 An overview of the theory and applications of metasurfaces: the two-dimensional equivalents of metamaterials *IEEE Antennas Propag. Mag.* **54** 10–35

[9] Glybovski S B, Tretyakov S A, Belov P A, Kivshar Y S and Simovski C R 2016 Metasurfaces: from microwaves to visible *Phys. Rep.* **634** 1–72

[10] Chen H-T, Taylor A J and Yu N 2016 A review of metasurfaces: physics and applications *Rep. Prog. Phys.* **79** 076401

[11] Li A, Singh S and Sievenpiper D 2018 Metasurfaces and their applications *Nanophotonics* **7** 989–1011

[12] Bukhari S S, Vardaxoglou J Y and Whittow W 2019 A metasurfaces review: definitions and applications *Appl. Sci.* **9** 2727

[13] Quevedo-Teruel O *et al* 2019 Roadmap on metasurfaces *J. Opt.* **21** 073002

[14] Munk B A 2000 *Frequency Selective Surfaces* (New York: Wiley)

[15] Yu N, Genevet P, Kats M A, Aieta F, Tetienne J-P and Capasso F 2011 Light propagation with phase discontinuities: generalized laws of reflection and refraction *Science* **334** 333–7

[16] Engheta N and Zilokowski R W (ed) 2006 *Metamaterials—Physics and Engineering Explorations* (New York: IEEE Press, Wiley Interscience)

[17] Ni X J, Emani N K, Kildishev A V, Boltasseva A and Shalaev V M 2011 Broadband light bending with plasmonic nanoantennas *Science* **335** 427

[18] Zhang X Q, Tian Z, Yue W S, Gu J Q, Zhang S, Han J G and Zhang W L 2013 Broadband terahertz wave deflection based on C-shape complex metamaterials with phase discontinuities *Adv. Mater.* **25** 4567–72

[19] Zhao Y and Alù A 2011 Manipulating light polarization with ultrathin plasmonic metasurfaces *Phys. Rev.* B **84** 205428

[20] Hao J, Yuan Y, Ran L, Jiang T, Kong J A, Chan C T and Zhou L 2007 Manipulating electromagnetic wave polarizations by anisotropic metamaterials *Phys. Rev. Lett.* **99** 063908

[21] Yu N, Aieta F, Genevet P, Kats M A, Gaburro Z and Capasso F 2012 A broadband, background-free quarter-wave plate based on plasmonic metasurfaces *Nano Lett.* **12** 6328–33

[23] Kar S 2022 *Microwave Engineering—Fundamentals, Design, and Applications* 2nd edn (Hyderabad: Universities Press)

[24] OBrien S and Pendry J B 2002 Photonic band-gap effects and magnetic activity in dielectric composites *J. Phys. Condens. Matter* **14** 4035

[25] Zhao Q, Zhou J, Zhang F and Lippens D 2009 Mie resonance-based dielectric metamaterials *Mater. Today* **12** 60–9

[26] Yang Y M, Wang W Y, Moitra P, Kravchenko I I, Briggs D P and Valentine J 2014 Dielectric meta-reflectarray for broadband linear polarization conversion and optical vortex generation *Nano Lett.* **14** 1394–9
[27] Khorasaninejad M, Aieta F, Kanhaiya P, Kats M A, Genevet P, Rousso D and Capasso F 2015 Achromatic metasurface lens at telecommunication wavelengths *Nano Lett.* **15** 5358–62
[28] Li J, Liu C, Wu T, Liu Y, Wang Y, Yu Z, Ye H and Yu L 2019 Efficient polarization beam splitter based on all-dielectric metasurface in visible region *Nanoscale Res. Lett.* **14** 1–7
[29] Shadrivov I V, Morrison S K and Kivshar Y S 2006 Tunable split-ring resonators for nonlinear negative-index metamaterials *Opt. Express* **14** 9344–9
[30] Chen H T, Padilla W J, Zide J M O, Bank S R, Gossard A C, Taylor A J and Averitt R D 2007 Ultrafast optical switching of terahertz metamaterials fabricated on ErAs/GaAsErAs/GaAs nanoisland superlattices *Opt. Lett.* **32** 1620–2
[31] Chen H T, OHara J F, Azad A K, Taylor A J, Averitt R D, Shrekenhamer D B and Padilla W J 2008 Experimental demonstration of frequency-agile terahertz metamaterials *Nat. Photonics* **2** 295–8
[32] Heyes J E, Withayachumnankul W, Grady N K, Chowdhury D R, Azad A K and Chen H T 2014 Hybrid metasurface for ultra-broadband terahertz modulation *Appl. Phys. Lett.* **105** 181108
[33] Chen H T, Padilla W J, Zide J M O, Gossard A C, Taylor A J and Averitt R D 2006 Active terahertz metamaterial devices *Nature* **444** 597–600
[34] Miao Z, Wu Q, Li X, He Q, Ding K, An Z, Zhang Y and Zhou L 2015 Widely tunable terahertz phase modulation with gate-controlled graphene metasurfaces *Phys. Rev.* **5** 041027
[35] Yao Y, Shankar R, Kats M A, Song Y, Kong J, Loncar M and Capasso F 2014 Electrically tunable metasurface perfect absorbers for ultrathin mid-infrared optical modulators *Nano Lett.* **14** 6526–32
[36] Chen H-T, Padilla W J, Cich M J, Azad A K, Averitt R D and Taylor A J 2009 A metamaterial solid-state terahertz phase modulator *Nat. Photonics* **3** 148–51
[37] Sautter J, Staude I, Decker M, Rusak E, Neshev D N, Brener I and Kivshar Y S 2015 Active tuning of all dielectric metasurfaces *ACS Nano* **9** 4308–15
[38] Ou J-Y, Plum E, Zhang J and Zheludev N I 2013 An electromechanically reconfigurable plasmonic metamaterial operating in the near-infrared *Nat. Nanotechnol.* **8** 252–5
[39] Chen K *et al* 2017 A reconfigurable active Huygens metalens *Adv. Mater.* **29** 1606422
[40] Radi Y and Alù A 2018 Reconfigurable metagratings *ACS Photon.* **5** 1779–85
[41] Lee J, Tymchenko M, Argyropoulos C, Chen P Y, Lu F, Demmerle F, Boehm G, Amann M C, Alu A and Belkin M A 2014 Giant nonlinear response from plasmonic metasurfaces coupled to inter-sub-band transitions *Nature* **511** 65–9
[42] Campione S, Benz A, Sinclair M B, Capolino F and Brener I 2014 Second harmonic generation from metamaterials strongly coupled to inter-sub-band transitions in quantum wells *Appl. Phys. Lett.* **104** 131104
[43] Huang J and Encinar J A 2007 *Reflectarray Antennas* (Hoboken, NJ: Wiley)
[44] Mesa F, Rodríguez-Berral R and Medina F 2018 Unlocking complexity with ECA. The equivalent circuit model as an efficient and physically insightful tool for microwave engineering *IEEE Microw. Mag.* **19** 44–65

[45] Asadchy V, Díaz-Rubio A, Kwon D-H and Tretyakov S 2019 Analytical modelling of electromagnetic surfaces *Surface Electromagnetics with Applications in Antenna, Microwave, and Optical Engineering* ed F Yang and Y Rahmat-Samii (Cambridge: Cambridge University Press) pp 30–65
[46] Asadchy V, Díaz-Rubio A, Tcvetkova S, Kwon D-H, Elsakka A, Albooyeh M and Tretyakov S 2017 Flat engineered multichannel reflectors *Phys. Rev.* X **7** 031046
[47] Díaz-Rubio A, Asadchy V, Elsakka A and Tretyakov S 2017 From the generalized reflection law to the realization of perfect anomalous reflectors *Sci. Adv.* **30** 1602714–23
[48] Radi Y, Sounas D and Alù A 2017 Metagratings: beyond the limits of graded metasurfaces for wave front control *Phys. Rev. Lett.* **119** 067404
[49] Epstein A and Rabinovich O 2017 Unveiling the properties of metagratings via a detailed analytical model for synthesis and analysis *Phys. Rev. Appl.* **8** 054037
[50] Epstein A and Eleftheriades G V 2016 Arbitrary power conserving field transformations with passive lossless omega-type bianisotropic metasurfaces *IEEE Trans. Antennas Propag.* **64** 3880–95
[51] Asadchy V, Díaz-Rubio A and Tretyakov S 2018 Bianisotropic metasurfaces: physics and applications *Nanophotonics* **7** 1069–94
[52] Holloway C L, Deinstfrey A, Kuester E F, OHara J F, Azad A K and Taylor A J 2009 A discussion on the interpretation and characterization of metafilms/metasurfaces: the two-dimensional equivalent of metamaterials *Metamaterials* **3** 100–12
[53] Vahabzadeh Y, Chamanara N, Achouri K and Caloz C 2018 Computational analysis of metasurfaces *IEEE J. Multiscale Multiphys. Comput. Tech.* **3** 37–49
[54] Kuester E F, Mohmed M A, Piket-May M and Holloway C L 2003 Averaged transition conditions for electromagnetic fields at a metafilm *IEEE Trans. Antennas Propag.* **51** 2641–51
[55] Chamanara N, Achouri K and Caloz C 2017 Efficient analysis of metasurfaces in terms of spectral-domain GSTC integral equations *IEEE Trans. Antennas Propag.* **65** 5340–7
[56] Tretyakov S 2003 *Analytical Modeling in Applied Electromagnetics* (Norwood, MA: Artech House)
[57] Francavilla M, Martini E, Maci S and Vecchi G 2015 On the numerical simulation of metasurfaces with impedance boundary conditions *IEEE Trans. Antennas Propag.* **63** 2153–61
[58] Oksanen M I, Tretyakov S A and Lindell I V 1990 Vector circuit theory for isotropic and chiral slabs *J. Electromag. Waves Appl.* **4** 613–43
[59] Scher A D and Kuester E F 2009 Extracting the bulk effective parameters of a metamaterial via the scattering from a single planar array of particles *Metamaterials* **3** 44–55
[60] Holloway C L, Kuester E F and Dienstfrey A 2011 Characterizing metasurfaces/metafilms: the connection between surface susceptibilities and effective material properties *IEEE Antennas Wirel. Propag. Lett.* **10** 1507–11
[61] Holloway C L, Mohamed M A, Kuester E F and Deinstfrey A 2005 Reflection and transmission properties of a metafilm: with an application to a controllable surface composed of resonant particles *IEEE Trans. Electromagn. Compat.* **47** 853–65
[62] Liu N, Mesch M, Weiss T, Hentschel M and Giessen H 2010 Infrared perfect absorber and its application as plasmonic sensor *Nano Lett.* **10** 2342–8
[63] Liu X, Fan K, Shadrivov I V and Padilla W J 2017 Experimental realization of a terahertz all-dielectric metasurface absorber *Opt. Express* **25** 191–201

[64] Yao Y, Shankar R, Kats M A, Song Y, Kong J, Loncar M and Capasso F 2014 Electrically tunable metasurface perfect absorbers for ultrathin mid-infrared optical modulators *Nano Lett.* **14** 6526–32
[65] Luo Z, Long J, Chen X and Sievenpiper D F 2016 Electrically tunable metasurface absorber based on dissipating behaviour of embedded varactors *Appl. Phys. Lett.* **109** 071107
[66] Lee Y P, Rhee J Y, Yoo Y J and Kim K W 2016 *Metamaterials for Perfect Absorption Springer Series in Material Science* vol 236 (Singapore: Springer Nature)
[67] Esquius-Morote M, Gomez-Dias J S and Perruisseau-Carrier J 2014 Sinusoidally modulated graphene leaky-wave antenna for electronic beam scanning at THz *IEEE Trans. Terahertz Sci. Technol.* **4** 116–22
[68] Wang X C, Zhao W S, Hu J and Yen W-Y 2015. Reconfigurable terahertz leaky-wave antenna using graphene-based high-impedance surface *IEEE Trans. Nanotechnol.* **14** 62–9
[69] Fong B H, Colburn J S, Ottusch J J, Visher J L and Sievenpiper D F 2010 Scalar and tensor holographic artificial impedance surfaces *IEEE Trans. Antennas Propag.* **58** 3212–21
[70] Mavridou M, Konstantinidis K and Feresidis A P 2016 Continuously tunable mm-wave high impedance surface *IEEE Antennas Wirel. Propag. Lett.* **15** 1390–3
[71] Huang Y, Wu L S, Tang M and Mao J 2012 Design of a beam reconfigurable THz antenna with graphene-based switchable high-impedance surface *IEEE Trans. Nanotechnol.* **11** 836–42
[72] Pfeiffer C and Grbic A 2015 Planar lens antennas of subwavelength thickness: collimating leaky-waves with metasurfaces *IEEE Trans. Antennas Propag.* **63** 3248–53
[73] Lu D and Liu Z 2012 Hyperlenses and metalenses for far-field super-resolution imaging *Nat. Commun.* **3** 1205
[74] Ma C B and Liu Z 2011 Designing super-resolution metalenses by the combination of metamaterials and nanoscale plasmonic waveguide couplers *J. Nanophotonics* **5** 051604
[75] Ma C B, Escobar M A and Liu Z 2011 Extraordinary light focussing and Fourier transform properties of gradient-index metalenses *Phys. Rev.* B **84** 195142
[76] Bosiljevac M, Casaletti M, Caminita F, Sipus Z and Maci S 2012 Nonuniform metasurface Luneburg lens antenna design *IEEE Trans. Antennas Propag.* **60** 4065–73
[77] Khorasaninejad M, Chen W T, Devlin R C, Oh J, Zhu A Y and Capasso F 2016 Metalenses at visible wavelengths: diffraction-limited focussing and subwavelength resolution imaging *Science* **352** 1190–4
[78] Holloway C L, Kuester E F and Novotny D 2009 Waveguides composed of metafilms: two-dimensional equivalent of metamaterials *IEEE Antennas Wirel. Propag. Lett.* **8** 525–9
[79] Gregoire D J and Kabakian A V 2011 Surface-wave waveguides *IEEE Antennas Wirel. Propag. Lett.* **10** 1512–5
[80] Bisharat D J and Sievenpiper D F 2017 Guiding waves along an infinitesimal line between impedance surfaces *Phys. Rev. Lett.* **119** 106802
[81] Engheta N 2002 An idea for thin subwavelength cavity resonators using metamaterials with negative permittivity and permeability *IEEE Antennas Wirel. Propag. Lett.* **1** 10–3
[82] Holloway C L, Love D C, Kuester E F, Salandrino A and Engheta N 2008 Sub-wavelength resonances: on the use of metafilms to overcome the $\lambda/2$ size limit *IET Microwaves Antennas Propag.* **2** 120–9
[83] Glaser P E 1968 Power from the Sun: its future *Science* **162** 857–61

[84] Alvarado U, Juanicorena A, Adin I, Sedano B, Gutiérrez I and de Nó J 2012 Energy harvesting technologies for low-power electronics *Trans. Emerg. Telecommun. Technol.* **23** 728–41
[85] Fowler C, Silva S, Thapa G and Zhou J 2022 High efficiency ambient RF energy harvesting by a metamaterial perfect absorber *Opt. Mater. Express* **12** 1242–50
[86] Iyer A, Alu A and Epstein A 2020 Metamaterials and metasurfaces—historical context, recent advances, and future directions *IEEE Trans. Antennas Propag.* **68** 1223–31
[87] Ali A, Mitra A and Brahim Aïssa B 2022 Metamaterials and metasurfaces: a review from the perspectives of materials, mechanisms and advanced metadevices *Nanomaterials* **12** 32
[88] The global market for metamaterials and metasurfaces to 2033, *Report*, Future Markets Inc. (ID: 5449076),152 pages, November 2022
[89] Sievenpiper D, Schaffner J, Song H, Loo R and Tangonan G 2003 Two-dimensional beam steering using an electrically tunable impedance surface *IEEE Trans. Antennas Propag.* **51** 2713–22
[90] Caloz C and Itoh T 2006 *Electronic Metamaterials: Transmission Line Theory and Microwave Applications* (New York: Wiley Interscience)
[91] Zheludev N I 2010 The road ahead for metamaterials *Science* **328** 582–3
[92] Ricci M, Orloff N and Anlage S M 2005 Superconducting metamaterials *Appl. Phys. Lett.* **87** 034102
[93] Caloz C and Itoh T 2003 Positive/negative refractive index anisotropic 2D metamaterials *IEEE Microw. Wirel. Compon. Lett.* **13** 547–9
[94] Balmain K G, Lüttgen A A E and Kremer P C 2003 Power flow for resonance cone phenomenon in planar anisotropic metamaterials *IEEE Trans. Antennas Propag.* **51** 2612–8
[95] Lin I-H, Caloz C and Itoh T 2005 Near-field focussing by a nonuniform leaky-wave interface *Microw. Opt. Technol. Lett.* **44** 416–8
[96] Zharov A A, Shadrivov I V and Kivshar Y S 2003 Nonlinear properties of left-handed metamaterials *Phys. Rev. Lett.* **91** 037401
[97] OBrien S, McPeake D, Ramakrishna S A and Pendry J B 2004 Near-infrared photonic band gaps and nonlinear effects in negative magnetic metamaterials *Phys. Rev.* B **69** 241101
[98] Zharova N A, Shadrivov I V, Zharov A A and Kivshar Y S 2005 Nonlinear transmission and spatiotemporal solitons in metamaterials with negative refraction *Opt. Express* **13** 1291–8
[99] Agrawal G P 2001 *Nonlinear Fiber Optics,* 3rd edn (New York: Academic)
[100] Yoo S and Park Q-H 2019 Metamaterials and chiral sensing: a review of fundamentals and applications *Nanophotonics* **8** 249–61
[101] Kim J, Rana A S, Kim Y, Kim I, Badloe T, Zubair M, Mehmood M Q and Rho J 2021 Chiroptical metasurfaces: principles, classification, and applications *Sensors* **21** 4381
[102] Ashalley E, Acheampong K, Vazquez Besteiro L, Yu P, Neogi A, Govorov A and Wang Z 2020 Multitask deep-learning-based design of chiral plasmonic metamaterials *Photon. Res.* **8** 1213–25
[103] Mahmoud A M, Davoyan A R and Engheta N 2015 All-passive nonreciprocal metastructure *Nat. Commun.* **6** 8359
[104] Li C, Yu P, Huang Y, Zhou Q, Wu J, Li Z, Tong X, Wen Q, Kuo H-C and Wang Z M 2020 Dielectric metasurfaces: from wavefront shaping to quantum platform *Prog. Surf. Sci.* **95** 100584

[105] Solntsev A S, Agarwal G S and Kivshar Y S 2021 Metasurfaces for quantum photonics *Nat. Photonics* **15** 327–36
[106] Kowerdziej R, Wróbel J and Kula P 2019 Ultrafast electrical switching of nanostructured metadevice with dual-frequency liquid crystal *Sci. Rep.* **9** 20367
[107] Shrekenhamer D, Chen W-C and Padilla W J 2013 Liquid crystal tunable metamaterial absorber *Phys. Rev. Lett.* **110** 177403
[108] Feng L, El-Ganainy R and Ge L 2017 Non-Hermitian photonics based on parity-time symmetry *Nat. Photonics* **11** 752–62
[109] El-Ganainy R, Khajavikhan M, Christodoulides D and Ozdemir S 2019 The dawn of non-Hermitian optics *Commun. Phys.* **2** 37
[110] Lazarides N and Tsironis G P 2013 Gain-driven discrete breathers in PT-symmetric nonlinear metamaterials *Phys. Rev. Lett.* **110** 53901
[111] Yang F, Hwang A, Doiron C and Naik G V 2021 Non-Hermitian metasurfaces for the best of plasmonics and dielectrics *Opt. Mater. Express* **11** 2326–34
[112] Silalahi H, Chen Y-P, Chen Y-S, Hong S, Lin X-Y, Liu J-H and Huang C-Y 2021 Floating terahertz metamaterials with extremely large refractive index sensitivities *Photon. Res.* **9** 1970–8
[113] Karvounis A, Gholipour B, Macdonald K F and Zheludev N I 2016 All dielectric phase-change reconfigurable metasurface *Appl. Phys. Lett.* **109** 051103
[114] Cui T J, Shuo Liu S and Zhang L 2017 Information metamaterials and metasurfaces *J. Mater. Chem.* C **5** 3644–68

IOP Publishing

Metamaterials and Metasurfaces
Basics and trends
Subal Kar

Appendix A

Harmonic oscillator model of materials—Lorentz and Drude model

A.1 Lorentz model

The harmonic oscillator model of the motion of electron in material media was developed at higher frequencies, especially to evaluate the material properties at optical frequencies. Here an electron in the atom of a material is modelled as a driven damped harmonic oscillator where the electron (e^-, m_e) is supposed to be connected to the nucleus (N) via a hypothetical spring (with spring constant C), see figure A.1. The driving force is the applied time varying electric field (E_L). The oscillations do not grow to infinity due to the restoring force ($P = -er$) giving rise to a damping force having damping coefficient γ. The model for electron motion in material is useful to obtain expressions for the frequency-dependent dipole moment, polarization, susceptibility, and dielectric constant of dispersive materials.

From the dynamic point of view, the charge oscillation will lead to the charge redistribution, which will create an additional induced electric field. The induced field will restore the charge to its equilibrium position. Newton's second law of motion states that the product of mass with the acceleration equals the sum of all the

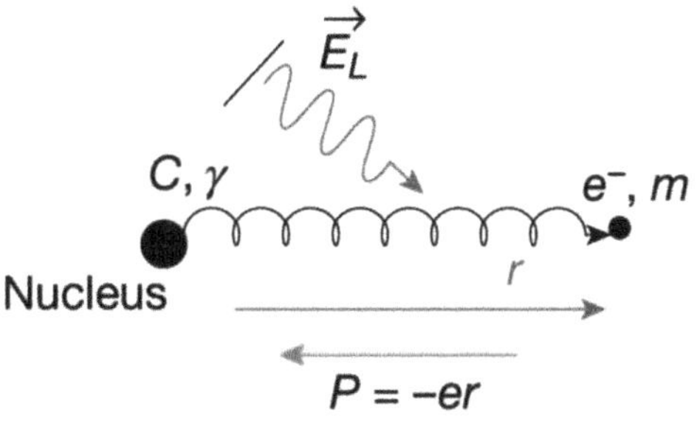

Figure A.1. The schematic diagram of spring-like classical driven damped harmonic oscillator system used for modelling the charges in a material.

doi:10.1088/978-0-7503-5532-2ch7

forces acting on the system. The governing equation of motion for the system in figure A.1 may thus be written as:

$$m_e \frac{\partial^2 \vec{r}(t)}{\partial t^2} = \sum_i F_i = F_E(t) + F_S(t) + F_D(t) \tag{A1}$$

where m_e denotes the effective mass of electron, $\vec{r}(t)$is the instantaneous distance deviation from its equilibrium position.

The first term on the right hand side of equation (A1), $\vec{F_E}(t) = q_e \vec{E}_0 e^{-j\omega t}$is the local force attributed to the driving electric field E (where we have assumed a harmonic time dependence of the driving electric field: $\vec{E}(t) = \vec{E}_0 e^{-j\omega t}$), q_e is the electronic charge, and $\vec{E}_0$ is the amplitude of the driving electric field. The second term in the right side of the said equation, $\vec{F_S}(t) = -q_e \times \vec{r}(t)$ is the restoring force proportional to the distance deviation from the equilibrium position with characteristic frequency $f_0 = \omega_0/2\pi$. The restoration forces are zero for free electrons within conductor materials (i.e. in metals) because the free electrons are not bound to a particular nucleus. The third term,$\vec{F_D} = -\gamma[\partial \vec{r}(t)/\partial t]$ refers to the damping force due to the collision energy loss, where γ is the damping coefficient in hertz. Due to the damping force, the oscillations don't go infinite when the driving force is at the resonant frequency. Note that the restoration and damping forces are negative because they are opposite to the direction of motion of the charge. Thus the differential form of equation (A1) becomes:

$$m_e \frac{\partial^2 \vec{r}}{\partial t^2} + m_e \gamma \frac{\partial \vec{r}}{\partial t} + m_e \omega_0^2 \vec{r} = q_e \vec{E}(t) \tag{A2}$$

In the frequency domain, equation (A2) can be written as:

$$m_e[(-j\omega)^2 \vec{r}(\omega) + \gamma(-j\omega)\vec{r}(\omega) + \omega_0^2 \vec{r}(\omega)] = q_e \vec{E}(\omega) \tag{A3}$$

On simplification, the instantaneous distance for a monochromatic electric field works out to be:

$$\vec{r}(\omega) = \frac{q_e}{m_e(\omega_0^2 - \omega^2 - j\gamma\omega)} \vec{E}(\omega) \tag{A4}$$

Now, the polarization P is the dipole moment per unit volume. For macroscopic material properties, if N_e is the density of charges (electrons) per unit volume, then considering that each electron is acting with the same dipole moment we have the frequency-dependent polarization as:

$$\vec{P}(\omega) = N_e q_e \vec{r}(\omega) = \frac{N_e q_e^2}{m_e(\omega_0^2 - \omega^2 - j\gamma\omega)} \vec{E}(\omega) \tag{A5}$$

while the electric susceptibility χ_e is related with the polarization P and electric field E with the following relation:

$$\chi_e(\omega) = \frac{P(\omega)}{\varepsilon_0 E(\omega)} \tag{A6i}$$

Thus, from equations (A6i) and (A5)

$$\chi_e(\omega) = \frac{\omega_p^2}{\omega_0^2 - \omega^2 - j\omega\gamma_L} \tag{A6ii}$$

The permittivity is then obtained as follows:

$$\varepsilon(\omega) = \varepsilon_0\Big[1 + \chi_e(\omega)\Big] = \varepsilon_0\left[1 + \frac{\omega_p^2}{\omega_0^2 - \omega^2 - j\omega\gamma}\right] \tag{A7}$$

where $\omega_p^2 = (N_e q_e^2/\varepsilon_0 m_e)$ is the *plasma frequency* of the bulk metal.

In terms of relative permittivity ($\varepsilon_r = \varepsilon/\varepsilon_0$) or dielectric constant of the material, the above equation (A7) may be written as:

$$\varepsilon_r(\omega) = 1 - \frac{\omega_p^2}{\omega^2 - \omega_0^2 + j\omega\gamma} \tag{A8}$$

The real and imaginary parts of the dielectric function may be written explicitly as:

$$\varepsilon_{r,\ \text{real[Lorenz]}}(\omega) = 1 + \frac{\omega_p^2(\omega_0^2 - \omega^2)}{(\omega_0^2 - \omega^2)^2 + \gamma^2\omega^2} \tag{A9}$$

$$\varepsilon_{r,\ \text{imag[Lorenz]}}(\omega) = 1 + \frac{\gamma\omega_p^2\omega}{(\omega_0^2 - \omega^2)^2 + \gamma^2\omega^2} \tag{A10}$$

It may be observed from equation (A8), at higher frequencies ε_r practically becomes equal to 1 while the DC value is: $\varepsilon_{r(DC)} = [1+ (\omega_P/\omega_0)^2]$. It is only in the neighbourhood of resonance ($\omega = \omega_0$) that the complex nature of dielectric function becomes important.

The plot of equations (A9) and (A10) are shown in figure A.2.

This resonant-type harmonic oscillator characterization of frequency-dependent dielectric function is known as the *Lorentz model*, the model was named after the famous Dutch physicist Hendrik Antoon Lorentz.

Similar magnetic response models follow immediately. The corresponding magnetization field components M_i and the magnetic susceptibility χ_m equations are obtained from the polarization and the electric susceptibility expressions with the replacents $E_i \rightarrow H_i$, $P_i/\varepsilon_0 \rightarrow M_i$. The permeability is given as: $\mu(\omega) = \mu_0[1 + \chi_m(\omega)]$.

The expression for the Lorentz model for effective permeability of the material medium is given by:

$$\mu_r(\omega) = 1 - \frac{\omega_{pm}^2}{\omega^2 - \omega_{0m}^2 + j\omega\gamma_m} \tag{A11}$$

where ω_{pm}, ω_{0m}, and γ_m are the plasma frequency, resonant frequency, and the damping coefficient in the magnetization of a magnetic dipole.

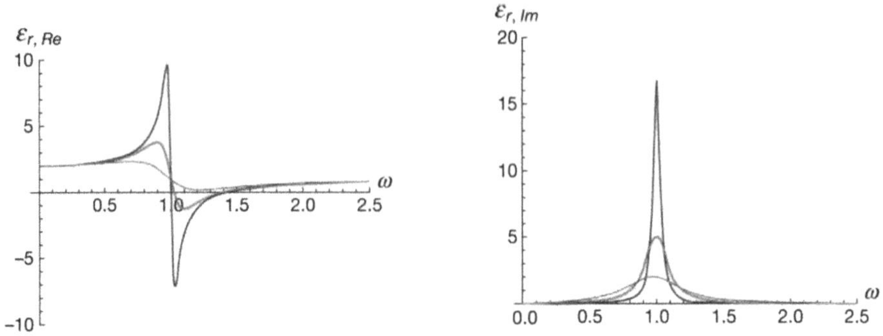

(i) (ii)

Figure A.2. Plots of (i) $\varepsilon_{r,\mathrm{Re}}$ and (ii) $\varepsilon_{r,\mathrm{Im}}$ for a material with an electronic resonance, plotted with $\omega_p = \omega_0 = 1$ and three different damping values (γ : 0.06, 0.2, and 0.5). As the damping decreases, the peaks get narrower and taller.

A.2 Drude model

Now, if the material is a pure metal then the electrons are free to move around the crystal lattice and are not bound to any particular nucleus and thus there is no question of the restoration force F_S of equation (A1). However, in metal the fixed (i.e. relatively immobile) background of positive ions obstructs the flow of electrons. The electrons oscillate in response to the applied electromagnetic field, and their motion is damped via collisions occurring with the fixed positive ions. Under this situation the equation of motion of free electrons under the influence of a time varying electric field may be described by (deleting the restoration force term in equation (A2)):

$$m_e \frac{\partial^2 \vec{r}}{\partial t^2} + m_e \gamma \frac{\partial \vec{r}}{\partial t} = q_e \vec{E}(t) \tag{A12}$$

where q_e is the electric charge of a free electron. The damping effect γ is proportional to the Fermi velocity, i.e.: $\gamma = v/l = 1/\tau$, where v denotes the Fermi velocity, and l is the mean free path of an electron between successive collision events. The relaxation time τ is the averaged time interval between subsequent collisions of an electron. In general, the relaxation time is about 10^{-14} s at room temperature, corresponding to $\gamma = 100$ THz.

The solution of instantaneous distances in equation (A12) for a monochromatic electric field is thus given by:

$$\vec{r}(\omega) = \frac{q_e}{m_e(\omega^2 + j\gamma\omega)} \vec{E}(\omega) \tag{A13}$$

Now, following the steps from equation (A4) till equation (A6) above it may easily be established that:

$$\varepsilon(\omega) = \varepsilon_0 \left[1 - \frac{\omega_p^2}{\omega^2 + i\gamma\omega} \right] \tag{A14}$$

When the frequency-dependent relative permittivity or dielectric function of the material can be written as:

$$\varepsilon_r(\omega) = 1 - \frac{\omega_p^2}{\omega^2 + j\gamma\omega} \tag{A15}$$

The corresponding relative permeability is given by:

$$\mu_r(\omega) = 1 - \frac{\omega_{pm}^2}{\omega^2 + j\gamma_m\omega} \tag{A16}$$

where ω_{pm}, and γ_{m} are the plasma frequency and the damping coefficient in the magnetization of a magnetic dipole.

The real and imaginary parts of the dielectric function may be written explicitly as:

$$\varepsilon_{\mathrm{r,real[Drude]}}(\omega) = 1 - \frac{\omega_p^2}{\omega^2 + \gamma^2} \tag{A17}$$

$$\varepsilon_{\mathrm{r,imag[Drude]}}(\omega) = \frac{\gamma\omega_p^2}{\omega(\omega^2 + \gamma^2)} \tag{A18}$$

The plots of equations (A17) and (A18) are shown in figure A.3.

This high-pass filter type harmonic oscillator characterization of frequency-dependent dielectric function is known as *Drude model*, proposed by Paul Drude in 1900 to explain the transport properties of electrons in materials (especially for metals).

A few observations may be made with respect to both the Lorentz and Drude models. It may be noted that both in Lorentz model and in Drude model, at high

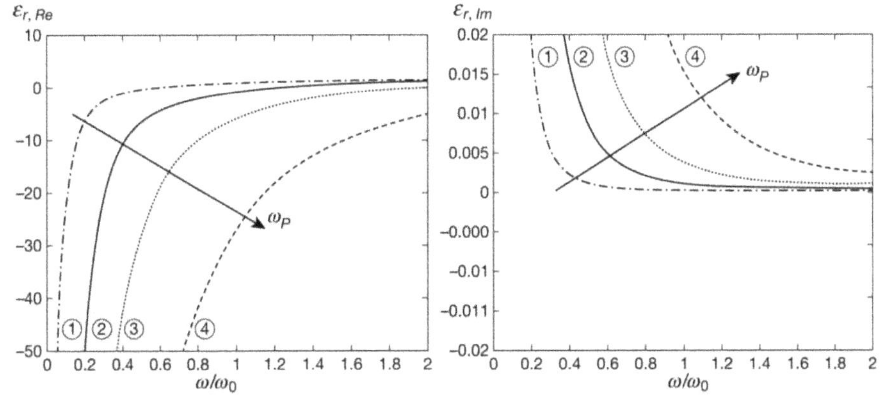

(i) (ii)

Figure A.3. Plots of (i) $\varepsilon_{\mathrm{r,Re}}$ and (ii) $\varepsilon_{\mathrm{r,Im}}$ for a conducting material (i.e. metal) plotted with four parametric values of plasma frequency: (1) 100 GHz, (2) 265 GHz, (3) 500 GHz, and (4) 1000 GHz.

frequency limit the permittivity of the medium reduces to that of the permittivity of free-space ε_0 and hence the relative permittivity or dielectric constant of the medium becomes practically 1. Because the Lorentz model is resonant in nature, the real part of the permittivity becomes negative in a very narrow frequency region immediately above the resonance. On the other hand the real part of Drude model can yield a negative real part of the permittivity over a wide spectral range.

IOP Publishing

Metamaterials and Metasurfaces
Basics and trends
Subal Kar

Appendix B

Derivations for f_{e0} and f_{ep}

B.1 Derivation for f_{e0}

The expression of the resonant frequency (f_{e0}) for the cut-wire (CW) structure has been derived [B1] by using the circuit theory approach where the capacitance and inductance of the wire structure has been separately calculated.

The cross-sectional area of each cylindrical shaped wire elements is πr^2. The two wire sections with gap d in between must behave as a parallel plate capacitor with area mentioned above. Thus, the capacitance between the CW elements is given by:

$$C = \varepsilon_{sub} \frac{\varepsilon_0 \pi r^2}{d} \tag{B1}$$

where ε_{sub} is the dielectric constant of the substrate.

As depicted in figure B.1 the entire volume of the periodic CW arrangement is divided into rectangular unit cells having a square face with sides h on the x–z plane (i.e. with horizontal lattice constant h is equal to the vertical lattice constant v), and length $(l + d)$ along y-axis. Let the square cross-section of the unit cells be approximated with a circular cross-section having the same area, i.e. with the radius $R_c = h/\sqrt{\pi} \approx h$. Each of the wire elements is located at the centre of each unit cell approximated as cylinders with radius h. The condition of current continuity holds good with conduction current along the wire elements and displacement current in the gap between them. At a given point, the magnetic field and the associated magnetic flux will have contribution only from the wire element in the unit cell within which the point lies. All other cells will have zero contribution.

The magnetic field for the wire element of length l is derived as:

$$B = \frac{\mu_0}{4\pi} \frac{I \; l}{R \, (l^2 + r^2)^{1/2}} \tag{B2}$$

where R is the perpendicular distance of the reference point of observation from the wire element. The associated total magnetic flux is thus given by:

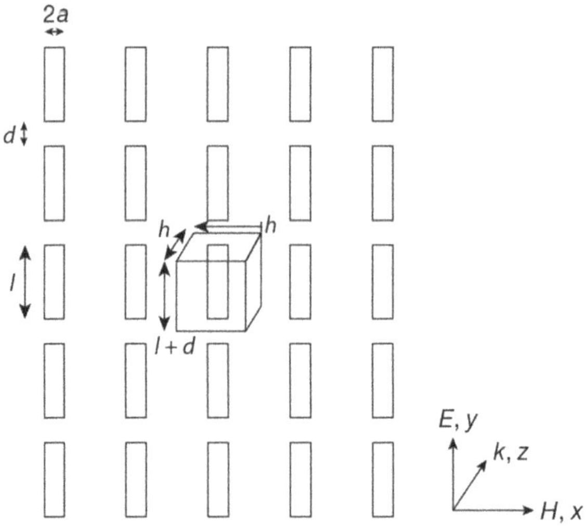

Figure B.1. CW array with unit cell indicated with dimensions labelled.

$$\varphi = \frac{\mu_0}{4\pi} I \ (l + d)\left[\log\left|\tan\frac{1}{2}\left(\tan^{-1}\frac{h}{l}\right)\right| - \log\left|\tan\frac{1}{2}\left(\tan^{-1}\frac{r}{l}\right)\right|\right] \quad \text{(B3)}$$

Since $\varphi = LI$, so the inductance of the each wire element is given by:

$$L = \frac{\mu_0}{4\pi} (l + d)\left[\log\left|\tan\frac{1}{2}\left(\tan^{-1}\frac{h}{l}\right)\right| - \log\left|\tan\frac{1}{2}\left(\tan^{-1}\frac{r}{l}\right)\right|\right] \quad \text{(B4)}$$

Let us consider that there are N_x, N_y, and N_z unit cells in the x-, y- and z-directions, respectively. The capacitors and inductors of a line along y-direction are in series with each other, whereas each of those in x- and z-directions are in parallel with each other. Thus the total impedance is:

$$Z_{\text{total}} = N_x N_z\left(jN_y\omega L + \frac{N_y - 1}{j\omega C}\right) \quad \text{(B5)}$$

By equating the imaginary part of the total impedance to zero, we get the expression for the electric resonant frequency as:

$$f_{e0} = \frac{1}{2\pi}\sqrt{\frac{4c^2(N_y - 1)d}{a^2\varepsilon_{\text{sub}}N_y(l + d)\left[\ln\left|\tan 1/2(\tan^{-1}h/l)\right| - \ln\left|\tan 1/2(\tan^{-1}a/l)\right|\right]}} \quad \text{(B6)}$$

where c is the velocity of light in free space.

Since for effective design of negative permittivity structures, the inequality $a, h \ll l$ must be obeyed, the expression for electric resonant frequency may be approximated as:

$$f_{e0} = \frac{c_0}{\pi a}\sqrt{\frac{(N_y - 1)d}{\varepsilon_{\text{sub}}N_y(l + d)\ln(h/a)}} \tag{B7}$$

From the above result we find that when $d = 0$, i.e., the case of a continuous wire or thin wire (TW) array, the resonant frequency vanishes. This is true because continuous TW structures are indeed non-resonant in characteristics.

B.2 Derivation for f_{ep}

The magnetic field around the wire element, as calculated in equation (B2) can also be expressed in terms of magnetic vector potential since $B = \nabla \times A$. Hence the magnetic vector potential may be written as:

$$A(r) = \frac{\mu_0}{4\pi}\,\pi a^2 n e v_m\left[\ln\left|\tan\frac{1}{2}\left(\tan^{-1}\frac{h}{l}\right)\right| - \ln\left|\tan\frac{1}{2}\left(\tan^{-1}\frac{a}{l}\right)\right|\right] \tag{B8}$$

where v_{m} is the mean electron velocity.

From classical mechanics, electrons in a magnetic field have an additional contribution to their momentum equal to $eA(\mathrm{r})$. Therefore, the momentum per unit length of the wire is given by:

$$m_{\text{eff}}\pi a^2 n e v_m = \pi a^2 n e \frac{\mu_0}{4\pi}\,\pi a^2 n e v_m \\ [\ln|\tan 1/2(\tan^{-1}h/l)| - \ln|\tan 1/2(\tan^{-1}a/l)|] \tag{B9}$$

Thus, the effective mass of the electrons is derived as:

$$m_{\text{eff}} = \frac{\mu_0}{4\pi}\,\pi a^2 n e^2\left[\ln\left|\tan\frac{1}{2}\left(\tan^{-1}\frac{h}{l}\right)\right| - \ln\left|\tan\frac{1}{2}\left(\tan^{-1}\frac{a}{l}\right)\right|\right] \tag{B10}$$

From plasma theory we have the plasma frequency given by:

$$\begin{aligned} f_{\text{ep}} &= \frac{1}{2\pi}\sqrt{\frac{n_{\text{eff}}\mathrm{e}^2}{\varepsilon_0 m_{\text{eff}}}} \\ &= \frac{1}{2\pi}\sqrt{\frac{4\pi c^2}{a^2\left[\ln\left|\tan\frac{1}{2}\left(\tan^{-1}\frac{h}{l}\right)\right| - \ln\left|\tan\frac{1}{2}\left(\tan^{-1}\frac{a}{l}\right)\right|\right]}} \end{aligned} \tag{B11}$$

Thus with the approximation $a, h \ll l$, as was used in the case of effective negative permittivity, the above expression for electric plasma frequency for CW may be written as:

$$f_{\text{ep}} = \frac{c_0}{2\pi h}\sqrt{\frac{4\pi}{\ln(h/a)}} \tag{B12}$$

The expression for f_{ep} derived by Pendry *et al* [B2] for continuous TW is given by:

$$f_{ep} = \frac{c_0}{2\pi h}\sqrt{\frac{2\pi}{\ln(h/a)}} \tag{B13}$$

A NOTE:

The plasma frequency expression given by equation (B13) was that derived by Pendry *et al* [B2]. Different formulation for plasma frequency has been given subsequently by Markos *et al* [B3], Maslovski *et al* [B4] and Kumar *et al* [B5], all of which tried to provide a more complete description of the phenomenon. The formulation given by Kumar *et al* [B5] is so far the most accurate formulation for the determination of plasma frequency of wire media.

References

[B1] Kar S, Roy T, Gangooly P and Pal S 2013 Analytical characterization of cut-wire and thin wire structures for metamaterial applications 2013 *Science and Information Conf. (London, October 7–9, 2013)* 665–9

[B2] Pendry J B, Holden A J, Stewart W J and Youngs I 1996 Extremely low frequency plasmons in metallic microstructures *Phys. Rev. Lett.* **76** 4773

[B3] Markos P and Soulkoulis C M 2003 Absorption losses in periodic arrays of thin wires *Opt. Lett.* **28** 846–8

[B4] Maslovski S I, Tretyakov S A and Belov P A 2002 Wire media with negative effective permittivity: a quasi-static model *Microwave Opt. Technol. Lett.* **35** 47–51

[B5] Kumar A, Majumder A, Chatterjee S, Das S and Kar S 2012 A novel approach to determine the plasma frequency for wire media *Metamaterials* **6** 43–50

Appendix C

Derivation of equation (2.7)

The generalized equation for the electromotive force (emf), following Faraday's law of electromagnetic induction, induced in the periodic array of a magnetic inclusion structure (MSRR/SR/LR) with the *H*-field of the electromagnetic wave appearing at right angles to the plane of the unit cells in the array, see figure C.1, is given by [C1, C2]:

$$-\pi r^2(N-1)^2\mu_0\frac{\partial}{\partial t}\left[H+i-\frac{\pi r^2}{h^2}i\right]-2\pi r(N-1)iR_c v+\frac{v\int idt}{C} \tag{C1}$$

where i is the induced RF current flowing in each ring caused by the magnetic field H of the incident electromagnetic wave, R_c is the resistance of a metallic ring per unit circumferential length of the ring, C is the distributed capacitance between the strips of the adjacent rings, N is the number of inclusions. The expression for capacitance C in the case of multiple split-ring resonator (MSRR), spiral resonator (SR), and labyrinth resonator (LR) (w.r.t. figure C.2) is tabulated in table C.1 [C2].

The circumferential length effective in each structure determines the multiplying factor to the capacitance between two rings while for LR an extra capacitance comes from the two gaps in each ring where t_M is the thickness of the metal strip of the ring and g is the cut of the gap.

Since the intensity of the time varying magnetic field incident on the structure is constant, the magnetic flux through the metallic rings does not change resulting in zero induced emf and thus from equation (C1):

$$j\omega\pi r^2(N-1)^2\mu_0\left[H+i-\frac{\pi r^2}{h^2}i\right]-2\pi r(N-1)iR_c v-\frac{jvi}{\omega C}=0 \tag{C2}$$

where we have used the fact that $\frac{\partial}{\partial t}=-j\omega$ and $\int dt=-\frac{j}{\omega}$, as the incident electromagnetic wave is time-harmonic in nature ($e^{-j\omega t}$).

From equation (C2) we can solve for the induced current as:

doi:10.1088/978-0-7503-5532-2ch9

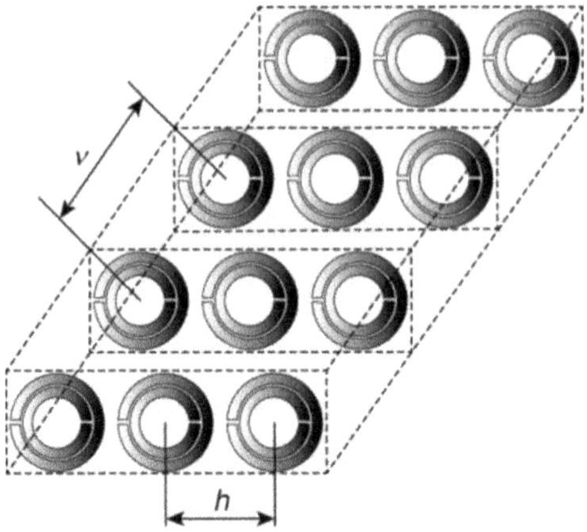

Figure C.1. SRR array showing the horizontal lattice constant h and vertical lattice constant ν.

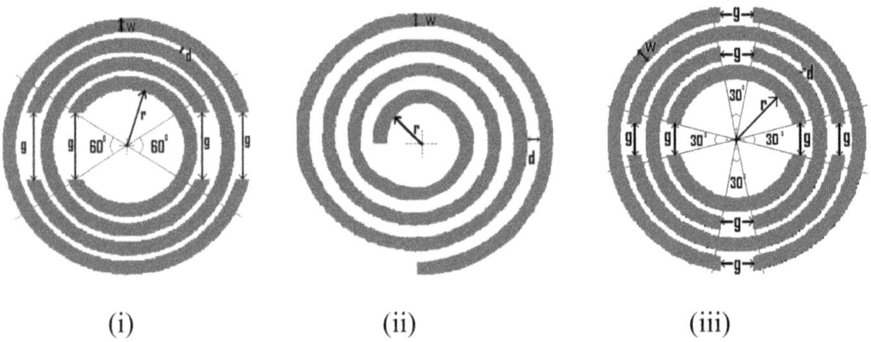

Figure C.2. Schematic representation of (i) MSRR, (ii) SR, and (iii) LR unit cell structure for deriving equation (2.7).

Table C.1. The capacitance C in case of MSRR, SR, and LR.

MSRR	SR	LR
$\frac{2\pi r}{6}\frac{\varepsilon_0}{\pi}\ln\left(\frac{wN}{d(N-1)}\right)$	$2\pi r\frac{\varepsilon_0}{\pi}\ln\left(\frac{wN}{d(N-1)}\right)$	$\frac{2\pi r}{24}\frac{\varepsilon_0}{\pi}\ln\left(\frac{wN}{d(N-1)}\right)+\frac{N}{2}\frac{\varepsilon_0 w t_M}{g}$

$$
\begin{aligned}
i &= \frac{-j\omega\pi r^2(N-1)^2\mu_0 H}{j\omega\pi r^2(N-1)^2\mu_0\left[1-\frac{\pi r^2}{a^2}\right]-2\pi r(N-1)R_c\nu-\frac{j\nu}{\omega}} \\
&= \frac{-H}{\left[1-\frac{\pi r^2}{a^2}\right]+j\frac{2(N-1)R_c\nu}{\omega r\mu_0}-\frac{\nu}{\omega^2\pi r^2(N-1)^2\mu_0 C}}
\end{aligned}
\tag{C3}
$$

Since, the effective relative permeability μ_{reff} in terms of the average induced magnetic field strength H_{av} and the average induced magnetic flux density B_{av} over the unit cell volume can be expressed as [C2]: $\mu_{reff} = B_{av}/\mu_0 H_{av} = H/H_{av}$, with $H_{av} = H - Fi$, where Fi is the depolarizing field; we have using equation (C3):

$$\mu_{reff} = 1 - \frac{F}{1 + j\frac{2(N-1)R_c\nu}{\omega r\mu_0} - \frac{\nu}{\omega^2\pi r^2(N-1)^2\mu_0 C}} \tag{C4}$$

When the values of the capacitance C from table C.1 is substituted in equation (C4) above we can get equation (2.7) of chapter 2 and also equation (2.6).

References

[C1] Pendry J B, Holden J, Robbins D J and Stewart W J 1999 Magnetism from conductors and enhanced non linear phenomena *EEE Trans. Microw. Theory Tech.* **47** 2075–8

[C2] Roy T, Banerjee D and Kar S 2009 Studies on multiple inclusion magnetic structures useful for millimeter-wave left-handed metamaterial applications *IETE J. Res.* **55** 83–9

Appendix D

The scattering parameters or *S*-parameters

NOTE: The students of RF and microwave engineering are familiar with *S*-parameters which are used for various designs at RF and microwave frequency domain. But for the benefit of all the other students of electronics and especially for students of physics, I am giving preliminary knowledge about *S*-parameters in the following.

D.1 Introduction

Devices and circuits in a network at low frequency are characterized in terms of *Z* (impedance), *Y* (admittance), or *h* (hybrid) parameters. These parameters are defined in terms of voltage and current in the device/circuit. This is because at low frequency, the device/circuit dimensions are small compared to the wavelength, and thus the devices and circuits may be considered as '*lumped*' components for which voltages and currents can be defined uniquely at any point in the network. However, at higher RF frequencies (microwave and above) the device/circuit dimensions are comparable to wavelength which makes them '*distributed*' components, i.e. the phase of the signal changes from one point to another in the network and this calls for field analysis of such networks. The voltages and currents are no longer measurable quantities (as they are difficult to measure) at microwave frequency and above, but are actually the derived quantities (voltage derived from electric field and current derived from magnetic field). The measurable quantities at higher RF frequency are the incident, reflected and transmitted power in the network (the reflected and transmitted powers are being generated due to scattering of the incident signal at a microwave junction). Again in view of the linearity of the field equations and most devices, the scattered wave amplitudes are linearly related to the incident wave amplitudes. The matrix describing this linear relationship between the reflected and transmitted waves with incident wave amplitude is called the *scattering matrix* and the microwave device/circuit is analyzed in terms of the scattering parameters or *S-parameters*.

It may be noted that unlike the *Z*, *Y*, *h*, and *ABCD* parameters (the last one being used at low frequencies for cascaded networks), the *S*-parameters do not use short

doi:10.1088/978-0-7503-5532-2ch10

and open circuit terminations. From a practical view point this has significance, as it is difficult to obtain perfect open and short circuit termination at RF and microwave frequency and also the impedance of short and open varies with frequency and thus defining parameters to characterize microwave device/circuit should not use them. Further, the presence of short/open eventually cause complete reflection of signal from the load terminal leading to possible oscillation and might even cause damage to the source. But use of matched load in defining S-parameters would cause absorption of the incident energy and thus cause no reflection back to the source.

From an RF designer's point of view it is also very important that the S-parameters can be easily imported and used for circuit simulations in electronic-design automation tools. S-parameters are the shared language between simulation and measurement. Since the 1970s, the popularity of S-parameters increased with the advent of the new generation network analyzers which could perform S-parameter measurement with ease.

D.2 *S*-parameter formulation of RF/microwave network

Let us consider an n-port microwave junction shown in figure D.1(i). Let a wave with an associated equivalent voltage V_1^+ be incident on the junction at terminal plane t_1. This will generate a reflected wave in line 1 being given by: $S_{11}V_1^+ = V_1^-$, where S_{11} is the reflection coefficient, or scattering coefficient, for line 1, with a wave incident at line 1. Waves will also be transmitted, or scattered, out to the remaining $(n - 1)$ lines from line 1 and will have amplitudes proportional to V_1^+. These amplitudes can be expressed as: $V_i^- = S_{i1}V_1^+$ where $i = 2,3,\ldots, n$, (S_{i1} is the transmission coefficient for signal transmitted to ith line from line 1).

When waves are incident in all the n lines, the wave voltage in a particular line has its own reflected wave voltage plus the scattered wave voltage contributions from all other $(n - 1)$ lines. Thus, in general, we can write:

$$\begin{aligned} V_1^- &= S_{11}V_1^+ + S_{12}V_2^+ + S_{13}V_3^+ + \ldots\ldots.. + S_{1n}V_n^+ \\ V_2^- &= S_{21}V_1^+ + S_{22}V_2^+ + S_{23}V_3^+ + \ldots\ldots.. + S_{2n}V_n^+ \\ &\ldots\ldots\ldots\ldots\ldots\ldots\ldots\ldots\ldots\ldots \\ &\ldots\ldots\ldots\ldots\ldots\ldots\ldots\ldots\ldots\ldots \\ V_n^- &= S_{n1}V_1^+ + S_{n2}V_2^+ + S_{n3}V_3^+ + \ldots\ldots.. + S_{nn}V_n^+ \end{aligned} \tag{D1}$$

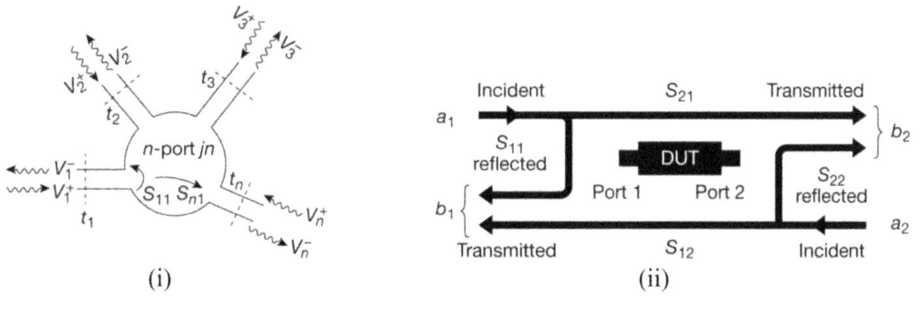

Figure D.1. (i) An n-port microwave junction. (ii) A two-port microwave network showing the incident and reflected wave variables and the S-parameters.

which can be written in matrix form as: $[V^-] = [S][V^+]$, where both $[V^-]$ and $[V^+]$ are column matrices and S is an $n \times n$ matrix called scattering matrix given by:

$$\begin{bmatrix} V_1^- \\ V_2^- \\ \vdots \\ V_n^- \end{bmatrix} = \begin{bmatrix} S_{11} & S_{12} & S_{13} & \cdots & S_{1n} \\ S_{21} & S_{22} & S_{23} & \cdots & S_{2n} \\ \vdots & \vdots & \vdots & \vdots & \vdots \\ S_{n1} & S_{n2} & S_{n3} & \cdots & S_{nn} \end{bmatrix} \begin{bmatrix} V_1^+ \\ V_2^+ \\ \vdots \\ V_n^+ \end{bmatrix} \tag{D2}$$

The coefficients S_{11}, S_{12}, …, S_{nn} are called scattering parameters (S-parameters) or scattering coefficients.

The scattering matrix definition so far was implicitly based on the assumptions that all ports have the same characteristic impedance. Though this may be the case in many practical situations, for general analysis we need to consider the case when each port has non-identical characteristic impedance. This will also bring out some significant meaning of the wave variables: the reflected wave variables (b_1, b_2) as dependent variables and the incident wave variables (a_1, a_2) as independent variables, see figure D.1(ii) (a two-port network), in terms of which we are going to define the S-parameters.

Taking each port's characteristic impedance into account, we define two normalized voltage waves:

$$a_i = \frac{V_i^+}{\sqrt{Z_{0i}}} \tag{D3a}$$

$$b_i = \frac{V_i^-}{\sqrt{Z_{0i}}} \tag{D3b}$$

where a_i and b_i represent the normalized incident and reflected voltages, respectively, at the ith port and Z_{0i} is the characteristic impedance of the ith port (which is a real number for loss-less line).

Thus the total voltage and current at each port can now be written as

$$V_i = V_i^+ + V_i^- = \sqrt{Z_{0i}}(a_i + b_i) \tag{D4a}$$

$$I_i = \frac{V_i^+}{Z_{0i}} - \frac{V_i^-}{Z_{0i}} = \frac{(a_i - b_i)}{\sqrt{Z_{0i}}} \tag{D4b}$$

The average net power delivered to the ith port can now be expressed in terms of a_i and b_i (with no concern of the fact that different ports have different characteristic impedance) as:

$$P_i = \frac{1}{2}\,\mathrm{Re}\,[V_i I_i^*] = \frac{1}{2}\,\mathrm{Re}\,[|\,a_i\,|^2 - |\,b_i\,|^2 + (a_i^* b_i - a_i b_i^*)]$$

$$= \frac{1}{2}[|\,a_i\,|^2 - |\,b_i\,|^2] \tag{D4c}$$

Since $(a_i^* b_i - a_i b_i^*) = 0$, when we take real part of it.

This equation is significant, as it clearly shows that the net power delivered to each port is equal to the normalized incident power less by the normalized reflected power. By defining the wave variables in this way it may be seen that the square of any wave variable becomes proportional to the average power in the wave, and these are now directly measurable quantities at microwave frequency.

The generalized S-matrix can now be defined in terms of the normalized voltage wave variables a and b as:

$$[b] = [S][a] \tag{D5a}$$

For a two-port network we may thus write:

$$b_1 = S_{11}a_1 + S_{12}a_2 \tag{D5b}$$

$$b_2 = S_{21}a_1 + S_{22}a_2 \tag{D5c}$$

where S_{11} and S_{22} are the reflection coefficients at ports 1 and 2, respectively, and S_{12} and S_{21} are the transmission coefficients to port 1 from port 2 and to port 2 from port 1, respectively.

Now we try to understand the physical significance of the S-parameters. Referring to equation (D5b), we find that:

$$S_{11} = \left.\frac{b_1}{a_1}\right|_{a_2=0} \tag{D6a}$$

(input reflection coefficient).

The variable a_2 is zero when there is no incident wave at port 2. The parameter S_{11} is, therefore, the reflection coefficient at port 1 when port 2 is terminated with a matched load.

Again from equation (D5c) we have:

$$S_{21} = \left.\frac{b_2}{a_1}\right|_{a_2=0} \tag{D6b}$$

(***forward transmission coefficient***).

This gives the ratio of a wave leaving port 2 for a wave entering port 1, i.e. the transmission coefficient from port 1 to port 2 under matched condition at port 2 (for a lossy transmission line or waveguide, S_{21} should be less than 1, and it indicates the amount of attenuation).

The other two parameters can also be found from equations (D5b) and (D5c) as follows:

$$S_{22} = \left.\frac{b_2}{a_2}\right|_{a_1=0} \tag{D6c}$$

(***output reflection coefficient***).

$$S_{12} = \left.\frac{b_1}{a_2}\right|_{a_1=0} \tag{D6d}$$

(***reverse transmission coefficient***).

It may be observed that S_{22} is the reflection coefficient at port 2 with port 1 matched and that S_{12} gives the transmission coefficient from port 2 to port 1 under matched condition at port 1. In general, the scattering parameters are complex numbers having both magnitude and phase.

Appendix E

Numerical techniques for metamaterial and metasurface characterization and design

E.1 Introduction

As is the case in many fields of engineering, besides the analytical approach (which will always remain essential for developing creative solutions), the role of numerical methods is becoming increasingly important in the characterization and design of metamaterials (MTMs) and metasurfaces (MSs). Indeed, they are useful to rapidly validate new concepts and also to efficiently optimize initial designs.

The starting point of all the numerical techniques is of course the source-free Maxwell's curl equations in isotropic medium which are given below for the time-dependent case and also for time-harmonic case [E1].

Time-dependent case:

$$\nabla \times \vec{E} = -\mu \frac{\partial \vec{H}}{\partial t} \qquad \nabla \times \vec{H} = \sigma \vec{E} + \varepsilon \frac{\partial \vec{E}}{\partial t} \tag{E1}$$

Time-harmonic case:

$$\nabla \times \vec{E} = -j\omega\mu\vec{H} \qquad \nabla \times \vec{H} = \sigma\vec{E} + j\omega\varepsilon\vec{E} \tag{E2}$$

There are several techniques, such as plane-wave expansion and spherical-wave expansion, finite-difference time-domain (FDTD) method, transfer-matrix method (TMM), finite-element method (FEM), method of moment (MoM) and so on, to solve the Maxwell's equations numerically. Finite differences and finite elements, as well as finite integration (which presents some similarities with finite differences), are generally used for solving equations in differential form. Regarding the integral-equation formulation, the method of moments (MoM) is clearly the most popular technique. It is generally confined to fields in the harmonic regime, since time-domain solutions remain a challenge from stability and memory points of view. Much of the commercial software is there to calculate various properties of MTMs,

doi:10.1088/978-0-7503-5532-2ch11

such as absorption, reflection and transmission spectra, field distributions, surface–current distributions, energy–density distributions, and so on. For the characterization of MTMs and MSs and also for various design problems we use simulation tools that use numerical techniques like FEM, FDTD, TMM, MoM and so on. Thus in this appendix we will discuss all these techniques in brief; emphasizing the FDTD technique which is used extensively for MTM/MS modelling. For thorough discussions of individual techniques one may refer to standard texts [E2, E3].

E.2 Finite difference time-domain (FDTD) method

The finite difference time-domain (FDTD) method was first introduced by Kane S Yee [E4] in 1966 and later developed by Taflove and others [E5]. FDTD is a time-domain method, the starting point of its implementation is the time-dependent Maxwell's equation, specifically equation (E1). The curl equations are discritized using a rectangular grid (which stores the field components of the material properties ε and μ) and central differences for the space and time derivatives. The procedure results in a set of finite-difference equations, which updates the field components in time. The FDTD equations for various types of materials, together with computational procedure, source incorporation procedure, stability criteria, and so on, are presented in a very clear and complete way in [E6]. We will discuss here Yee's algorithm (which treats the irradiation of scatterer as an initial value problem) and associated issues [E3].

FDTD method is the most popular method in MTM simulations. In FDTD the numerical differentiation is based on the finite difference in both space and time. The starting point of FDTD is the Taylor expansion of any function as:

$$f(x_0 + \Delta x) = f(x_0) + \Delta x f'(x_0) + \frac{1}{2!}(\Delta x)^2 f''(x_0) + \frac{1}{3!}(\Delta x)^3 f'''(x_0) + \ldots.. \tag{E3}$$

To apply FDTD the computational domain should be defined, first. The computational domain is a spatial domain where the simulation will be carried out. The computational domain is divided into small meshes, at which finite differences are calculated. From the Taylor expansion we can get the *central-difference equation* at the ith spatial mesh and at the nth time step. Using Taylor series expansions for: $f(x_0 \pm \Delta x)$ we can write:

$$f(x_i + \Delta x)|_{t_n} + f(x_i - \Delta x)|_{t_n} = 2f(x_i)|_{t_n} + (\Delta x)^2 f''(x_i)|_{x_i,t_n} + \frac{(\Delta x)^4}{12} f''''(x_i)\Big|_{x_i,t_n} + \ldots\ldots \tag{E4a}$$

$$f(x_i + \Delta x)|_{t_n} - f(x_i - \Delta x)|_{t_n} = 2\Delta x f'(x_i)|_{t_n} + \frac{(\Delta x)^3}{6} f'''(x_i)\Big|_{x_i,\, t_n} + \ldots\ldots \tag{E4b}$$

where i and n are integers, and Δx is the length of mesh. The first and second derivatives are approximately given by:

$$f'|_{x_i,t_n} = \frac{f(x_i + \Delta x)|_{t_n} - f(x_i - \Delta x)|_{t_n}}{2\Delta x} + \Theta[(\Delta x)^2] \quad \text{(E5a)}$$

$$f''|_{x_i,t_n} = \frac{f(x_i + \Delta x)|_{t_n} + f(x_i - \Delta x)|_{t_n} - 2f(x_i)|_{t_n}}{(\Delta x)^2} + \Theta[(\Delta x)^4] \quad \text{(E5b)}$$

where $\Theta[(\Delta x)^2]$ and $\Theta[(\Delta x)^4]$ are the error introduced by the truncation of the series. If the infinite Taylor series were retained, an exact solution would be realized for the problem. However, for practical reasons, the infinite series is usually truncated after the second-order term. This imposes an error (which is practically negligible) that exists in all finite difference solutions.

To solve the Maxwell's equation through FDTD, the fields are calculated in the so-called *Yee cell* (see figure E.1).

From equation (E1) we can write the equation for electric field as:

$$\frac{\partial \vec{E}}{\partial t} = \frac{1}{\varepsilon}[\nabla \times \vec{H} - \sigma \vec{E}] \quad \text{(E6)}$$

where σ is the electrical conductivity, the x-component of electric field in the Yee cell is given by:

$$\frac{\partial E_x}{\partial t} = \frac{1}{\varepsilon}\left(\frac{\partial H_z}{\partial y} - \frac{\partial H_y}{\partial z}\right) - (\partial E_x) \quad \text{(E7)}$$

Following Yee's notation, we define a grid point in the solution region as:

$$(i, j, k) \equiv (i\Delta x, j\Delta y, k\Delta z) \quad \text{(E8)}$$

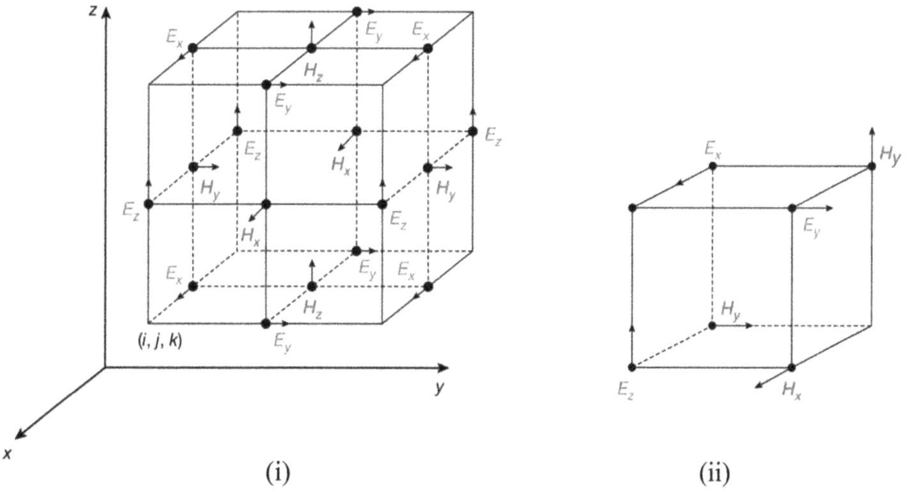

Figure E.1. (i) Positions of the field components in a unit cell of the Yee lattice. (ii) Typical relations between field components within a quarter of a unit cell.

and any space function and time as:

$$F^n(i, j, k) \equiv F(i\delta, j\delta, k\delta, n\Delta t) \tag{E9}$$

where $\delta = \Delta x = \Delta y = \Delta z$ is the space increment, and Δt is the time increment, while j, j, k, and n are integers. Using central finite difference approximation for space and time derivatives that are second-order accurate:

$$\frac{\partial F^n(i, j, k)}{\partial x} = \frac{F^n(i + 1/2, j, k) - F^n(i - 1/2, j, k)}{\delta} + \Theta(\delta^2) \tag{E10a}$$

$$\frac{\partial F^n(i, j, k)}{\partial t} = \frac{F^{n+1/2}(i, j, k) - F^{n-1/2}(i, j, k)}{\Delta t} + \Theta(\Delta t^2) \tag{E10b}$$

Applying equation (E10b) to equation (E7) we get the central-difference equation for x-component of the E field as:

$$\begin{aligned} E_x^{n+1}(i + 1/2, j, k) &= \left(1 - \frac{\sigma(i + 1/2, j, k)\delta t}{\varepsilon(i + 1/2, j, k)}\right). E_x^n(i + 1/2, j, k) \\ &+ \frac{\delta t}{\varepsilon(i + 1/2, j, k)\delta}[H_z^{n+1/2}(i + 1/2, j + 1/2, k) \\ &- H_z^{n+1/2}(i + 1/2, j - 1/2, k) \\ &+ H_y^{n+1/2}(i + 1/2, j, k - 1/2) - H_y^{n+1/2}(i + 1/2, j, k + 1/2)] \end{aligned} \tag{E11a}$$

The x-component of the magnetic field can be similarly obtained as [E3]:

$$\begin{aligned} H_x^{n+1/2}(i, j + 1/2, k + 1/2) &= H_x^{n-1/2}(i, j + 1/2, k + 1/2) \\ &+ \frac{\delta t}{\mu(i, j + 1/2, k + 1/2)\delta}[E_y^n(i, j + 1/2, k + 1) - E_y^n(i, j + 1/2, k) \\ &+ E_z^n(i, j, k + 1/2) - E_z^n(i, j + 1, k + 1/2)] \end{aligned} \tag{E11b}$$

Other electric and magnetic field components for y and z components may also be derived to get the complete set of equations for Yee's finite difference algorithm. It may be noticed from equations (E11a) and (E11b) and figure E.1(i) that the components of electric and magnetic fields are interlaced within the unit cell and are evaluated at alternate half-time steps. All the field components are present in a quarter of a unit cell as shown typically in figure E.1(ii).

It may be mentioned that to ensure the accuracy of the computed results, the spatial increment δ must be small compared to the wavelength (usually $\leqslant \lambda$) or minimum dimension of the scatterer. For practical reasons, it is best to choose the ratio of time increment to spatial increment as large as possible, however, without affecting the stability condition [E3]. Although the FDTD method is very versatile and spectrum of broad frequency range can be obtained at once, it requires a lot of computing resources.

E.3 Transfer-matrix method

One of the common techniques for numerical characterization of dispersive material like left-handed material (LHM) is the transmission matrix method (TMM). This method is able to calculate the stationary scattering properties, that is, the complex transmission and reflection amplitudes of finite length samples. Such calculations are extremely useful in the interpretation of experimental measurements of transmission and reflection data. The calculation of the transmission and reflection coefficients for the LHM slab is performed by assuming that the slab, which is finite in the direction of the incoming incident signal (z-direction), is placed between two semi-infinite slabs of vacuum and employing the time-harmonic Maxwell's equations (equation (E2)). By imposing periodic boundary conditions, the slab is considered infinite along the directions perpendicular to that of the propagation direction of the incident signal. The approach used with the TMM consists of the calculation of the electromagnetic field components at a specific z-plane (e.g., after the slab) from the field components at a previous z-plane (e.g., before the slab). For implementation of this procedure the time-harmonic Maxwell's equations are discritized, employing a rectangular grid on which the fields and material parameters are defined. TMM method has been applied to characterize LHM composed of SRRs of various shapes and metallic wires [E7]. The agreement between the theoretical calculations via TMM technique was found to match very well with the experimental results. TMM method has also been used extensively to characterize photonic crystal band structure [E8].

E.4 Finite element method (FEM)

FEM is a numerical technique to solve the boundary-value problems of partial differential equations (PDEs). To apply the FEM technique, the spatial region of interest is discritized or divided into a finite number of sub-regions or elements (hence the name: *finite element*), i.e. into properly-chosen meshes or grids. The shape of mesh can be a triangle in 2D space; however, it can be of any form. In 3D space it can be tetrahedral, hexahedral, prism, or pyramidal shape. All meshes should be connected to adjacent meshes. Then, those connected meshes produce nodes, which are usually points at which three or more meshes meet. Meshes should be chosen in such a way that there should be no empty space between meshes; however, the total outer surface of all outermost meshes should not necessarily be identical to the real boundaries.

The next step is to choose basis functions which can be used to set up the approximate solution at each node. The basis function can be piecewise linear in a mesh. In these days, the basis functions are usually piecewise polynomial. At each node the coefficients of polynomial are set to satisfy the boundary conditions. The field quantities are interpolated in the mesh. The numbers of equations are finite. If the size of meshes is infinitesimally small these approximate solutions can be exact, however, the degrees of freedom would be infinite, which is impossible to take care of. If the size of mesh is macroscopically large, but small enough to make approximate solution reliable, the degrees of freedom become finite.

Although the finite difference method (FDM) and MoM are conceptually simpler and easier to program than the FEM, FEM is a more powerful and versatile numerical technique for handling problems involving complex geometries and inhomogeneous media. The systematic generality of the method makes it possible to construct general-purpose computer programs for solving a wide variety of problems.

E.5 Method of moments

The MoM is essentially the method of weighted residuals applicable for solving both the differential and integral equations. The method owes its name to the process of taking moments by multiplying with weighing functions and integrating. The method has been successfully applied to a wide variety of electromagnetic problems of practical interest such as radiation due to thin-wire elements and arrays, scattering problems, analysis of microstrips and lossy structures, propagation over an inhomogeneous Earth, and antenna beam patterns, just to mention a few.

Most of the electromagnetic problems can be stated in terms of an inhomogeneous equation:

$$L\Phi = g \tag{E12}$$

where L is an operator which may be differential, integral or integro-differential, g is known as excitation or source function, and Φ is the unknown function to be determined. MoM is a general procedure for solving equation (E12).

E.6 TMM versus FDTD—advantages and disadvantages

Both TMM and FDTD can model finite-size LHM samples with arbitrary internal structures and arbitrary material combination (e.g., metallic, lossy), giving complex transmission and reflection amplitudes, that is, magnitude, phase, and polarization information. The simultaneous amplitude and phase knowledge can be used in the inversion of transmission and reflection data to obtain the effective material parameters (ε and μ) provided that the effective medium approach is valid (which is the case with LHM or MTM).

Among the drawbacks of the TMM is the necessity of the discritization of the unit cell, which introduces some numerical artefacts and some constraints into the shape and size of the components inside the unit cell. For example, to simulate tiny components, as is usually the case with the LHM study, one needs very fine discritization, practically possible only within a nonuniform discritization scheme. Otherwise large calculation times and large memory requirements are unavoidable.

The main advantage of FDTD compared to TMM is that it can give the transmission properties over a wide spectral range with just a single calculation. It also can give time-domain pictures of the fields and currents over the entire computational domain. However, FDTD also has the disadvantage, like TMM, that stems from the inherent discritization required.

Like TMM, the FDTD method can be used to calculate scattering properties of LHM, and it also involves discritization of Maxwell's equations. The difference between the two methods is that TMM is employed for steady-state solutions while FDTD method is used for general time-dependent solutions. The steady-state solutions are then obtained by employing Fast-Fourier Transform (FFT) of the time-domain results. This permits the study of both the transient and steady-state response of a system. An additional advantage of FDTD is the possibility of obtaining a broadband steady-state response with just a single calculation, as the excitation signal can be a pulse rather than a monochromatic wave.

E.7 FEM versus FDTD—advantages and disadvantages

There are several differences between FEM and finite difference method (FDM), such as FDTD. FEM approximates the solution, while FDM approximates the PDEs. FEM can handle the very complex shape of the boundary of structure, while FDM can handle a relatively simple shape only. Most of all, FEM is easy to implement. There are, however, several disadvantages. Because FEM approximates the solution itself with a piecewise polynomial at each node, it is not possible to find a general closed-form solution, from which the dependence of the investigated system's response on the variation of parameters is studied systematically. FEM inherently contains errors; therefore, it is fatal if users make mistakes.

All numerical methods inherently contain errors because of the so-called truncation error of the computer. Because the computer can handle only a finite magnitude of number itself and the number of digits is also limited. Furthermore, if the number of digits is infinite (irrational numbers) or larger than the allowed number of digits, it is inevitable to truncate the rest of the digits. If the truncations are accumulated during the process the results can be fatally inaccurate. Besides this truncation error, FEM contains a few sources of inherent errors. They deal simplified geometry, polynomial interpolation of field quantities, simple integration technique, such as Gaussian quadrature, and so on. There are also common mistakes made by users, such as wrong choice of type of elements, distorted elements, insufficient support which prevents, for instance, all rigid-body motion, inconsistent units, etc. In spite of these drawbacks, FEM is one of the most popular numerical methods in numerous fields of engineering because of ease of implementation and versatility in handling very complex surfaces.

References

[E1] Kar S 2022 *Microwave Engineering—Fundamentals, Design, and Applications* 2nd edn (Madison, CT: Universities Press)

[E2] Davidson D B 2010 *Computational Electromagnetics for RF and Microwave Engineering* 2nd edn (Cambridge: Cambridge University Press)

[E3] Sadiku M N O 2011 *Numerical Techniques in Electromagnetics with MATLAB* 3rd edn (New York: CRC Press, Taylor and Francis Group)

[E4] Yee K S 1966 Numerical solution of initial boundary-value problems involving Maxwell's equations in isotropic media *IEEE Trans. Antennas Propag.* **14** 302–7

[E5] Taflove A *et al* 1988 Detailed FDTD analysis of electromagnetic fields penetrating narrow slots and lapped joints in thick conducting screens *IEEE Trans. Antennas Propag.* **36** 247–57

[E6] Taflove A and Hagness S C 2000 *Computational Electrodynamics: The Finite Difference Time Domain Method* (Boston, MA: Artech House)
[E7] Markos P and Soukoulis C M 2003 Transmission properties and effective electromagnetic parameters of double negative metamaterials *Opt. Express* **11** 649–61
Markos P and Soukoulis C M 2003 Absorption losses in periodic arrays of thin metallic wires *Opt. Lett.* **28** 846–8
[E8] Pendry J B and McKinnon A 1992 Calculation of photon dispersion relations *Phys. Rev. Lett.* **69** 2772–5
Pendry J B 1994 Photonic band structures *J, Mod. Opt.* **41** 209–29

Milton Keynes UK
Ingram Content Group UK Ltd.
UKHW051505270224
438568UK00003B/11

9 780750 355308